The Convergence of Science and Belief

Georges Mallet

The Convergence of Science and Belief

Georges Mallet

Academica Press
Washington

Library of Congress Cataloging-in-Publication Data
Names: Mallet, Georges (author)
Title: The convergence of science and belief | Mallet, Georges.
Description: Washington : Academica Press, 2025. | Includes references.
Identifiers: LCCN 2021952081 | ISBN 9781680534184 (hardcover) |
9781680534191 (e-book)

In Loving Memory of My Mother

Acknowledgments

I am deeply grateful to my daughter, Laure, whose unwavering encouragement, invaluable technical advice, and wholehearted dedication have been instrumental in promoting my work.

My heartfelt thanks also go to my partner, Joëlle, and my son, Frédéric, whose steadfast support and thoughtful guidance were a constant source of strength throughout the writing process.

The infectious enthusiasm of my grandchildren, Maxence and Tristan, who championed me as a *"writer"* among their acquaintances, has been inspiring and motivating.

I am equally thankful to the readers of my previous essay *"Les scientifiques et Dieu"* for their insightful and constructive feedback, which has helped me grow as a writer.

A special thank-you to Christopher Rex for his generosity in spontaneously offering me to create my first cover.

Contents

To our esteemed readers!

You will make a captivating journey across science and spirituality within these pages. We've dedicated a significant portion of this work to delve into the most groundbreaking scientific discoveries. Why, you may ask? Because, just as materialists cannot confirm or deny God's existence, neither can scientists. However, scientific breakthroughs serve as objective building blocks, empowering us to engage our intellect and craft our perspective on this profound question. These insights can complement and enrich our understanding of religious texts, such as the Scriptures.

The initial section, comprising chapters 2 to 4, is tailored to kindle the curiosity of high school students entering the fifth year in England or 10th grade in the United States. However, you don't need to embark on a linear reading journey. If you are well-versed in the realm of science and are eager to delve into the opinions of esteemed scientists you have encountered before, feel free to skip ahead. You can always return to these chapters through the convenient Author Name Index (Pp. 331-359). Doing so will allow you to savor the text anew, akin to a novice discovering the artistry of a new discipline.

Welcome to a world where science and spirituality intertwine, inviting you to explore the profound mysteries that lie at the heart of our existence.

Happy Readings!

Prolegomena

Why is there something rather than nothing?
G. Leibniz[Lz2]

From the dawn of time, humanity has been fascinated by the origins of life, leading to passionate debates and countless theories. Every civilization has sought to unlock the mysteries of nature, which once terrified our ancestors, prompting them to associate gods with celestial bodies. This curiosity sets humans apart from other creatures, driven by our pursuit of knowledge and rational understanding.

The Genesis story, foundational to the three **Abrahamic religions**, attributes the creation of the universe and humanity to God. This belief has persisted, aided by religious leaders who have preserved these teachings. Yet, ancient Greek philosophers, known as *"pre-Socratics,"* sought to explain the natural world by physical and mechanical means, without dismissing the existence of deities. **Aristotle**, often called the *"Father of Logic,"* acknowledged the importance of Physics for understanding natural phenomena but valued Metaphysics, the study of *"being as being,"* even more.

Pope **John Paul II**[1] explains in his Encyclical *"Fides et Ratio"* [JP] (p. 37) that:

"...deep in the heart of man are sown the desire and longing for God,"

opinion shared by most religions.

So, **Leibniz**'s question remains pertinent as our scientific understanding continues to evolve, although a definitive answer remains elusive.

Faith is not a cultural inheritance but a personal choice based on trust in God's love for us. Rooted in Christ's message and the stories of early Christians, martyrs, and saints, our faith teaches us *"moral law,"*[2] and the concepts of good and evil.

---

[1] ***Wojtyla** Karol, J. (Wadowice, POL 1920 - 2005 VAT). After studying philology, he joined an anti-Nazi theater group in 1942. Ordained a priest in 1946 (religion was forbidden by the communist regime), and he was appointed bishop in 1958. He opposed the materialistic policy, which was the rule in his country, and defended the workers against the communists. Elected Pope in 1978, he took the name of **John Paul II**. He contributed to the fall of the communist bloc, and organized in 1986 the 1ˢᵗ international interreligious meeting (194 heads of religions) in Assisi. Beatified in 2011 by **Pope Benedict XVI**, he was canonized by **Pope Francis** in 2014.*
[2] The Moral Law: *"Act in such a way that the maxim of your will may always be valid at the same time as the principle of a universal legislation." It is not dictated to us by anyone, it is within us, it derives from the impartiality and universality of Reason. Such as it was formulated* [Gra] *by* **Kant** *Emmanuel (Konigsberg, East Prussia 1724-1804). German philosopher, his ideas*

Blaise **Pascal,**[3] argues that faith doesn't need science to prove God's existence, as Scripture states:

"...God is a hidden God... it is the heart that feels God, not reason...."

So, as a scientist and *Christian* believer, why do I choose to show that faith and science can coexist and even complement each other, using examples from some of history's most famous scientists; convinced that truth is not based on the reputation of the one who states it, whether he is a Nobel Prize winner (a prize that has only existed since 1901) or not?

I aim to nourish, inspire, and challenge the faith of those around me, helping them see that even *"rationalists"* and *"materialists"* can appreciate the profound connection between science and belief in a higher power. In this exploration, I won't exclude agnostics or atheists whose work has significantly contributed to the progress of science. St. **Thomas Aquinas** once said:

"What one man can bring through his genius
is little compared to the whole of science."

I recognize that this essay may draw criticism, both constructive (which I'll appreciate) and baseless (which I'll ignore). However, if it helps even one person see the harmony between faith and science, I'll consider my mission accomplished. In this way, I hope to engage young readers in a meaningful conversation about the interconnectedness of Faith and science in our quest to understand the world around us.

Atheists of good faith, hope that scientists will prove the non-existence of God. In the meantime, let them meditate on the aphorism of the philosopher Sir Anthony **Kenny** (Liverpool 1931):

"After all, if there is no God, then God is in an incalculable way the
greatest creation of the human imagination...."

---

are developed in "Critique of Practical Reason" *and* "Foundation of the Metaphysics of Morals" *in particular. According to him, our conscience is infallible and instinctively makes the difference between good and evil. A follower of* G. **Leibniz** *in the first part of his life, he thought a lot about the possible links between science and metaphysics. Although educated by a pietistic protestant mother (a Lutheran sect), he was nonetheless an* agnostic.
[3] *See [Pa] pp. 242 & 781. & 278-424.*

Chapter 1.

Overview of the interplay between Science and Religion

*"For a long time, we thought that science was
going to chase away the religious function, that was a mistake."*
Hubert Reeves[4]

The interplay between science and faith has sparked countless debates with materialist thinkers claiming rationality as their stronghold. These are the same *"common sense advocates"* who cried foul when Albert **Einstein** an assistant examiner in Bern at the Swiss patent Office, proposed his Theory of Relativity, inviting us to review our relationship with time and space. However, reason and logic are not exclusive to science and philosophy; they have played pivotal roles in establishing the doctrines of monotheistic religions, particularly *Christianity*. In this journey of discovery, it is essential to acknowledge that Asia, has contributed significantly to shaping our understanding of these questions.

In Chapter 2 we will embark with the great scientists on a journey through the main steps that have led us to our current understanding of the Universe. It is important to note that our knowledge remains incomplete, yet it should not undermine the Faith of those who hold it. Nor, should it be used to establish faith solely on scientific foundations, as some have erroneously attempted with the emergence of the Big Bang theory (cf. chap.3).

We will see that, as far back as **the 7th century B.C.**, Hellenistic philosophers embarked on a quest to unravel the mysteries of nature, seeking mechanistic explanations instead of relying on poetic and mythological interpretations. These philosophers, known as *"the Ionians,"* delved into the enigmatic workings of the natural world while still acknowledging the existence of divinities. However, their

---

[4] **Reeves** Hubert *(Montreal, CAN 1932-2023 Paris, FRA). Astrophysicist and popularizer of science. In 1953, he was awarded a BSc degree in Physics from the University of Montreal, an MSc in 1956 on* "Formation of Positronium in Hydrogen and Helium."
In 1960, he was awarded a Ph.D. degree at Cornell University on "Thermonuclear Reaction Involving Medium Light Nuclei." *Then he taught until 1964 at the University of Montreal and was appointed as an advisor to NASA. Since 1965 he has been a Director Of Research at the CNRS (French National Centre for Scientific Research).* Author of numerous articles, he was honored by various Medals.

radical ideas were met with resistance by some Athenians, who accused them of denying gods' powers, and calling them disdainfully: *"the physicists."* This clash between philosophical inquiry and traditional religious beliefs highlights the contrasting perspectives between different cultures and religions and their approach to faith.

In parallel, the Persians embraced *Zoroastrianism,*[5] a monotheistic religion centered around Ahura **Mazda** as the supreme god. Meanwhile, India was already practicing *Buddhism* and *Brahminic Hinduism*, with *Hinduists* drawing wisdom from the Vedas, ancient Sanskrit texts. The Vedas dating back 5,000 years before our era, encompassed various branches of knowledge and became the foundation for the development of Vedāntasūtras,[6] revealing the essence of the Vedas.

The destruction of *"Milesian School"*[7] by the Persians in 494 B.C., and the emergence of the *"Eleatic School"*[8] paved the way for the flourishing of Greek scientific philosophy. Although the proposed solutions were rudimentary, these philosophers laid the groundwork for future philosophical masters, who would shape the history course.

At the beginning of the **4th century B.C**. philosophical systems were already advanced but with many unanswered questions. However, the decline of Hellenistic philosophy coincided with geopolitical shifts; as Rome's dominion expanded over Greek cities, leading to a transformation in philosophical themes. During the prosperous period of the **Roman Empire**, philosophy evolved toward materialism, skepticism, and mysticism, intertwining with *Judaism* and *Christianity* in Alexandria.

Neoplatonism emerged under the guidance of influential philosophers like **Philon of Alexandria**, contributing to the development of *Christian* philosophy.

---

[5] **Zoroaster**, the *Greek form of* **Zarathustra**, *a Persian prophet and poet, who probably lived around the 10th century BC. He is the founder of Zoroastrianism, a monotheistic religion, derived from Mazdeism (the official Iranian religion having* Ahura **Mazda** *as its main god, from the 7th century BC until the Arab invasion in 651 AD). Zoroastrianism was imposed by* **Hystapes**, *father of the Achaemenid king* **Darius I** *(Persia - 521 to - 486 BC), on his people.*

[6] The Vedanta-Sūtras *are constituted from Upanishads, a set of 7 philosophical texts, the basis of the Hindu religion, written in the poetic form at the end of the Vedic period between the 9th and 6th centuries B-C. They were transcribed for the first time by* **Srila Vyāsadeva** *the seventeenth incarnation of Krishna (or Bâdarâyana, a group of authors).*

[7] Milesian School *(VI-Vth century BC). Miletus, a city located on the coast of Anatolia (Asian part of Turkey), bordering the Ionian Sea, was renowned for its philosophers, called: "the* Physicists," *who advanced Geometry, Physics, Astronomy, and Cosmogony (mythical or scientific theory, explaining the formation of the universe).*

[8] The Eleatic School *was probably founded by* **Xenophanes of Colophon** *(ref. 36-10). Following the capture of Phocaea by* **Harpagus**, **Cyrus**'s *general, in the 6th century BC, some Phocaeansb settled in Elea, a small city of Campania (Southern Italy) on the Tyrrhenian coast, making it one of the most important Hellenistic philosophy centers.*

a Phocaea: now Foça in the southeast of Turkey on the Aegean Sea.

b Phocaeans: at the origin of the creation of Massalia (Marseille, 13 FR), Nikkea (Nice, 06 FR) ...

Early **Church Fathers**[9] and thinkers such as **Origen** and **Plotinus**, alongside **St Augustine of Hippo**, achieved a significant synthesis of philosophical and theological thought. Their contributions fostered the Hellenization of the Muslim empire during the **Middle Ages**, as Arabic spread from India to Spain, blending Indian, Sassanid,[10] Jewish, Christian, Berber ad Muslim economic challenges, the emergence of the Mongolian Empire, and a significant turning point occurring in Hispania with the *Reconquista*,[11] culminating in 1492 when Granada was captured by the "*Catholic Kings*," profound influence continued to shape Western thought until the 17th century.

The **Renaissance**, which began in the middle of the **15th century**, coming from Italy and spreading rapidly throughout Europe, brought about an artistic and cultural renovation. This movement challenged the predominant focus on God. Besides, with the advent of Johannes **Gutenberg**'s movable metal type press, allowing the printing of both sides of a sheet of paper, knowledge became more accessible, allowing the dissemination of biblical texts (the Bible was 1455 the 1st book printed by this process), Greek and Roman literature, and scientific discoveries. The diffusion of cartography allowed great explorers to discover the world. The shift of focus, from God to humanity sparked a separation between theology and philosophy, leading to more radical positions that dismantled the philosophical edifice built by patristic and medieval thinkers. However, for St **John Paul II** the twentieth century saw "*a revival of Thomistic philosophy.*"

The scientific revolution of the **20th century** revolutionized humanity's understanding of the Universe. Geniuses such as Galileo **Galilei** a convinced *Catholic*, Johannes **Kepler** a *Lutheran Protestant*, Nicolas **Copernicus**, a

---

[9] The term "Church Fathers" *appeared during the ecumenical council of Nicaea in 325, which brought together all Christian churches. It referred to the three hundred or so bishops who took part in it and evolved over the centuries. On Pp. 16-19 we will look at the founding fathers who, according to* Henri **Sonier de Lubac** *(ref. 69 - p.22), satisfy the four requirements of the Church: 1. saints, 2. having lived before the 8th century, Patristic period, which will be followed by medieval scholasticism (from the Greek word for idleness, free time), 3. whose work, free of doctrinal errors, allows for a rational defense (inspired by platonic thought) of the Christian faith, and 4. approved by the Church, which limits their number to four. Later, others such as* **St. Thomas Aquinas***, will become so. Orthodox Christians have another classification.*
Doctors of the Church: *36, including 4 women: personalities whose exceptional authority in the field of theology is recognized by the Church. All have been canonized.*

[10] Sassanides. *Iranian dynasty (founded by* **Ardashir I** *(IRN: Banvan 180-242) reigned over Great Iran (Iran, Iraq, Armenia, Caucasus, Dagestan, Afghanistan, Turkey, Persian Gulf...) from 224 until the invasion by Muslims in 651. Baghdad was then the city of culture and capital of one of the most powerful empires (empire of the Aryans = noble people initially) of western Asia.*

[11] Reconquista *(1st half of the 8th century to the end of the 15th century), the Middle-Age period during which Iberian Christian kingdoms tried to take back their territories from Muslims who, benefiting from the weakness of the Visigoths and the warm welcome they received from the Jews, who had been mistreated by the natives for more than a century, had invaded the south of the peninsula in 711. It started at the beginning of the invasion but, was not effective until the 12th century with the formation of chivalric and religious orders. It was interspersed with long periods of status quo.*

Catholic canon, René **Descartes** *a Catholic*, and Isaac **Newton** an *Anglican* played pivotal roles in this transformation. **Galilei**'s observations through the telescope challenged the geocentric model, while **Kepler**'s laws of heliocentric motion unveiled the harmony and mathematical precision underlying celestial bodies. **Descartes**' rationalism and **Newton**'s laws of motion provided a robust philosophical and scientific framework for understanding the physical world. These visionaries laid the foundation for the Enlightenment which advocated reason, and scientific inquiry.

The subsequent centuries witnessed remarkable advancements in physics, with Albert **Einstein**'s theories of relativity reshaping our understanding of time, space, and gravity; allowing scientists to reach the infinitely large (from a few km to the limits of the known universe, that is of the order of 10^{24} km),[12] the quantum mechanics delved into the peculiar nature of the real world, revealing the fundamental uncertainties and interconnectedness of the subatomic world allowing the interpretation of the infinitely small (dimensions smaller than 10 thousandths of a micron meter (1 μm)). One of its fathers, the 1932 Nobel Prize for Physics, Werner **Heisenberg**, a *Christian*, wrote [He2]:

> "*Modern atomic physics has pushed the natural sciences off the materialistic path they were on in the nineteenth century.*"

In recent years the detection of gravitational waves has provided further evidence of cosmic events that occurred billions (10^9) of years ago.

While Science has made tremendous progress in unraveling the mysteries of the universe, allowing us for example, to know that the history of the universe began 13.8 billion ($13.8.10^9$) years ago: "*instant 0*," and that the Big Bang model can explain scientifically the sequence of events that occurred since Planck's time τ_P (= 10^{-43} s),[13] to the present day, some fundamental questions still elude our grasp:

-     What preceded the Big Bang? Although I have great respect for **St. Augustine** I don't agree with his provocative answer:

---

[12] *The magnitudes involved can be very large (the universe) or very small (many parts of 1, the elements that make up matter), so we have adopted the mathematical representation, which is very simple.*
The number 1 followed by x zeros is written: 10^x. Examples: $1.000 = 10^3$, 1 million = 1.000.000 = 10^6, 1 billion = 1000.000.000 = 10^9. Thus, 10^{24} km = one million of billions of billions of km. When we talk about the atom, x is preceded by the sign -. Examples: 1 thousandth = 1/1.000 = 10^{-3}, 1 millionth = 1/1.000.000 = 10^{-6}, 1 billionth = 1/1000.000.000 = 10^{-9}. One micron meter (1 μm) is equal to a thousandth of an mm = 10^{-3} mm = 10^{-6} m. 1 millionth of a m = 10^{-6} m.
[13] *Planck time τ_P = 10^{-43}s, the time interval between the "instant 0" and the triggering of the Bang, below it our measurements and physical interpretations are currently meaningless. Some cosmologists even wonder if below Planck time the notions of time and length have a meaning. For some, continuity gives way to a discrete structure (cf. [Gr] p. 469). It is the smallest duration of time having a physical signification.*

"...before God created heaven and earth, he made nothing," [Aug1]

An answer to a question he asked himself [Aug], echoing the one he was often asked:

"...what did God do before he created heaven and earth"?

- What were the initial conditions of the universe? Here, the religious belief steps in, offering answers rooted in faith, spirituality, and the quest for transcendence. Religion continues to provide a source of solace, purpose, and moral guidance, addressing questions that science alone cannot answer.

We will meet some, for whom Faith and Science are inseparable. This is the case of the *Jewish* scientist, Rabbi Carl **Feit**:

"Religion and science are two ways of studying the world, and each guides our search for understanding."

As well as the 1964 Nobel Prize for Physics, the *Protestant* Ch. **Townes** professes that there are:

"convergence between science and religion," [To]

although, the former is based on rationality and the latter on revelation. He demonstrates that the gap between scientific intuition and revelation is narrow or even non-existent.

We cannot scientifically prove God's existence, nor that He is the Universe's Creator because God does enter in the scientific parameters which are only concerned with the matter. Also, I adhere to the following proposals of A. **Einstein**:

"...science can only affirm what is, but not what ought to be...Religion, on the other hand, deals only with the values of human thought and action."

This idea, that humanity needs both the *"wings"* of science and religion to progress and find balance, is consistent with *Baha'i* teachings, a collection of talks and writings of **Abdu'l-Baha** [14] the eldest son of Abdu'lBaha, [15] the founder and

---

[14] **Abdu'l-Baha** *(Tehran, Persia 1844-1921 Haïfa, Mandatory Palestine) born 'Abbas, Eldest son of Baha'u'llah, head of the Baha'i faith from 1892 to 1921.*

[15] **Baha'u'llah** *(Nur, Persia 1817-1892 Acre, Mandatory Palestine) born Mirza Husayn-Ali Nûri from a noble family. Adept in 1847 of Bâb, he was a defender of the poor, sick, and oppressed. In 1850 he was banned, imprisoned, and then exiled for much of his life due to his beliefs. In Irak, in 1863, he claimed to be the last prophet. Exiled for his beliefs he wrote the basis of the Baha'i teachings and was the founder of the Baha'i faith.*

central figure of *Baha'i faith*,[16] a distinct faith that emerged from *Islam* in the 19th century in Persia (now Iran), and is characterized by its belief in the unity of all religions, the oneness of humanity, and the promotion of world peace. According to it, **Baha'u'llah** is the latest *"messenger"* sent by God after Abraham, Moses, Jesus, Muhammad, and others, each bringing a message tailored to the needs of their time. Abdu'l-Baha wrote:[17]

> *"Religion and science are the two wings upon which man's intelligence can soar into the heights, with which the human soul can progress. It is not possible to fly with one wing alone."*

More recently, **St. John Paul II** had a similar point of view and stated in the preamble to his Encyclical *"Fides et Ratio"* [JP]:

> *"FAITH AND REASON are like the two wings that enable the human spirit to soar toward the contemplation of truth. It is God who has placed in the heart of man the desire to know the truth and, in the end, to know Himself so that, knowing and loving Him, he can attain the full truth about himself."*

These two last quotes underscore the idea, shared by other religious leaders, that science and religion can work together, rather than being in conflict.

---

[16] Baha'i faith. *This monotheist and Abrahamic religion emphasizes the fundamental unity of all religious teachings, viewing them as progressive stages in the spiritual evolution of humanity. It advocates for the establishment of world peace and the unity of all nations and peoples.*
[17] "Le Courage d'Aimer," Fascicule 4, p. 46, *is part of a collection of writings and talks of* **Baha'u'llah**.

Chapter 2.

The Minds that shaped our understanding of the universe

"The heavens tell of the glory of God;
the firmament publishes the work of his hands"
David Psalm 19:2

"By faith, we understand that the universe was ordered by the word of God,
so that what is visible originates from the invisible"
St Paul's Letter to the Hebrews 11: 3

"There can be no opposition between faith and science."
St. John Paul II[18]

Caution. In this chapter, I aim to distill the key scientific breakthroughs that, much like assembling pieces of a puzzle, have brought us to our current understanding of the Universe. While this understanding is far from complete, it should not undermine the Faith of those who hold it, nor should it be misconstrued as an attempt to validate faith through science, as some have mistakenly done with the advent of the Big Bang theory.

This chapter invites the average reader to appreciate the remarkable qualities inherent in the laws governing the Universe and the fundamental particles. Without Life as we know it, these qualities could not exist; they are revealed through scientific progress. Moreover, it highlights how scientific advancements can prompt profound religious inquiries, paving the way for a meaningful dialogue between science and religion.

Belief in a higher power has deep roots in human history, serving as an innate and compelling need for individuals who have continually pondered the purpose, of their existence, their origins, and what lies beyond this life. As the philosopher E. **Kant** [Ka2] once aptly noted:

---

[18] **John Paul II** *on March 30, 1979, receiving in the Vatican the members of the European Physical Society who had participated in a seminar, reaffirmed the researcher's freedom in his search for the understanding of the world.*

"Our reason has this peculiar fate, concerning one class of its knowledge, it is always troubled with questions which cannot be ignored because they spring from the very nature of reason, and which cannot be answered, because they transcend the powers of human reason."[19]

In the beginning, men believed in the existence of more or less important gods that they had to honor in order not to suffer their wrath, then, of a Trinity, or even of a single God. Thus, archaeologists have found[20] in the caves of Skhul and Qafzeh (in Israel) burials and funerary furniture dating back to the ancestors of Homo Sapiens and Neanderthals, that is to say, more than 100,000 years BC; similarly, rock paintings associated with shamanism (mediation between Man and the world of the spirits) dating back to -33,000 years BC have been discovered. Visiting the caves of Lascaux for example, or the Valley of Wonders closer to us, everyone can see that twenty to thirty thousand years ago our ancestors had an animistic vision of the universe. Nature, sun, moon, stars, vegetation, and animals were all part of a whole and had a soul. For some, everything was under the control of a *"Great Spirit"* whose kind was unknown. Hinduists had the Vedas (rejected by Bouddha), ancient Sanskrit texts dealing with all branches of knowledge. Vedas were transcribed for the first time, by **Srila Vyäsadeva** (or Bâdaräyana, a group of authors) 5,000 years before our era. They added aphorisms revealing the essence of the Vedas giving the Vedänta-sütras (ref.582, p.192).

The first traces of Greek mythology and Mycenean civilization go back to the 18th century before our era. The Universe and more particularly the Sky and the Earth, respectively assimilated to **Ouranos** (Saturn) the father and **Gaia** (Ge) the mother, would have generated the **Titans**, the first gods, monsters with superhuman forces. Following a titanic battle, the most powerful of them: **Cronos**, is sent to Tartarus[21] by his son **Zeus**, who becomes the king of the gods. Although stronger than all his brothers, Zeus nevertheless shares the responsibilities with them, reserving the sky for himself, the rain, and the clouds.[22] He thus gives his children particular missions.

The inhabitants of ancient Egypt (-3,150 to -50 B.C.), the union of Upper and Lower Egypt, respectively in the South and the North, adored gods who were eternal, like the soul, but did not believe in reincarnation. The cycles were very important: that of the Sun (24 hours), that of the great flood of the Nile (annual), and that of births and deaths. At that time, they shared the feeling that in the beginning there had been a *"noun,"* a great primordial ocean, of which the Nile was a part, at the origin of Life and then of Death, and from which, according to

---

[19] *Cf [Ka2] p.63, from the prolegomena of the 1st edition in 1781.*
[20] *Science et Vie N° 1055, 08.2005.*
[21] Tartarus: *a place where the greatest criminals are gathered.*
[22] Brothers of Zeus: **Poseidon** *the god of the seas was very important to the Greeks who were sailors,* **Hades** *his elder brother god of the underworld, and the dead, nicknamed* "master of hell."

the Heliopolitan myth, the demiurge [23] **Atum** emerged from himself, on the Benben (the mound of the primordial ocean), who shaped the world from the inert matter at his disposal. Atum is originally the Sun god, hence the name Re-Atum. First, of the **Ennead,** [24] he begat the first divine couple: **Tefnut** (the goddess of Waters) and **Shu** (the god of the Air). Their descendants: **Geb** (the god of the Earth) and **Nut** (the goddess of the Sky) begat: **Osiris** *"the King,"* initially the god of the plants whose role increased over time becoming the god of the dead, **Set** (the god of the Evil), Nephthys (the goddess of the dead) as well as Isis *"the Queen,"* nicknamed *"the great sorceress,"* to whom one lent more extensive powers than those of **ReAtum**. She was the wife of her brother **Osiris**. The Egyptologist, and *Dominican friar*, Richard **Beaud** [25] qualifies [Be] this period as henotheistic rather than polytheistic because, at that time the divine is ONE, it is hidden and manifests itself in various forms. It was only around 536 BC that a small group of Hebrews returning from exile discovered monotheism.

In 69 **Caligula** had a temple built on the capitol in honor of **Isis** who then became the universal goddess: *"the mother."* A short parenthesis opened in the middle of the 14th century B.C. under the influence of **Amenhotep IV** who, completing the work of his predecessor, imposed **Aten** as the only God of Egypt, assimilating him to the Sun, thus making him a descendant of **Atum-Re**. He took advantage of this to attribute to himself the name of **Akhenaten** (the brilliance of Aten). The high priests, generally astrologers, were the intermediaries between humans and the gods.

For the Chinese, the universe is then the result of the interaction between Yin (Earth) and Yang (Heaven). This holistic [26] vision with its alternating cycles was enough for them.

The Torah [27] *"Jewish Law"* - comprising 613 commandments compiled in the

---

[23] Demiurge*: [derived from dêmiourgos (people - creation)] the god, architect of the universe common to the Egyptians, the platonists, ...*

[24] Ennead*: group of 9 Egyptian deities:* **Atum** => **Tefnut** *and* **Shu** => **Geb** *and* **Nut** => **Osiris**, **Set**, **Isis** *&* **Nephthys** *wife of his brother* **Set**. *The latter, jealous of* **Osiris**, *killed him and threw his body into the Nile from where he was resurrected.*

[25] **Beaud** Richard *(Albeuve, CHE 1942). Egyptologist, he taught at the French Biblical and Archaeological School of Jerusalem, gave courses in Ptolemaic writing (used from the 4th to the 1st century BC by the Ptolemaic dynasty) at the Ecole Hebrew of Jerusalem. An archaeologist, he participated in excavations at Karnak; a professor of philosophy, he is a theologian and a member, among others, of the Nice section of the International European Academy of Sciences.* Dominican.

[26] Holistic or globalist*: the universe is an indivisible whole that cannot be explained by its different components.*

[27] The Torah *is initially the written teaching transmitted to* **Moses** *by God (Pentateuch). Nevertheless, the existence of an oral Torah transmitted by Moses from generation to generation is admitted by all. The writing of these accounts began at the time of the second destruction of the temple and gave rise to a text: the* Mishna *which, with the contributions of the rabbis of the second century AD, led to the* Talmud. *The whole Talmud + Torah is the basis of the* Halakha *(Jewish Law).*

Pentateuch[28] - initially attributed to **Moses**,[29] the second *"man of God"* (the first being Adam), includes Genesis teaching that God, infinite power, the perfectness form of existence, Creator of the Earth and Life, is eternal and elusive to human reason, is then law for the Hebrews. This may explain why they did not seek, in the beginning, to know scientifically the origins of the world and left the field open to the Greek, Ionian, Milesian, and Eleatic scholars who were at the crossroads of the known world. In the same way, the Septuagint[30] will be sufficient for the *Christians, Jews,* or converted *Samaritans,*[31] of the 1st century.

---

[28] Pentateuch. *Name given by the Christians to the whole of the 5 books:* Genesis *(1st written account of the creation of the world, without scientific basis)*, Exodus *(exit from Egypt)*, Leviticus *(law of the priests of the tribe of Levi)*, Numbers *(census)* and Deuteronomy *(the second law). If the first draft can be attributed to* **Moses** *following the revelations made to him by God at Mount Sinai (Ex 19-40), the definitive version has been questioned by specialists such as Rabbi M.* **Maimonides** *(ref.114-p.33), known as the "second Moses," the Jewish philosopher B.* **Spinoza** *(ref.172-p.49), and the Protestant theologian Andreas* **Boderstein** *[1486-1541]) who have observed additions. It is attributed to* **Ezra** *[5th century B.C., Levite descendant of* **Aaron,** *brother of* **Moses***].*

[29] **Moses**, *birth situated between the 16th and 12th centuries BC, according to the authors. Some scholars refute the authorship of Moses in the Pentateuch. The official Christian doctrine attributes to Moses the initial writing which would have undergone modifications and additions at various times, which would explain certain duplications. The Decalogue (the 10 commandments) was enunciated to him by God, in Egyptian, on Mount Sinai.*

[30] Septuagint. *The Greek translation of the Hebrew Bible was made in the 3rd century BC by 72 Jewish scholars (6 from each of the 12 tribes of Israel). It was a commission from* **Ptolemy II** *[Isle of Cos -309/-308 to -246, king of Egypt who was the first Greek to be crowned pharaoh and to proclaim himself god]. Two legends have been retained: 1. that of the Letter of Aristaeus* * *to* Philocrates, *according to which these scholars would have been in charge of translating together as philologists. 2. That of* **Philo of Alexandria** *(ref. 64-p.21), according to whom the translators worked separately. The similarity of the 72 translations, done separately, would have aroused the wonder of the readers and led them to think that, God inspired the translators. It was very well received by the Christians who translated it into Syriac, Coptic, Armenian, Latin, etc., as well as by the Hellenistic Jews of the time. On the other hand, the rabbis of the third century of our era found it too messianic. Slight discrepancies between an initial text and its translation can appear because there is always a part of interpretation due to the translator. For example, in the Jewish text, Mary (Christ's mother) is qualified as "a young girl" whereas for us Christians as "the Virgin." Note that this Greek translation is in agreement with some of the "Dead Sea Scrolls" discovered in 1947 in caves near Qumram (northwest of the Dead Sea).* * Aristaeus *is the pseudonym given by* **Flavius Josephus** *([Jerusalem 37/38 to 100 Roma], Roman historiographer of* Jewish *origin) to an unknown Jew.*

[31] Samaritan: *Following the death of Solomon in 930 BC, the Kingdom of Israel unified by David split into two kingdoms. The kingdom of Israel in the north with its capital "Samaria" (currently "Naplouse") and the kingdom of Juda has Jerusalem as its capital. They are considered a sect by Jews because they reject all the books of the Bible after the Pentateuch and in particular the oral tradition that became the Torah.*

2.1. Axial Era: From the dawn of the 8th century to the beginning of the 2nd century BC.

"This, in practice, is the faith of all physicists: the deeper we search, the greater our wonder, the greater our wonder at what we see."

Sura 67, 3-4. Quoted by A. Salam on receiving the Nobel Prize in Physics.

This period, as defined by the *Christian existentialist*[32] philosopher Karl **Jasper**[33] [Ja], seemed interesting to me because it is the temporal crossroads where humanity questioned the myths and customs, ... that gave rhythm to its life and gave rise to spirituality and the main religions that we know:

-   Judaism in Palestine with the prophets Joshua, Samuel, Elijah, Isaiah, Jeremiah, Ezekiel, ... leading to the Bible,

-   Zoroastrianism in Persia,

-   Buddhism and Brahminic Hinduism in India with the Upanishads (ref.6-p.2), - Confucianism in China.

Considering the purpose of this essay, the pre-Socratic period, although pagan, is interesting because it gave birth to Greek scientific philosophy. The Ionians, who were at the *"center"* of the world, traveled a lot and were open to the ideas of other peoples, especially the Egyptians, whom they liked to visit to meet the scientists of the time, who were none other than the high priests. The first ones searched for the primordial substance[34] that would explain the laws of Nature and free man from false superstitions. Thus, they had successively as *"Arche"* or *"Principle"*:

---

[32] Christian Existentialism. *According to "Le Robert": existentialism emphasizes the philosophical importance of individual existence, with its irreducible characteristics. According to the same source, for the **1957 Nobel Prize** in Literature,* Albert **Camus** *"Existentialism has 2 forms: one with* Soren **Kierkegaard**, Karl **Jasper**... *leads into the divinity by the criticism of reason, the other atheist existentialism ... with* Edmund **Husserl**, Jean-Paul **Sartre**...

[33] **Jasper** Karl *(Oldenburg, DE 1883-1969 Basel, CH), psychiatrist and philosopher. He studied law and defended his doctoral thesis in medicine in Heidelberg in 1909, then in 1913 a thesis on* "General Psychopathology." *In 1919, he became a lecturer, and in 1921 he was appointed to the chair of philosophy in Heidelberg. Married to a Jewish woman, the Nazi regime deprived him of his functions, which he did not regain until 1945. In 1948 he succeeded* Friedrich **Nietzche** *in Basel and became a Swiss citizen in 1968.*

[34] Substance *did not only have the material meaning we attribute to it today; it was the carrier of its own qualities; and life was inherent to it. Its indestructible character leads us to attribute to it, in the present language, the two forms: matter and energy. This does not allow us to compare these Egyptian philosophers and their Ionian followers, as some do, to the particle physicists of the 20th century.*

-   Water for **Thales of Milet.**[35] Mathematician, he was trained in geometry in Egypt (we all studied in high school the *"theorem of Thales"* from which we can deduce, for example, the height of an object according to the length of its shadow, ...). When he wrote: *"everything is full of gods"* he went further than his contemporaries and assumed the presence of gods in certain inanimate objects. According to **Aristotle**, Thales was the first of the Ionian philosophers to believe that the universe was a big air bubble in the middle of an ocean of water. Water is not only at the origin of Life but is essential to it, it is found in all food.

-   The Apeiron, for **Anaximander,**[36] the initiator of meteorology, and a **Thales'** disciple. This indefinite substance could be assimilated to the ether, which had been conveyed until A. **Einstein**. Apeiron was [Bt] for its designer *"immortal, imperishable and engendered,"* all the divine attributes of the time.

-   Fire for **Heraclitus of Ephesus,**[37] symbolized constant change. Fire is omnipresent (like the Sun), it was the divine symbol of Zoroastrianism, the first known monotheistic religion (ref.5-p.2). Heraclitus' principle is that:

"What exists is not being but becoming: there is the only reality in change. The real is only the being, it is absolute, the One, the Truth, the Goodness...."

Plato admired his philosophy. On the other hand, **Aristotle** attacked it because too much based on the field of the senses.

-   The Air for **Anaximenes.**[38] It is also essential to Life, and it can be transformed into water by condensation. Let's note that **Diogenes of Sinope**[39] assimilated God (Zeus) into the Air.

---

[35] **Thales** *(Milet-Greece -625 to -547 BC). Initiator of the Greek philosophy. Mathematician (especially in geometry), Physicist, Astronomer, and Politician. His method based on observation and demonstration is the basis of scientific reasoning.*
[36] **Anaximander** *(Milet -610 to < -546 BC). Philosopher and scholar, he did not mix the divine and philosophy.*
[37] **Heraclitus** *(Ephesus, TR -576 to - 480 BC) nicknamed* "the obscure." *Philosopher, he tries to develop a non-conformist concept of the divine, attacking the cult of the images, the rites in use as that of the purification by the blood of those who are contaminated by blood. He compares this ritual to washing with mud a body covered in mud, it would be insane.*
[38] **Anaximenes** *(Milet -550 to -480 BC), Philosopher, disciple of Anaximander.*
[39] **Diogenes of Sinope** *(Sinope, TUR - 413 to - 327 Corinth, GRC). Philosopher representing the Cynic school. Was accused of atheism by his contemporaries.*

- Fire and Earth for the **Eleate Parmenides,**[40] a pupil of **Xenophanes of Colophon.**[41] It remained of him a famous work: *"Of the Nature"* which treats the eternity of the being. With **Zeno of Elea**,[42] his pupil, they made the distinction between the world of the senses which is subjective, material, thus degradable, and the Intelligible world, of *"the universal Being,"* that of the Spirit, which is eternal, immutable, revealed by the Reason.

- Then, the 4 elements: Water, Air, Fire, and Earth were gathered by **Empedocles of Agrigento**.[43] Let's note that at this time the Hindu currents directed by Brahmanas,[44] had arrived at the same conclusion. He believed in the metempsychosis of the souls, in the expectation of a better future but, did not grant the gods any power in the organization and the conduct of the world [Bt].

The atom[45] concept, an unbreakable particle, was probably introduced by **Leucippus**[46] and his disciple **Democritus;**[47] nevertheless, it will be necessary to

---

[40] **Parmenides of Elea** *(Elea, GRC -530 to -444), Pythagorean philosopher and then Eliotian, introduced logic, which he opposed to experience, into Greek philosophy. Because, according to him, it was the only way to access the truth. He developed Pythagorean physics, affirming that the Earth is spherical and at the center of the universe.*

[41] **Xenophanes of Colophon** *(Colophon, Ionia -570 to -475 Elea or Colophon), Ionian storyteller. Polymath according to Heraclitus, a supporter of an infinite flat Earth, all things are eternal according to him. The theologian rejected all the divination of his time and doubted man's ability to know the world's reality because sensations vary from one being to another and according to situations.*

[42] **Zeno of Elea** *(Elea -490 to -430 BC). Parmenides' disciple, was according to Socrates a skilled dialectician.*

[43] **Empedocles** *(Agrigento, GRC~-495 to~-435 BC). Poet, Philosopher, Scholar, and a Mystic, tried to make a synthesis of the ideas of the time on the* "Nature of the universe." *His gods were not immortal but endowed with greater longevity.*

[44] Brahmanas**,** *Hindu sages, and scholars recognizing religious authorities were followers of the Upanishads (ref.6-p.3).*

[45] **Pullman** Bernard *(Wloclawek, POL. 1919-1996 Paris, FRA). Chemist and quantum biochemist. Professor at the Sorbonne, Paris, director of the laboratory of the Institute of Physical and Chemical Biology. Co-founder, and president of the International Academy of Quantum Molecular Sciences. Created in Nice in 1990, together with J-Pierre **Changeux***, *neurobiologist, Professor at the Collège de France, and Dan Vasilescu, Professor in Biophysics at the University of Nice, the International Center of Reflection on Molecular Biophysics and Biology, to which they honored me by associating me as Treasurer. This center was sponsored by the Global Network of Cell and Molecular Biology of UNESCO. B. Pullman wrote a very interesting and well-documented book:* "The Atom: in the History of Thought" [Pu], *in which these aspects are very well described.*

[46] **Leucippus** *(Miletus -460 to -370). Some authors doubt the genus or even his existence.*

[47] **Democritus** *(of Abdera-Thrace -460 to –370°) and* **Leucippus** *are considered co-inventors of the atom concept. They exclude the intervention of gods in the creation of the universe and admit two principles: the vacuum and the corpuscles to which they attribute equal importance to that of the atoms: elementary corpuscles of the matter, unbreakable, compact, full,*

wait 22 centuries so that the first modern atomic theory is worked out by John **Dalton** (1801). For Leucippus, the Earth was at the center of the world. Democritus sought, among other things, "*the origin of the idea of gods*" [Bt] and already envisaged the possibility that there were several universes, an idea taken up by **Descartes** and then at the end of the 20th century (cf. § 4.2, pp. 183-186).

Let's note that among the 6 great philosophical currents [48] coming from Brahmanism, two: the *Vaisésika*[49] and the *Nyaya*[50] were the followers of 9 basic substances, 4 of which were common to the 4 elements of Empedocles. These 9 substances have the properties of the atom; namely, eternity, uniqueness, and great number, B. **Pullman** [Pu] calls them "*Indian atoms.*" For the followers of this religion, God *omniscient*, is one of the 2 variants of the soul which is one of the 5 other basic substances[51] that they recognize. Among the two great currents of Buddhism (the Hinayana or Small Vehicle, and the Mahayana or Great Vehicle) which were born in Nepal, only the Hinayana defended the atomic theory which is somewhat similar to the Hindu theory in that it distinguishes four different types of atoms, associated with the four elements of **Empedocles** (without having had any knowledge of them). Let's note that Buddha rejected the *Upanishads.*

Philolaos of Crotone.[52] Scientist, he is the first of the Greek philosophers to affirm that the Earth is not immobile but turns around a "*central fire*" (center of the world) distinct from the Sun, fire that he calls: "*Hestia.*" Twenty centuries later, Copernicus would admit having been influenced by his work!

---

indestructible, unalterable, in infinite number, animated of an eternal and constant movement. Their atoms are differentiated by their shape, their arrangement, and their position. When they collide, they group together, forming compounds that can be found everywhere in the cosmos; in the matter as well as in the soul (the atoms contained therein are however particularly fine, round, and therefore easily mobile) which is perishable. The senses (smell, color, ...) are purely subjective. For **Democritus,** *there is an infinite number of universes, an idea that some cosmologists share (cf. chap. 4.).*

[48] The six main currents of Brahmanism form six pairs: *the Nyaya and the Vaisheshika, the Vedanta and the Mimamsa, and the Samkhya and the Yoga.*

[49] the Vaisesika or Vaiheshika-sutra, the *founding text of the Vaisésika (one of the 6 orthodox Hindu philosophical schools) was written between the 2nd and the 1st century B.C. by* **Kanada**, *an Indian philosopher and scientist.*

[50] The Nyaya - sutra *was written mainly by* Akshapada **Gautama** *and other collaborators including scientists between the 6th century BC and the 2nd century AD.*

[51] The other substances are ether, space, time and, manas (organs of thought, of atomic nature; each manas being associated with a soul). The soul is divided into two categories: that of God and the other, grouping all the souls associated with living beings. *The properties of the 4 types of atoms are respectively solidity, viscosity, movement, and heat for those associated with earth, water, air, and fire.*

[52] **Philolaos of Crotone** (Metaponto, ITA. -470 to - 390 Thebes), Philosopher and politician. Mathematician, he was interested in the works of Pythagoras and in the theory of numbers in particular. Astronomer, he evaluated among other things, the solar year to 365,25J. He was interested in natural sciences and medicine. According to I. **Newton** ([Ne2], T2 p.2), the idea that the Earth was not the center of the World goes back to the Babylonians, to Pythagoras and his followers.

2.2. The Masters [Th]

*"The fundamental creative principle is found in mathematics. Therefore,
in a certain sense, I consider it true and possible for pure thought to
apprehend reality, as the ancients revered."*
Albert Einstein ([Ei1] p. 134)

Socrates[53] is an icon of philosophers. His intellectual journey in Physics led
him to reject it, as well as the physicists. He said:

*"... their physical science is impossible because the field of the world
exceeds us and its study generates only contradictions; it is even
impious, the gods having reserved this field to themselves: if it were
possible moreover it would be useless to make the men good and happy."*

Deeply troubled by the moral decay within Athenian society, Socrates
attributed the blame to the sophists and the physicists who:

*"... had applied themselves entirely to the study of Nature and, moved by
a truly scientific spirit, had tried to reduce the moving multiplicity of
things to the stable unity of a principle."*

Yet, Socrates, demonstrating his merits, applied a scientific methodology to
his philosophical quest for Wisdom. Despite his disdain for prevailing immorality,
he embarked on a noble mission: the moral transformation of his fellow citizens.
His unique approach involved direct dialogues, incorporating an *"ironic"* phase
where he feigned ignorance to prompt his interlocutors to unveil their
understanding. The *"maieutic"* aspect, reminiscent of his mother's midwifery and
father's medical background, engaged participants into introspective
contemplation, exposing their internal contradictions.

While his condemnation of impiety met resistance, Socrates emerged as one
of the most religious philosophers of his era. He proposed that the multitude of
gods revered by his fellow citizens served a singular, imperceptible deity, not a
creator, but an organizer of the universe. This divine force, encompassing the
entire world, desired the well-being of its creations. Socrates urged veneration and
beseeched humanity to seek guidance from this benevolent force to reach virtue
and goodness.

In the captivating world of **Plato's**[54] intellectual exploration, the realms of

---

[53] **Socrates** *(Alopèce, near Athens about -469 to -399 Athens). Condemned to death for impiety
(not respecting the official deities, which many, including Plato, contested) asked to be able to
swallow hemlock to die with more dignity. The spiritual master of **Plato** among others, he is
described as physically courageous (he participated, as an infantryman, in three campaigns
during the Peloponnesian wars) and politically, which certainly earned him some enmity.*
[54] **Plato** *(Athens -428/-427 to -348 B.C.), whose real name was **Aristocles***, is said to be
descended by his father from **Codros**, the last king of Athens according to Greek mythology, and*

Physics and Astronomy intertwine with his profound philosophical musings. Plato a luminary of ancient thought, delved into the celestial mysteries with a fervor for meteors and a deep connection to **Empedocles of Agrigento's** four elements. Elevating his cosmology to new heights, Plato introduced a fifth element, sculpted by the hands of the demiurge, a mathematician of divine proportions. Plato, a devotee of **Pythagoras,** [55] the trailblazer who unveiled the Earth's spherical nature, drew inspiration from Pythagoras' geometric marvels. Imagine, Plato's Academy adorned with the inscription:

"Let no one enter here who is not a geometer,"

a testament to his awe for the timeless dance between numbers and geometric wonders.

In his magnum opus: the Timaeus[Pla], Plato orchestrates a symphony of intellect through the voices of Socrates, Timaeus, Critias, and Hermocrates. Their dialogue weaves a tapestry of Plato's scientific and political vision, anchored by five elements embodying the essence of matter.

Each of his 5 elements, which would be the smallest parts of the matter, [56] corresponds to one of five regular polyhedra whose faces are identical, regular polygons:

- **Fire**, the fiery essence, symbolizing vision, is pictured by a tetrahedron:

---

by his mother from **Solon** *(statesman, poet, Athenian 6th century B.C.). Student of* **Socrates**, *he is the founder of the modern Philosophy centered on the Dialectic, relying on the Dialogue that he was the first to introduce. Like* **Parmenides** *(ref.40-p.13) before him, he practiced ontology (discipline related to the questioning of being, what it is); on the other hand, he did not share his idea of Being, which he replaced by a myriad of Ideas (ref.58–p.17), that F-J. Thonnard assimilates to intelligible atoms. B.* **Pullman** *wrote [Pu (p.74)] that Plato shows* "a thought with a definite atomic character" *while noting that he does not have the same conception as* **Democritus** *and* **Epicurus**. *For him, God and matter are* eternal, in the same way, that mathematical truths are eternal. Astronomer, he was interested in the meteors. According to B. Pullman [Pu], all his cosmology is based on the theory of **Empedocles**. He influenced **Aristotle**, the **Church Fathers**, and the following generations. Would have been nicknamed Plato because of his large forehead. As for **Socrates**, it is very difficult to define his relationship to religiosity. * Not to be confused with **Aristocles**, a peripatetic of the 2nd century B.C., born in Messina, nor with the tutor of **Septimius the Severus**.
[55] **Pythagoras** *(Island of Samos, Aegean Sea -580 to - 495 Metapontum-south of Italy, before J-C). Philosopher, and mathematician (arithmetic), no one is unaware of his famous theorem on the relationship between the sides of a right-angled triangle. For him everything was a number, these are the principles of nature. He established a correspondence between geometric figures and numbers (1 is the point, 2 is the line, 3 is the surface (triangle), and 4 is the three-dimensional object (pyramid), ...). Was according to* **Hegel** *(ref. 154-p.44) the first universal Master. An accomplished athlete, he participated in the 57th Olympic Games where he would have won all the pugilistic competitions. Very interested in religious questions he is the author of the transmigration of the souls. Founded his school (philosophical, scientific, political, religious) in Crotone in -532.*
[56] *He does not speak about atoms but does not close the door to such a possibility.*

a pyramid with four identical faces in the form of equilateral triangles,

- **Earth**, grounded in materiality, takes the form of a cube with 6 identical square faces; each consisting of 4 half-equilateral triangles.

- **Air** dances in the octahedron, 8 identical faces, similar to those of fire,

- **Water** flows gracefully in the icosahedron: 20 identical faces, similar to the previous ones,

- Its **fifth** element is the dodecahedron, with 12 identical pentagonal faces, also made up of the basic elements of Pythagoras, mirroring the universe's foundation, an animated and organic whole, of spherical form.

Plato, the philosopher-pioneer, championed the Dialectic and metempsychosis, embracing the concept that learning is akin to remembering:

"Learning by reason is more like remembering previous knowledge than discovering entirely new truths."

He envisioned an ideal world beyond the senses and championed a Craftsman benevolent to all. Stating the timeless existence, Plato unfolded the canvas of eudemonism.[57]

Shaped by the unjust trial of his mentor Socrates, Plato's quest crystallizes in defining the conditions of a just state.

Plato was:

- *deist* by some sides: believes in an eternal God, the Demiurge; efficient cause of the world who took hold of the five elements, eternal too, and shaped them (did not create them) in the most beautiful and best possible way, thus giving them a precise, particular form. But:

- *Pagan*, because kept all the gods of Greek mythology that were subordinate to God, making them faithful servants, emanations of the Royal Soul.

Nevertheless, he rejected from mythology all that seemed to him absurd in particular the *"lying fables"* of **Homer**, while recognizing his ignorance of the nature of the gods.

François-Joseph **Thonnard** [Th], a philosopher par excellence, ordained an *Augustinian* priest in 1922, encapsulated Plato's essence:

"... his philosophy is dominated and unified by the theory of IDEAS:[58]

---

[57] Eudemonism: *doctrine according to which happiness is the goal of life.*
[58] IDEA: *essence (intimate nature of things) eternal and purely intelligible of sensible things. Some ideas are compatible with each other (snow is compatible with whiteness but not with heat); they possess 4 properties: Spirituality (invisible to the eye, they are perceptible only by*

the proper object of our science (dialectics) is the world of Ideas of which the sensible world is only the shadow or the copy."

Aristotle,[59] the Master Mind of Metaphysics, often hailed as the "*Prince of philosophers*," was a towering intellect whose influence on the world of thought remains immeasurable. While physics was the science of nature's universal causes in his eyes, it paled in comparison to the grandeur of his magnum opus, /M*Metaphysics*. We will explore Aristotle's world, where he delved deep into metaphysical realms and expanded our understanding of the natural world.

a)   Eudonism and Beyond: Aristotle a fervent follower of eudemonism, pursued the ultimate human good and happiness. Though he was a student of **Plato,** he did not share his Master's ideas. Instead, Aristotle charted his unique path in philosophy, making a profound impact that still resonates today.

b)   The Four Elements and the Fifth Essence (fig. 1). Aristotle's elemental theories, based on the four "*qualities*" two active (hot and cold) and two passive (dry and humid), gave rise to the four classical elements: earth, water, air, and fire. To these, he introduced the enigmatic fifth element, the "*ether*," contributing to a richer understanding of the cosmos.

c)   Convinced Anti-Atomist. In a world filled with countless particles, Aristotle rejected both the concept of the Atom and that of the Vacuum, believing it would hinder all forms of motion. Instead, he introduced the notion of *Ylem*, the fundamental substance from which all matter originates, containing infinite potential within itself. Today, physicists of particles call it: "*the string*" cf. § 4.2.1.

---

the Intelligence), *Reality (endowed with life and thought, constituting the real world), Immutability (eternal) and Purity (perfect and distinct from each other). See [Th pp. 51 & 52].*
Thonnard *François-Joseph (BE: Barvaud-sur-Ourthe 1896 - 1974 Saint-Gérard),* Catholic priest, *professor of Philosophy. Novice in 1913 he studies Philosophy from 1915 to 1917, Theology until 1922.*
[59] **Aristotle** *called "the Stagira," because born in - 384 in this city of Greece called today Stavros. He died in 322 BC in Chalcis (Euboea, a Greek island in the Aegean Sea) where he had taken refuge after the death of* **Alexander the Great** *because, thinking and speaking too much, according to some members of the Free Republic of Athens, he had been accused of impiety and felt threatened. In addition to philosophy, he was interested in letters, poetry, arts, sciences, metaphysics, and politics. Real encyclopedia he founded, in -335, the Lyceum in Athens. His observations led him to believe that the earth was round, contrary to the first Greek philosophers, and contemporaries of Thales, for whom it was flat. See his treatise "On Heaven" which is in itself a true cosmology. Just like* **Plato**, *his Master, he thought that everything that moved was endowed with life, which made them assimilate the stars into gods. For him, all the laws governing the world of his time could be deduced from our only reflection, observation was useless.*

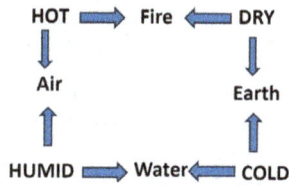

Fig.1. Elements and qualities

d) A Different Universe: Aristotle, drawing on the astronomical knowledge of his time, defended the *"geocentric"* model where the Earth (round), stood still at the center; while the Sun, the 5 Planets he had observed (Mercury, Venus, Mars, Jupiter, and Saturn), the Moon and the stars turn around it. This was a radical stance, five centuries ahead of **Ptolemy** (ref.68-p.17), reflecting his deep-rooted beliefs and interpretation of ancient texts, like Genesis.

e) The First Motor: In Aristotle's world, rest was the natural state of all things. He postulated the existence of:

"a first, motor, immobile, pure (immaterial) Force and Thought, unalterable, omniscient, omnipotent, eternal" [Ro1],

Responsible for setting the world into motion. This concept would later inspire many philosophical and theological discussions.

f) A Universe without a Creator: Aristotle challenged the prevailing idea of a god as a creator. He rejected the notion of a temporal beginning of the world, envisioning a universe that had always existed. His ideas had profound implications for the understanding of time[60] and matter.

g) The Unity of Body and Soul: In Aristotle's philosophy, the soul was the essence of all living beings. He posited that human souls possessed not only the faculties of plants and animals but also the intellectual capacity for reasoning, distinguishing us as unique creatures.

h) Dynamics and the Four Causes: Aristotle's *"Physics"* (eight books) delves into the dynamics of the natural world, guided by his theory of the four causes. He explored the natural places of the elements and made distinctions between natural and violent motion, providing valuable insights into the concept of motion.

---

[60] *This notion of* time *is still debated within the scientific community (which distinguishes the course of time (which is familiar to us) from the arrow of time (cf. the work of* **Sir A. Eddington** [Ed]*) linked to the causality and irreversibility of macroscopic phenomena. With the advent of cosmology (cf. [Gr], [Pr]...) thermodynamics intervenes at the same time as the Theories of General Relativity and Quantum Physics, making appear a simultaneous creation of entropy and entropic time.*

i)   The Legacy of Impetus Theory: to address specific cases of motion, Aristotle introduced "*Impetus theory*," which evolved over centuries, finding its way into the works of philosophers and scientists like Philoponus (490-570 A-C), and **Avicenna** (ref.101, p.30) This theory eventually made room for **Newton**'s groundbreaking first law of motion, also named "*Inertia law*," solidifying the framework for understanding how objects behave in the absence of forces.

In conclusion, Aristotle's profound philosophical insights and scientific inquiries continue to inspire and shape our understanding of the world. His pursuit of metaphysical truths, his theories on the elements, and his unique perspectives on motion and the cosmos ensure that he remains a pivotal figure in the history of human thought.

Epicurus,[61] like **Democritus**, thought that only the concepts of vacuum and atoms - particles: unbreakable, impenetrable, with different dimensions, sizes, and weights - could explain the real world; excluding any causality from the influence of gods and exempting man from revering their wisdom. The atomists of that time shared the idea of an uncreated, eternal world and time, the fruit of chance. He was one of the few (**Plato** was one of them) who wondered about the possibility of other living beings, plants, existing in other worlds (§.IV.3. p.186).

Aristarchus of Samos.[62] Astronomer, he is one of the first to have had the idea that the Earth rotates on itself and around the Sun, which made little noise at the time because it was contrary to the Aristotelian ideas as well as to that of the great mathematician and astronomer **Hipparchus.**[63] He also developed methods of calculating the distances between the Earth and the Sun and between the Earth and the Moon by taking the diameter of the Moon for a unit.

Eratosthenes,[64] the inventor of Geography, imagined two systems of projection of the geographical area known at the time and compiled all available data. His work, rediscovered in the West in the 15th century, revived scientific geography.

---

[61] **Epicurus** *(Samos -341 to -270 Athens). Philosopher and astronomer, interested in Physics* ("letter to Herodotus") *and meteors* ("letter to Pythocles") *founded various philosophical schools. He is at the origin of Epicureanism, a doctrine misunderstood by many. For him, pleasure is the beginning and the end of a happy life. By pleasure he meant knowing how to be satisfied with little, leading a simple, wise, almost monastic life, based on friendship; far from the debauched life that some people attribute to him.*

[62] **Aristarchus of** *(Samos -310 to -230), mathematician, astronomer.*

[63] **Hipparchus** *(Nicaea, TUR -190 to -120 Rhodes, Dodecanese). Astronomer, Mathematician, and Geographer, he established the foundations of the theory of the epicycles and the bases of the trigonometric tables. The origins of trigonometry date back to 4,000 years before our era.*

[64] **Eratosthenes** *(Cyrene, LBY -276 to -194 BC Alexandria, EGY). Mathematician, astronomer, historian, writer, musicologist, and geographer, he was the first to calculate, with a precision of 1%, the circumference of the Earth. Director General of the Library of Alexandria.*

Philo of Alexandria,[65] a *Jewish philosopher*, sketches Neoplatonism linking Aristotelian philosophy to biblical works. By using, even abusing, allegorical interpretations, he concludes that there is no discrepancy between **Plato**'s idea of God, the creator of the universe, and the one stated by the Bible, which is surprising because, for Plato, the matter was eternal. According to **Philo,** God by his transcendence, cannot be known by non-philosophers; the angels are his intermediaries with men, and more particularly the Logos, the divine Word, situated in **Plato**'s ideal world, which he places on a pedestal. Plato did a great work of rereading the Septuagint and wrote 34 treatises that would inspire all the *Church Fathers.*

2.3. Birth of Jesus Christ: Year 0 of the Gregorian calendar[66] of the so-called *"Christian Era."*

A revolution began, as Jean-Jacques **Rousseau** would point out [Ru]: **Jesus** established a spiritual kingdom on earth, separating the theological and political systems, which was not the case with the pagans, where each people had its gods, or even with **Moses**, who proposed the God of Israel.

Christianity, in line with Judaism, adopted the Old Testament and particularly Genesis, leading the official Christian philosophy to reject, for several centuries, **Aristarchus'** heliocentrism.

St. Paul the **Apostle**[67] was one of the first Christians to give a rational basis to his statements when he argued notably in Athens (Acts 17: 16-23), with Stoics and Epicureans.

Claudius **Ptolemy,**[68] is at the origin:

-   of the *"geocentric"* model, an error that prevailed for 14 centuries (until

---

[65] **Philo of** *(Alexandria, EGY -20 BC to 45 AD). Hellenized* Jewish *philosopher. He is one of the representatives of the Jews to the Romans. Had little influence on the Jewish tradition.*

[66] The Gregorian calendar, *which accompanies us since 1582, responds to a request from* **Pope Gregory XIII** *(Bologna 1502-1585 Rome, ITA) to correct the drift of 10 days from the Sun of the Julian calendar established by* Julius **Caesar** *(Rome -100 to -44 BC,) in 46 BC or 708 years after the founding of Rome.*
It is **Christophorus Clavius** *(Bamberg 1538-1612 Rome), known as* "the Euclid of the 16th century," *who was in charge of it.* German Jesuit, *astronomer (a crater of the Moon was named Clavius in his honor), mathematician (translated and completed the Latin version of the* "Elements of Euclid."

[67] **St Paul the Apostle** *(born in Tarsus, TUR around the year 0, executed in Rome in 65). A* roman Jew *from the tribe of* **Benjamin***, he spoke Aramaic, Greek, and Hebrew. He persecuted the Christians until his conversion around 33 following a* "mystical" *encounter with Christ on the road to Damascus. From there he converted many pagans around the Mediterranean.*

[68] **Ptolemy** Claudius *(Upper Egypt around 90 - 168 Canopy, EGY). Greek Geographer, Astronomer, and Astrologer, author of numerous works including the* "Almagest," *meaning* "the greatest," *which was the encyclopedia of mathematical and astronomical knowledge from the 2nd century A.D. until the publication (discreetly for fear of reprisals from the Church) of the heliocentric theory of* **Copernicus** *16th century.*

Copernicus); it made **I. Newton** say ([Ne2]-T.2-Intro. §III): *"There is a great appearance that the authority of* **Aristotle***, which was almost the only rule of authority at the time of* **Ptolemy***, is what led this great astronomer into error...."* In this model, the Earth, spherical, is at the center of 8 spheres carrying the Moon, the Sun, the 5 known Planets traveling epicycles [69] around their spheres, and the 8th: the *"celestial vault"* limit of the system, housing the stars. This complicated model according to I. **Newton**, who was quite critical of **Ptolemy** and especially of the Aristotelian influence, made him write ([Ne2], T.2, p. 2):

"...there is nothing so difficult as to put an error in the place of truth."

-       of the *"Geography,"* a treatise on cartography, compiling geographical knowledge around 150 BC (Roman Empire).

But it is with **Origen,** [70] a *Christian* theologian, that Christian theology begins. He knew, like his successors, how to apply in a critical way the rational philosophy of **Plato** to interpret the texts of the Septuagint. His gigantic work, called *Origenism*, was a source of inspiration for the Church Fathers. At the end of the 19th century, the *Catholic* Cardinal Giovanni **Mercati**[71] found an ancient work (stored in the National Library of Religions in Paris) on which one can see **Origen**'s Hexapla,[72] a synopsis written to settle the differences between *Jews* and *Christians* on the Scriptures. Origen was the first *Christian*, according to Henri **Sonier de Lubac,**[73] to distinguish 4 meanings or 4 types of levels of reading of

---

[69] Epicycle. *Each planet describes a circle whose center moves on a larger circle having the Earth as the center.*

[70] **Origen** *(Alexandria, EGY 185-253 Tyre, LYB). His father, a Christian victim of* **Septimius Severus** *([Leptis Magna, LY 146-211 Eboracum-Yorkshire, emperor from 193) was beheaded before his eyes, leaving the family in deepest destitution. He provided for his mother and his brothers and sisters. He was a true ascetic, very rigorous, and would have castrated himself to avoid succumbing (he was undoubtedly influenced by a too-literal reading of* **Matthew** *19-12:* "...there are eunuchs who have made themselves..." *and of* **Mark** *9-43:* "if your hand causes you to fall, cut it off...." *He later regretted this gesture). He was declared a heretic in 553, during the 2nd Council of Constantinople, because his mutilations made him unfit for the priesthood. He was imprisoned and tortured in 250, under the reign of* **Decius.***
Decius (Budalia-ALB 201-251 Rome, ITA), emperor in 249, he took the name of **Trajan Decius**.

[71] **Mercati** Giovanni *(Reggio Emilia, ITA 1866-1957 VAT), cardinal. His pears nicknamed him:* "the most cultured priest of the 20th century."

[72] Hexapla: polyglot edition commented, in 6 columns, on various versions of the Hebrew Bible.

[73] **Sonier de Lubac** Henri *(Cambrai 1896-1991 Paris). Jesuit, theologian, cardinal. Born into a wealthy noble family, he entered the Society of Jesus in 1913 and spent his novitiate on the island of Jersey. Seriously wounded in the First World War, life in the trenches alongside non-believers showed him the difficulties priests face in sharing their faith. Ordained a priest in 1927, he became a professor of Theology in 1929 at the Catholic Faculty of Lyon. During the Second World War, he participated in 1941 in the creation of* "Témoignage chrétien," *a weekly letter denouncing nazi ideology. Appointed ambassador to the Vatican in 1945. Following the*

the Bible: the historical, the allegorical, the tropological,[74] and the anagogical.[75] Let's take the example of the word Jerusalem; it can be translated as City of the Hebrews, Church of Christ, Human Soul, or Celestial City depending on the meaning we choose. But he uses only three: the literal, the moral, and the spiritual in his commentaries on Scriptures (corresponding respectively to the three parts of man: body, soul, and spirit). If some of his theses were contested after his death and in particular during the Constantinople synods of 543 and 553, he was rehabilitated in 1893 by **Pope Leo XIII**[76] in his Encyclical *"Providentissimus deus."* Recalling the work of the exegetes, he wrote [Le]:

"Among those of the East, the first place belongs to Origen, a man admirable for the promptness of his mind ... It is from his numerous works ...... that almost all his successors have drawn."

This was confirmed by Pope **Benedict XVI**[77] in 2007, during a tribute he paid to him at a seminar organized in his honor.

Plotinus,[78] making a synthesis of **Plato**'s work, interpreted allegorically the biblical texts by applying the method of this last one, creating the *"Neoplatonism,"* outlined by **Philo of Alexandria**. This will lead to a mystical philosophy crossed by a great religious breath. The ONE as he calls God, transcendent, is at the top of the pyramid and addresses his creatures through his intermediaries: Intelligence – the source of Truth, made up of a multitude of Ideas - which he creates, which in turn produces the Soul of the World containing all the corporeal souls.

---

publication in 1946 of "Surnaturel, Etudes historiques" *he was forbidden to teach by the Jesuit General and transferred to Paris. Authorized to resume his courses in 1958, he became interested in the theory of the four senses and was elected to the Academy of Moral and Political Sciences. Appointed expert of the Vatican II Council (1962-65) by* **John XXIII** *he became a recognized theologian. Refused in 1969 to be named a cardinal by Pope* **Paul VI** *because he wanted to continue to be a priest, but accepted in 1983 to be named Cardinal-deacon by Pope* **John Paul II.**

[74] Tropological: *figurative interpretation of the reading of the Bible. The allegorical is more mystical.*

[75] Anagogical: interpretation of the reading of the Bible tending towards a spiritual meaning.

[76] Pope **Leo XIII** born **Pecci** Vincenzo G. *(Carpineto Romano (French Empire 1810 - 1903 VAT). Elected Pope in 1878.*

[77] Pope **Benedict XVI** born Joseph A. **Ratzinger** *(Marktl-Bavaria 1927). Elected Pope in 2005, retired to the Vatican in 2013.*

[78] **Plotinus** *(Lycopolis, EG 205 - 270 Naples). Greco-Roman philosopher, a student of Ammonius Saccasil, taught until he died in the school he opened in Rome in 244. Although he started writing only in 255, he left considerable work. He placed* **Plato** *above all Greek thinkers.*

2.4. Church Fathers: Reason as an ally of Faith.

"If we submit everything to reason, our religion will
have nothing mysterious or supernatural about it.
If we shove the principles of reason, our religion will be absurd."
Blaise Pascal ([Pa] 273–173)

St. **Augustine**[79] *"the African,"* a nickname he gave himself to remind his belonging to the Latinized Berber intelligentsia, was one of the greatest rhetors and *Christian* thinkers of the West. Theologian and Philosopher, the reading of the *"Hortensius"* by **Cicero**[80] *the eclectic* incited him to study philosophy and then teach rhetoric. He returned to Christianity at the age of 33, thanks to the Neoplatonic preaching of **Ambrose of Milan**[81] who gave him new tools to rethink the faith of his youth, and especially thanks to *"divine"* intervention, in the gardens of Milan, who *"told"* him to reread the Bible and in particular the epistle of **Paul** to the Romans. He then decided to abandon the world, dedicate himself to God,

---

[79] **Augustine of Hippo**, *(DZ: Thagaste* [Souk Ahras] 354-430 Hippo [Annaba]), from a well-to-do Berber family he is better known as* **St. Augustine**. *Beatified in 1298 by* **Pope Boniface VIII** *who named him Doctor of the Church and Church Father. Formed in Carthage, raised in the Christian religion by his mother St. Monica, but not understanding the hidden meaning of the Bible, he moves away from it while being internally faithful to Christ. Until he read* **Cicero**'s *Hortensius - at 19 - he led a rather dissolute and idle life among pagans. He then thought he had found the answer to the question that had always nagged at him: how to find the Truth? and turned to Manichaeism (the Persian religion of Good and Evil); then, in 383, toward* **Cicero**'s *probabilism, derived from skepticism (one cannot be sure of knowing the truth). After having taught rhetoric (the art of speaking eloquently and forcefully on any subject) in Carthage, he was offered a position in Rome in 383 and then in Milan in 384. This was to be a revelation, and* **St Ambrose of Milan** *confirmed his baptism. The Neoplatonist texts, including Plotinus'* *"Enneads," provided him with doctrinal elements and a dialectical process, that of* **Plato**. *A monk in 388, he returned to Hippo with some disciples. He was ordained priest in 391 and acclaimed bishop in 395. Author of numerous works of which the 3 most important are: The Confessions, The City of God, and On the Trinity. Influenced the West until the advent of scholasticism. * Thagaste, now Algeria, was a roman province.*
[80] **Cicero**, Marcus Tullius (ITA: Arpinum -106 to -43 Gaeta, murdered). After studying rhetoric and law, he became a lawyer and then consul. Was a remarkable orator. Philosopher, he wrote in particular "The Hortensius," a dialogue between 4 roman consuls (work having disappeared) to encourage the Romans to study philosophy. Roman representative of Eclecticism, a doctrine derived from skepticism which consists in taking what is best, according to its followers, in the previous philosophies but to which they do *not adhere in their totality. Retains from* **Plato** *his ideas on the existence of God, the Soul being his emanation. On the other hand, God does not intervene in human affairs.*
[81] **St. Ambrose of Milan** *(Trier, DEU 340-397 Milan, ITA), Doctor and Father of the Western Church. A high Roman official from 365. He studied in Rome and became a lawyer and then administrator of the province of Liguria-Emilia (Milan). Bishop of Milan in 374 against his will, he defended the rights of Christians to the emperors* **Valentinian I & II**, *and* **Theodosius the Great**. *Author of numerous works and hymns.*

and generate a new way of thinking about the relationship between man's problems and the Mystery that surrounds the Being. Having received an excellent education and although not a scientist (theology is not a hard science, but one of faith), he eloquently demonstrated the necessary combination of reason and faith to find the truth. From an allegorical reading of the Bible, he tried to demonstrate the humanistic character of the Christian philosophy according to which the goal of the creation is the union of Man with God, in the image of that of Christ with God, affirming the goodness of the Nature assimilated to God. The central point of his philosophy is the search for Truth, which is the supreme good. Not only his work but, also his truth philosophical search is impressive. He is a proponent of virtue ethics allowing for self-realization. Addressing God, he writes ([Aug1]-Book 13,33): *"matter you made out of nothing, the form you took out of formless matter."*

St. **Augustine** adhered to the four Cardinal Virtues concept enunciated by **Plato**; namely Prudence or Wisdom, Courage, Moderation, and Justice. Until St. **Thomas Aquinas** took up the torch, 850 years later, he was the reference for the *Christian* Western and *Catholics*. Although he never doubted the existence of God, he proposed a rational demonstration of his existence and deduced the attributes of God [Th]. For him, as for St. John **Duns Scotus** (ref.128, p.36) and **William of Ockham** (ref.581, p.188), there is nothing in Genesis to indicate that God used different materials to create heaven and earth. Among his errors which, unfortunately, were emulated by the Church Fathers, let us cite his total adherence to the anti-atomic theories of **Plato** and **Aristotle**. If his refutation of the vacuum and multi-universes has long been reproached to him, now some eminent cosmologists seriously consider, but without experimental proof, that the vacuum is a state of matter and that the multiverse exists (cf. chap. 5). In his defense, let's recall that he was not a scientist and that his opposition to the ideas of **Epicurus**, who was an atomist and whose epicureanism was wrongly assimilated to libertinism, led him to reject the theory of **Democritus** and to rally to the theory of the 4 elements of **Empedocles** [Aug2]. On the other hand, he supported the thesis that time was created at the same time as the world and, anticipating the work of Ch. **Darwin** introduced the concept of evolution of the cosmos (§5.4pp.234-254). According to **John Paul II** ([JP] - p.57) St **Augustine** *"succeeded in producing the first great synthesis of philosophy."*

From the 5th to the 6th century, the barbarians slaughtered the Greco-Roman civilization, dealing a heavy blow to Philosophy, which struggled to get back on track.[82] The Arab Muslim Empire will take over.

---

[82] Let's note that St **Augustine** died during the siege of Hippo by the Vandals.

2.5. 7th to 13th century.
Hellenization of the Arab-Islamic world

"I feel like saying: it doesn't matter what the religion is as long as there is God"
George Smoot

We often overlook the rich set of influences that shaped what we refer to as the *"Arab sciences"* or the *"Islamic Golden Age"* (5th to 13th c.). These remarkable developments were the fruit of collaborative contributions from Hindu, Chinese, Persian (particularly from the Academy of Gondishapur[83]), and Greek thinkers. Following the closure of the pagan schools of Athens in 538 by **Justinian I**,[84] these brilliant minds sought refuge in Alexandria, Harran (S-E of present-day Turkey, near Syria and Iraq), and Antioch (Turkey), finding homes among *Jews, Syriac Christians*,[85] and Berbers. This intellectual fusion occurred within the expansive Arab Muslim empire that emerged after the passing of **Mohammed** (632-635), stretching from India to Spain. The transformative conquest of the Sassanid empire by the Caliph **Oma**, in 636 of our era (year 14 of the Hegira[86]), led to the mass conversion of the conquered peoples to Islam, often under pressure, facing a grim choice between death and conversion. The enlightened caliphs of that era, like **Al-Ma'mûn**,[87] displayed wisdom by

---

[83] Academy of Gondishapur. *Created by Nestorian Christians (ref. 96) deported from Antioch and Athenian philosophers banished from their country, it was renowned for its medical center which was the most important in the world, as well as its mathematicians and astronomers.*

[84] **Justinian I** or **Justinian the Great** *(Tauresium, ALB 482-565 Constantinople, TUR). Eastern Roman Emperor, he is considered by many as the greatest Byzantine emperor as his work (religious, legislative, military: territorial expansion, the influence of the Roman Empire, ...) is important.*

[85] The Nestorians, *followers of Nestorius (Syria 380-451 Kargeth, EGY). Nestorius was the Patriarch of Constantinople from 428 to 431 year, in which his doctrine was declared heretical by the Council of Ephesus. Stating that the 2 natures of Christ (divine and human) are distinct he concludes that Mary was Christ's Mother but not God's Mother. Millions of Nestorians in the 12th century were scattered in Persia, India, China, and Central Asia.*

[86] Hegira = immigration. *In 622 Muhammad and his companions left Mecca for the Oasis of Yathrib to create a more fraternal community.*

[87] **Al-Ma'mûn** *(Baghdad, IRQ 786-833 Tarsus, TUR. Hârûn ar-Rachîd's son and a Persian slave. From 827 he tried to impose on all his subjects the Mu'tazili doctrine, according to which the Koran has a created and not eternal character, advocating a purified concept of deity, making man responsible for his acts and placing* **Ali** *(600 - 661, eldest son of Muhammad's uncle) above all the prophets. In 829, he created the Baghdad Observatory, and in 832 the House of Wisdom in Baghdad, where he brought together scholars of all faiths and introduced Greek philosophy...* **Muhammad** *is said to have taken his cousin* **Ali** *at birth in his hands and to have given him his name. In 622* **Al-Ma'mûn** *married* **Fatima**, *Muhammad's daughter. Let's note that the division between Sunnis and Shiites dates back to 632 when the question of Muhammad's successor arose. The future Shiites designated* **Ali**, *while the future Sunnis (the majority) chose* **Abu Bakr** *(Mecca 573-634 Medina) a rich merchant, father-in-law, and close advisor of Muhammad as the first Rashidun caliph.*

orchestrating the translation of diverse texts into Arabic, primarily undertaken by local scholars, including **Hunayn ibn Ishaq,**[88] the Persian *Nestorian Christian* affectionately known as *"the Master of the translators,"* and **Thâbit ibn Qurra,**[89] a *Sabaean* with roots in the *Judaeo-Christian Baptist* current. They meticulously translated Greek and other manuscripts from scholars of the conquered countries. **Al-Ma'mûn**'s visionary leadership extended to building numerous observatories and advocating for a rationalist form of Islam, bridging divides between Shiites and Sunnites. The Arabic language, acting as a unifying force, facilitated the dissemination and enhancement of knowledge throughout the empire. Muslims didn't just transport information, they refined it, greatly enriching not only their own culture but also transmitting Aristotelian thoughts, which profoundly influenced Western intellectualism until the 17th century. Within this community of scholars and exegetes (ulama), were atomists like the **Mu'tazilites**, the pioneers of Kalâm;[90] an intricate dogmatic science and rational theology inspired by peripatetic philosophy while diverging on fundamental philosophical points. In Appendix I on p. 141, I have highlighted the initial four propositions out of twelve from the Kalâm and provided two illustrative instances of its misuse.

Among the main rationalists we have retained in chronological order:

- **Al Khwarizmi,**[91] Mathematician, Astronomer, Astrologer, and Persian Geographer, was a member of the Houses of Wisdom whose role, in the transmission of Persian, Syriac, Indian, Chinese, and Greek knowledge, is considerable because he translated, and annotated in some cases, into

---

[88] Hunayn **ibn Ishaq** *(Al-Hira, IRQ around 808-877). Physician, son of a Nestorian pharmacist, he followed courses in medicine in Baghdad, traveled, and then returned to Baghdad where he translated into Syriac and Arabic many authors including* **Hippocrates***, *the "father of Medicine" and* **Galen**** Claudius, *the two main founders of the bases on which Western medicine is settled... It is said that* Hippocrates *was the sower and* Galen *the harvester.*
*Hippocrates *(Greece: Island of Cos -460 to -377 BC Larissa). According to him, nature has a divine character but is not under the influence of the whims of the gods.*
**Galen Claudius *(Pergamon, TUR 129-216 Rome, ITA)* Agnostic[(a)]. *Physician, he appealed observation.* Surgeon, *being the appointed physician of the gladiators could observe the interior of the human body,* Biologist, *and prolific writer he commented on the writings of Hippocrates. He was the last of the great physicians of Greco-Roman antiquity.*
[(a)] According to Antoine Pietrobelli "Galien agnostique: un texte caviardé par la tradition," Revue des études grecques 2013.
[89] Thabit **ibn Qurra** *(Harran, TUR 826-901) was a Sabean. He was a polymath: Musicologist, Philosopher, Mathematician, and Astronomer from a family of Neoplatonists. Considered a pagan by* **Justinian I,** *(the Byzantine emperor from 527 to 565) he was exiled from Athens. He studied at the House of Wisdom, Bagdad. Author of numerous works in Musicology and mathematics (translated by Archimedes, Euclid, Ptolemy, ...). He made important discoveries in algebra, conics, theory of numbers, ... A crater on the Moon bears his name.*
[90] Kalâm or Calam: *meaning word attributed to God. Muslim doctrine that appeared in the 8th century, based on Reason.* Mu'tazilite: *members of the "Party of Divine Justice and Unity." They withdraw and stand apart from the dispute between* **Ali** *and his opponents.*
[91] **Al-Khwarizmi,** born Muhammad Ibn Musa al-Khwarizmi *(Khiva, UZB 780-850 Baghdad, IRQ). Persian mathematician, astronomer, and geographer. His name gave "algorithm."*

Arabic and then into Latin the main texts of the schools of Philosophy, Medicine, Mathematics, Astronomy, ... of these countries. He introduced into the Muslim empire the Indian decimal system and algebra, hence his appellation: *"the father of algebra"* shared with **Diophantus of Alexandria,**[92] because although the works of the latter were before his, some think that Al Khwarizmi had not been aware of them.

-   **Al-Kindi**[93] *Muslim*, nicknamed *"the father of Arab philosophy,"* was commissioned by various caliphs, including **Al-Mamûn** to translate Greek works. Founder in 830 of the House of Wisdom in Baghdad, he was one of the greatest Hellenising philosophers of the Arab world. Rationalist, follower of Aristotelianism, he studied mathematics, physics, astronomy, morals, and metaphysics, the first source of knowledge of the reason for things according to him. Physics came after because he considered it as the simple knowledge of things. According to the Koran, God is for him the First Principle, the *"One,"* the Creator.

-   **Muhammad al-Fazari**[94] *Muslim,* Astronomer, and Astrologer built the first astrolabe (device to measure the height of the stars) in the Muslim world.

-   **Abbas ibn Firnas**[95] *Muslim,* Poet, Physician, and Chemist. He invented a planisphere, a water clock, and various instruments to work quartz and glass; but is also known to have tried to fly by himself wearing feathers, the landing was reportedly painful.

-   **Al-Razi**[96] *Persian* scholar. An atomist, opposed to Aristotelianism, Al-Razi read a lot of Greek and Hindu works that influenced him, as did his

---

[92] **Diophantus of** (Alexandria between the 3rd and 2nd centuries BC). *Wrote, among other things, the "Arithmetic" which influenced Arab and Renaissance scholars.*

[93] **Al-Kindi** *(IRQ: Kufa 801-873 Baghdad). Kufa was the first capital of the Sunni Abbasid caliphate. His father, who came from a tribe in southern Arabia, was governor of Basra. He went to Baghdad, the third intellectual place with Kufa and Basra in the Muslim empire in that time. He believed that time and space are finite. A true encyclopedia, he wrote numerous works on mathematics, particularly citing Persian knowledge. He was interested in geometric optics and the study of burning mirrors (cf. the feat attributed to Archimedes, but very controversial, of having burned the Roman fleet with the help of mirrors), hydrostatics, the chemistry of essential oils, pharmacy, music, astronomy, ... that is to say more than 250 works.*

[94] Muhammad **al-Fazari** (Persian or Arab died in 796 or 806).

[95] Abbas **ibn Firnas** *(ESP: Ronda 810-887 Cordoba). Berber ancestors participated in the conquest of Spain.*

[96] **Al Razi** *(Rayy (near Tehran, IRN) 865-925 or 935). Mathematician, Philosopher, Physician,* Muslim. *He did most of his research in Rayy, which did not prevent him from working in Syria, Egypt, and Andalusia. He first devoted himself to Music and then to Alchemy, which led him, when he became a doctor in 895, to introduce iatrochemistry, the origin of modern chemical pharmacopeia. He was the best physician of his time thanks to the medical knowledge of ancient*

contacts with the Chinese free thinkers and scientists, who were numerous in Tehran. He was one of the greatest Iranian scholars, who contributed to developing disciplines such as Alchemy, Medicine, and Philosophy. Defender of the animal cause, he advocated vegetarianism. Officially *Muslim*, he was critical of Islam; rejecting the idea of a single god, he drew much criticism from his colleagues.

- **Al Fârâbi,**[97] *Persian Muslim* philosopher. He studied in Baghdad the political Philosophy of **Plato**, Mathematics, Arts ... He frequented the Nestorians and all the great thinkers of the time. Called the "*Second Master*" by **Maimonides**, for whom the first was **Aristotle**, he was **Avicenna**'s teacher and contributed, like **Al-Kindi**, to the spread of Aristotelian and Neoplatonic thought in the Muslim world while integrating his data.

- Saadja **Gaon**[98] was the first *Rabbi-philosopher* to attempt to reconcile his rabbinic Jewish tradition with Aristotelian philosophy. This opposed him to the Caraïsts, who rejected the Oral Law.

- It should be noted that the Christians had not remained inactive. From the 7th century onwards, a set of doctrines was born, interpreting the known universe in the light of reason. Their method is intellectual and dialectical. They rethink Greco-Roman philosophy, and initiate scholasticism - independent philosophy and perfect science in its order - subject to faith. If the period extending from the 7th to the 12th century was that of the development of these new doctrines, the 13th century was more fertile and dominated, as we shall see later, by St **Thomas Aquinas,** a *Dominican theologian.* **Gerbert d'Aurillac,** [99] **Pope Sylvester II,**

Greece. He was notably influenced by the Persian Christian physician Sâbûr **Ibn Sahl** *(Khuzestan, IRN born ? - died in 869 in Samarra (Iraq)). As a physician to the prince of Rayy, he ran a hospital that was the first to have an unit dedicated to the mentally ill. He was interested in neurology, and described smallpox as well as allergic asthma, ...*

[97] **Al Fârâbi** or **Fârâbî** *(Fârâb-Turkestan 872-950 Damascus, SYR) is said to be the "Second Teacher of Intelligence" because he tried to appropriate all the sciences of the time as well as the arts including music.*

[98] **Gaon** *Saadia (Faiyum, EGY 882-982 Bagdad, IRQ). Jewish Philosopher, Rabbi: only the written law prevails. At the origin, towards the end of the 9th century, of the first rational Jewish works on the Bible.*

[99] **Gerbert of Aurillac** *(Belliac, 15 FR between 945 and 950-1003 VAT),* **Pope Sylvester II.** *Scientist, and philosopher, he taught rhetoric. Of modest origin, probably a shepherd, he was admitted at the age of 12 to the Benedictine monastery where he began his studies. The monks noticed his abilities and entrusted him to* **Count Borrell II of Catalonia** *who introduced him to the Catalan scholars of the time. He studied astronomy in Spain and would have invented the 1st mechanical clock [Th]; The emperor of Occident* **Otton 1st** *to whom* **Pope John XIII** *presented him, enrolled him as the tutor of his son. Appointed archbishop of Ravenna he was elected Pope in 999.*

Philosopher, Mathematician, Physicist (Mechanics), and French Astronomer was one of the major actors in the renewal of the medieval West. During his pontificate, Sylvester II promoted the scientific renaissance and imposed the Indian decimal numbers, known as "*Arabic numbers*" on the Christian world.

- **Alhazen.**[100] Raised in the *Shiite* movement, he abandoned the study of Islamic texts to turn to the sciences because he was disillusioned with religious quarrels of a theological nature. The specialists say that he is the heir of the Greek and Indian scientists. He was the first to:
  - base his research on experimentation rather than reasoning,
  - realize that an object with continuous motion only stops when acted upon by a force. His work influenced Johannes **Kepler**.

- **Avicenna,**[101] the "*Prince of Scholars*" according to his students, was a *deistic Muslim* of Iranian origin. A student of **Al-Fârâbi**, he was influenced at an early age (21) by the ideas of **Aristotle**, the philosopher par excellence, and the Koran. A mystical philosopher (the third Master - after **Aristotle** and **Al-Fârâbi** - according to his students), he tried to reconcile the Aristotelians and the Neoplatonists. According to him, God created the world but does not take care of human problems; their lives are subject to the laws of Nature. He proposed the "*Theory of the Hierarchy of Beings and Causes*" according to which:
  - At the top God, who is not only the pure Act and first Motor of **Aristotle** but above all the Necessary Being, the source of all possibilities, the only one whose intimate nature (essence) is identical to existence. First Cause, in relation with his creatures through the intermediary of the First Caused who possesses "*the Holy Spirit.*" For some, Aristotle was a pantheist.
  - Below, the Celestial world, whose management God entrusts to a perfect Intelligence: the First Cause. This latter is an intermediary between its Creator and the 1st sphere of which it creates the soul and the secondary Intelligence, playing the same role for the 2nd

---

[100] **Alhazen,** or **Ibn al-Haytham** *(EGY: Bassora 965-1039 Cairo) one of the promoters of experimentation. Mathematician, Physicist (physiological optics), Astronomer (a crater of the Moon and an asteroid are named after him), Physician (first to realize that light can damage the eye), and Philosopher.*

[101] **Avicenna** or **Ibn Sina** *(Afshana, near Bukhara in UZB 980-1037 Hamadhan, IRN). A Persian Muslim with deistic tendencies, introduced to experimental sciences at a very young age. Physician, his "Canon of Medicine" was a reference work for Eastern and Western physicians until the 7th century. For him, Metaphysics and Astronomy are intertwined and complement each other; the soul, spiritual and immortal, is unique in each living being. He was in the service of various sultans and emirs.*

internal sphere; this is until one arrives at the last celestial sphere moved by the human Intelligence or human Soul managing the:

The terrestrial world contains, in its center, the immobile Earth and its satellite the Moon according to the model of Ptolemy.

- Michel **Psellos,**[102] a *Christian* monk, was called: "*consul of the philosophers*" by **Constantine IX.**[103] Writer, a Neoplatonic philosopher, he taught at the University of Constantinople. He was very erudite: History, Rhetoric, Theology, Geometry, Medicine... He was the author of numerous works and contributed to the scientific and literary Renaissance of the 11th century.

- **Avicebron**[104] an Andalusian *Rabbi*, was also a follower of Neoplatonism, but he influenced the Jews less than Christian scholasticism. According to his teaching: from the ineffable God comes the universe, then the angels, the souls, and finally the men.

- Abu'l-Barakât **al-Baghdadi,**[105] *a Jewish* physician was forced to convert to Islam. A physicist, he was interested in kinematics, linking force and acceleration, and defending the theory of vacuum and experimentation in science. As a philosopher, he introduced a concept of metaphysical time breaking the link between space and time, opposing **Aristotle's** conception of time, for whom time is a measure of movement, existing only if there is a change. Al-Baghdadi professed the immortality of the soul, independent of the body. At the same time, the *Christian:*

- **Guillaume de Conches**[106] *Catholic theologian,* a humanist, and physicist

---

[102] **Psellos** Michel *(Constantinople, TUR 1018-1078). Coming from a family of little fortune he had to occupy very quickly a job of a secretary near a civil judge but lived in contact with strong personalities who helped him to climb the ladder. He was a very influential politician. Disgraced in 1054 he became a* Christian *monk on Mount Olympus, until his return to business in 1057.*

[103] **Constantine IX** Monomachos *(Antioch 1000-1055 Constantinople, TUR),* Theodosius **Monomachos'** *son (judge, nobleman), married in 1042 the empress* Zoe **Porphyrogenes** *(64 years old, widow for the second time) and became Byzantine emperor. The same year he adopted the* Christian *religion which had been recognized in 313 (Edict of Milan) by* **Constantine I** *before being proclaimed the State's religion in 380 (Edict of Thessalonica) by* **Theodosius I.**

[104] **Avicebron,** *real name:* Solomon **ibn Gabirol** *(ESP: Malaga 1020-1057 Valencia). Andalusian* Rabbi, *Poet, Theologian, Metaphysician.*

[105] Abu'l-Barakât **al-Baghdâdî** *(IRQ: Balal, near Mosul 1080-1164 or 65 Baghdad).* Avicenna's student *was a physician of* Jewish *origin who converted to* Islam *at the end of his life for unclear reasons, but probably to avoid trouble. His work had a great influence on Islam, but little on Judaism.*

[106] **Guillaume de Conches** *(Conches-en-Ouche, 27? FR: 1080-1154).* Catholic theologian, *grammarian, and philosopher, commented on* **Plato'**s *"Timaeus" and knew the works of the main Greek and Arab physicians. Defended dialectics.*

- of renown, and an influential member of the School of Chartres,[107] defended atomism, the soul of the world, and the use of dialectics.

- Zahir al-Dîn **al-Bayhaqî**.[108] A *Muslim* and, according to some, a descendant of one of **Mohammed**'s companions. He dedicated himself to the study of sacred texts, Mathematics, and Astrology. He authored numerous works across various disciplines of his time and was a precursor to the Encyclopedia.

- **Al-Ghazali**.[109] *Sufi*[110] *Theologian*, a defender of the orthodoxy of the Koran, opposed **Avicenna**. He was a philosopher who also tried to reconcile faith and reason, his principle was that reason alone is incapable of discovering the truth and that mystical enlightenment must be added to it.

- **Averroes**,[111] of his true name **Ibn Rochd**, a Muslim Andalusian of Berber origin, received a strong religious education in Hadith and Fiqh[112] from his father, he then studied Physics, Astronomy, Philosophy, Mathematics, and Medicine before becoming a physician at the Almohad court (a Berber dynasty that ruled the Maghreb and Al Andalus from 1120 to 1269). In 1166, at the request of the Caliph, he studied and translated the works of **Aristotle**, striving to distinguish between science and faith. According to Averroes, the Quran[113] could be interpreted in three ways: for the general populace (pure faith), for theologians (search for mystical meaning), and for the philosopher (true meaning, scientifically proven). He believed that evidence should prevail over

---

[107] School of Chartres. *One of the four components of the University of Paris, recognized in 1200 by* Philip **Augustus**, King of France, *and in 1215 by His* **Holiness Innocent III**, *played an important role in the Middle Ages through its humanism and the high quality of its teachers. It is at the origin of the baccalaureate (contraction of baccalarius: the young man who aspires to become a knight and laureare: crowned with laurels), of the Licence, of the Institutes and Schools.*

[108] Zahir al-Dîn **al-Bayhaqî** *(Sabzevar, IRN 1097-1169). Coming from a family of judges and imams he studied with the great scholars of the time and was appointed Cadi of Bayhaq.*

[109] **Al-Ghazâlî** Abû Hamid Mohammed **ibn Mohammed Ghazâlî** *(Tus, N-E IRN 1058-1111). Initiated in law and philosophy in Baghdad. A Muslim of Persian origin, he turned away from the philosophers whom he accused of incoherence and moved to Sufism. He influenced many theologians of the 3 monotheistic religions.*

[110] Sufi. *Muslim ascetic, follower of Sufism, mystical Muslim doctrine.*

[111] **Averroes** *(Cordoba, ESP 1126-1198 Marrakech, MAR). Philosopher, Scientist, and Muslim, Cadi of Cordoba. He commented on the works of* **Aristotle** *and developed certain rationalist and materialist aspects of them, distinguishing, even opposing, rational and revealed truths, which earned him attacks from certain theologians.*

[112] Hadith: *a collection of traditions relating to the words and deeds of* **Muhammad**. *A major source of guidance for Muslims with the Koran.*
Fiqh: *Islamic jurisprudence, temporal interpretation of the sharia rules.*

[113] *[Th] p. 270.*

faith in case of conflict, necessitating an allegorical interpretation of the Quran.

Averroes was one of the greatest thinkers of Muslim Spain, legitimizing Philosophy according to Muslim law. His work remained a reference in the West until the end of the 16th century. His Aristotelian conception of God as the Creator of the Earth, rather than as a craftsman, was opposed to traditional interpretations of the Quran, leading to unjust suspicions of heresy.

- At that time Moses **Maimonides**, [114] born Moses ben **Maimon**, the greatest *rabbi* of the Middle Ages, hence his nickname: *"the second Moses,"* sought to strengthen *Jews' faith*. Physicist, [115] metaphysician, philosopher, and theologian, believed that the scientific search for God would bring him closer. To do this, he applied Aristotelian philosophy to the reading of the Bible. Although, he adhered to the five elements of **Aristotle**, to the absence of emptiness, ... he differed from him. For **Maimonides** God, eternal, created the universe, while for **Aristotle**, God eternal would have only modeled the materials, also eternal, which he had at his disposal. He wrote and commented on numerous religious books, including *"Mishnah Torah"* or *"The Strong Hand"* (a reference work on *Jewish* law consisting of 14 books). His main work is *"The Guide for the Undecided"* [Ma1&2], also called *"The Guide for the Lost."* Although difficult to read, it is interesting because **Maimonides** gives keys to his student and disciple *Rabbi* Joseph ben **Judah of Ceuta** [116] (among others) to understand the metaphors, homonyms, and numerous allegories encountered in the Scriptures. His basic principle is [Th]:

---

[114] **Maimonides** Moses, *(Cordoba, ESP 1138-1204 Fostat-Cairo, EGY), Sephardic Jew. His father, rabbi Maimonides, was a member of the rabbinical court of Cordoba like his ancestors who took the name of* **Ibn Abdallah** *after the capture of Cordoba by the Almohads (a Berber dynasty established in Mauritania and Senegal). He was interested in the sciences and Greek philosophy from an early age. In 1351, when the Almohads took power and gave him the choice between conversion or death, his family went into exile and joined Fez. He left then for Egypt. He was appointed Raïs al Yahoud (leader of the Jews in 1171 in Fostat, near Cairo. Around 1180 he became the appointed physician of the vizier* **Al Qadi al Fadil**, **Saladin**'s *secretary, and then of the sultan's family. He was an astronomer, physician, philosopher, and theologian renowned for his writings which attempted to bridge the gap between philosophy and Judaism, notably the* "Guide for the Undecided." *He had a good knowledge of the works of* **Al-Ghazali**.

[115] *Like* **Avicenna**, **Maimonides** *placed Physics before Metaphysics,*

[116] Rabbi Joseph ben **Judah of Ceuta** *(Ceuta, Maghreb 1160-1226). Physician in Alexandria devoted his free time to poetry. He left for Fostat (incorporated into Cairo) where* **Maimonides** *taught him Logic, Mathematics, and Astronomy. For two years, he followed Maimonides' commentaries on the prophets and then moved to Aleppo, Baghdad while following Maimonides' teaching at a distance (Book of the Lost). He became a great rabbi* "the light of the West" *according to* **Al-Harizi** *and defended Maimonides against the rabbis of Baghdad.*

"(that) *there is necessarily agreement between faith and philosophy... in case of disagreement, although the Bible is "One" and expresses the truth it is necessary to have an allegorical reading,"*

because for him, reason does not contradict the Bible but explains it.

Maimonides' thought radiated far beyond the Jewish community and crossed the centuries; exerting an influence not only on medieval philosophers such as **St. Albert the Great, St. Thomas Aquinas** who fought against him, **John Duns Scotus,** ... on those of the Enlightenment: Baruch **Spinoza,** Moses **Mendelssohn,** but even today. The scientist and *Orthodox Jewish* philosopher Yeshayahu **Leibowitz,** a controversial figure in contemporary Judaism, was one of his followers.

We cannot avoid the fact that he had opponents, the best known of whom, Solomon **Ben Abraham of Montpellier**[117] (Middle of the 13th century), found contradictions between the Talmud and the tendency of **Maimonides** to allegorize biblical narratives, downplaying the role of miracles. His incitement to have Maimonides' works publicly burned by *Dominican monks* led to his excommunication by **Maimonides'** horrified followers. The respected rabbi Moses **ben Nachman,**[118] nicknamed "**Ramban,**" an admirer of Maimonides but not a supporter of all his ideas, attempted a conciliation rejected by both sides.

According to B. **Pullman** ([Pu] ref.3-p.68), for **Maimonides,** the only limit of God is his inability to do evil.

From the 13th century onwards, most of the thinkers of the Muslim empire attacked rational thought.[119] The Persian *Muslim theologian,* mathematician, physicist, astronomer, physician, and philosopher, Nasir al-Din **al-Tusi**[120] was one of the last *Muslim* masters of thought adept at rationalism. This is perhaps one of the reasons (in addition to the "*sacking*" of **Averroes**) that made Abdus **Salam,** an *Ahmadi Muslim* winner of the 1979 Nobel Prize in Physics, state:

"... *of all the great civilizations of the planet, the Islamic community is the one that has given science the most restricted place.*"

Among the precursors of Thomism who tried to merge philosophy and theology, to correct **Aristotle**'s opposition to the dogma of God the Creator, in particular,

---

[117] Rabbi Solomon **Ben Abraham of Montpellier** *(first half of 13th century). Provencal prominent Talmudic authority, he did not have an overview of Maimonides.*

[118] Rabbi **ben Nachman** Moses *(Girona 1194–1270 Acre, kingdom of Jerusalem).* Sephardic *philosopher, physician, and a* kabbalist *he had great authority over* Jews.

[119] *Sapi Inès* "Sciences médiévales et islam" *La recherche N° 518-2016. Quantum physicist at CERN, Muslim specialist of sufism.*

[120] Nasir al-Din **al-Tusi** *(Tus, IRN 1201-1274 Baghdad, IRQ). A theorem bears his name. With* **Al-Urdi** *(Urd, SYR 1200 - 1266 Maragha, IRN), they realized the Maragha observatory (N-W Iran).*

by relying on **St. Augustine**, let's quote the first great secular[121] scholastic of the 13th century:

- **Guillaume d'Auvergne**,[122] a *Catholic theologian*. He was influenced by **Avicenna** and **St Augustine** and achieved a synthesis between Christian concepts and Aristotelian ontology reviewed by the Neoplatonists. Partisan of a hierarchy: the angels, immaterial and without form, being above the men made of matter, thus subjected to corruption.

- Defending the Neoplatonist eclecticism one could quote Dominicans like **Roland of Cremona**, **Hughes of Saint-Cher**, ...; as well as *Franciscans:* **Alexandre de Hales** (Halesowen, UK 1180 - 1245 Paris), **Jean de la Rochelle** (La Rochelle 1200-1245 Paris) ... but the Master was:

- **St. Albert the Great**.[123] *Dominican theologian* he taught at the University of Paris where his most prestigious pupil was **St Thomas Aquinas**. St Albert the Great was a Philosopher, passionate about Greek and Arabic science, he made many experiments to try to reconcile science and faith. It is not for his scientific contribution that I mention him, even though it is generally agreed that it was important, but because he made the Arab thinkers known in the West and introduced into his teaching Peripatetics[124] and Neoplatonic ideas, showing that it is possible to reconcile science and faith. His main goal[125] was to *"remake Aristotle for the Latins' use."* For that, he took again scientifically all the work of **Aristotle** by commenting on it, purifying it, and *Christianizing* it. He corrected it, when he considered it necessary, using his knowledge of most of the known sciences of the time,[126] sciences that he experimented with. It should not be forgotten either that he did everything possible to allow his favorite pupil, the most gifted, **Thomas Aquinas**, to flourish.

---

[121] Secular: *who belongs to the century, layman living in the world, as opposed to ecclesiastic*[Ro].
[122] **Guillaume d'Auvergne** *(FR: Aurillac around 1190 - 1249 Paris). Canon of Notre-Dame de Paris in 1223, professor of theology in 1225, founded in 1226 the convent of the* "Daughters of God;" *Bishop of Paris in 1228 he was entrusted with various missions by* **Pope Honorius III**. *He was the confessor of* **Louis IX of France***, revered as* **St. Louis**.
[123] **St. Albert the Great**, *whose real name was* **Albrecht von Bollstädt** *(Lauingen-Bavaria 1193-1280 Cologne, DEU) better known as* Albert of Cologne *or* Albertus Magnus. *He came from the nobility (counts) and entered the Dominicans in 1223. Gifted for studies, he attended the greatest universities of the time. He became known for his commentaries on Aristotle. Appointed bishop of Regensburg in 1260 he resigned two years later. From 1263 to 64 he preached the crusade through Germany. In 1271 he returned to Paris to support the doctrine of* **St. Thomas Aquinas**. *Was the most fruitful of all theologians.*
[124] Peripatetic, *a word derived from the Greek peripapetikos (who likes to walk around discussing) was given to the students of* **Aristotle** *because he liked to dialogue while walking.*
[125] *[Th] p. 315.*
[126] *He wrote about Zoology, Botany, Astronomy, Alchemy, Medicine... sciences that he practiced ([Th]-p.316).*

St Thomas Aquinas. [127] *Dominican theologian, a* specialist in oriental religions, established a dialogue with the Jewish and Arab thoughts of his time. A philosopher, he criticized Averroism but was an undeniable peripatetic. His **Holiness Urban IV** entrusted him with the revision of **Aristotle**'s texts, *"resurrecting"* him on occasion for the *Christian* community, imposing the application of his principles, but distancing himself on certain fundamental points, incompatible with the *Christian* faith. His work considered the most important of the Middle Ages, aims at reconciling Faith (according to the dogmas of Christianity) and Reason (according to the philosophical views of **Aristotle**), two visions of Truth that appear to him as dual. His school, *"Thomism"* made the most perfect synthesis of all anterior scholastic schools. He was nicknamed the *"Angelic Doctor"* because his life was exemplary and his response to the numerous attacks, he was the target of, was so benevolent. Except during the Renaissance (15th and 16th centuries), a period during which Europe significantly abandoned the austere mores of *Catholicism* to return to a pagan Hellenism, and thanks to the perennial fidelity of Dominicans, this theological Science imposed itself into the *Christian* world. H.H. **Pius XI** proclaimed him: *"Guide to Studies,"* cf. the Encyclical *"Studiorum ducem"* 29/06/1923. **St. John Paul II** says of him (cf. [JP] p.106): *"... **St. Thomas** is an authentic model for those who seek the truth. Indeed, the demands of reason and the strength of faith have found the highest synthesis that thought has ever achieved...."*

On the other hand, the Franciscans felt closer to the theses of St John **Duns Scotus**.[128] *Franciscan, theologian* philosopher, nicknamed the *"Subtle Doctor,"* the founder of *"Scotism,"* a school characterized by a metaphysical formalism, the

---

[127] **St. Thomas Aquinas** *(ITA: Aquino-Lazio 1224-1274 Priverno-Lazio). Born into an aristocratic family (Counts) in the castle of Roccasecca, he was educated from the age of 5 by the Benedictine monks of the Abbey of Monte-Cassino. From the age of 10 to 18 he studied Didactics, Metaphysics, and Morals at the University of Naples. At the age of 18, he opted against the advice of his family, for a monastic life and entered the Dominicans where he was ordained in 1244. Studying in Paris from 1245 to 48, he met* **Albert the Great** *(course on Aristotle) whom he followed for a time in Cologne as a disciple and assistant. He taught at the University of Paris from 1252 to 1259 and from 1268 to 1272. In the meantime, he taught at the Vatican stadium. Canonized in 1323 by* **Pope John XXII,** *he was declared Doctor of the Church in 1567 by* **Pope Pius V.** *Initiator of the Thomist School. Very charismatic, he had many followers and aroused a lot of jealousy among the secularists, especially in Paris. In response to their violent attacks, he instituted* "disputes" *at Christmas and Easter; meetings during which he answered all questions. He left an immense work, his masterpiece being, according to the connoisseurs, the two* "Summas": *the philosophical and the theological which he will not have time to complete. A master in the search for truth, he adhered to the formula of* **St. John the Evangelist** *(8:32):* "Then you will know the truth and the truth will set you free."
[128] **St John Duns Scotus** *(Duns, Scotland 1266-1308 Cologne, DEU). He entered the Franciscans in 1290, was ordained a priest in 1291, completed his education at Oxford (where* **Thomas Aquinas** *was not appreciated) until 1302, then in Paris. In 1305 he obtained his doctorate, was director of studies at the Franciscan Stadium in 1306 and 1307, then went to Cologne. Author of numerous works. He was beatified in 1993 by* **Pope St John Paul II.**

"*univocity*[129] *of the being*" and voluntarism (an autonomous force that allows us the possession of the real from which it deduces the individual freedom of the man), also proposes to defend the Faith by the Reason, attaching himself to demonstrate that God created the world but directs it in all freedom. Freedom of action as well on his creatures as in the border that he creates between the Good and the Evil. Professor Thonnard distinguishes [Th] two "*remarkable*" applications of his doctrine:

1. The limit which separates the natural from the supernatural depends on the free determination of God,[130] justifying his skepticism towards the absolute immortality of the soul, who would deny God the power to abandon some of them.

2. The three first commandments, concerning the duties of men towards God, are alone of absolute natural right;[131] the others[132] are of natural right consecutive to the free will of God. If he often arrives at the same conclusions as **St Thomas Aquinas**, the spirit is different, the attacks are all the easier as the latter is no longer of this world to defend his theses. Despite everything, the "*subtlety*," at the limit fallacious of his argumentation, served him well. He will be himself the target of the nominalists.[133]

2.6. 14th to 17th century: Introduction to heliocentrism

It was a time of exciting discoveries and debates that changed our understanding of the universe. We will meet:

---

[129] Univocity: *characteristic of a word that keeps the same meaning in two different uses. For J.* **Duns Scotus** "Being" *means both God and his creatures, but the difference between them comes from the fact that God is infinite whereas Man is finite.*

[130] F.-J. Thonnard *([Th]-p.424) deduced that God could justify a sinner and give grace without removing sin because according to J.* **Duns Scotus,** *there is no essential relationship between nature and grace.*

[131] *Recall the* three first commandments *(cf. Exodus):* 1. I am the Lord your God, who brought you out of the land of Egypt, out of the house of slavery. 2. You shall have no other gods but me. You shall not make any idols, any images, ... 3. You shall not call upon the name of the Lord your God for evil, for the ....

[132] *The* 7 other commandments *(cf. Exodus):* 4. You shall make the Sabbath a memorial, a holy day. Six days you shall work and do all your work; but the seventh day is a day of rest... 5. Honor your father and your mother, that your days may be long upon the land which the LORD your God is giving you. 6. Thou shalt not commit murder. 7. Thou shalt not commit adultery. 8. Thou shalt not commit theft. 9. Thou shalt not bear false witness against thy neighbor. 10. You shall not covet your neighbor's house; you shall not covet your neighbor's wife, ...
According to [Th] p. 424. God could make lying a good act, and declare the hatred of the neighbor a virtue. I think he is considering the case where the hated individual is a demoniac.

[133] Nominalists: *followers of* **William of Ockham** *(ref. 581-p.188) for whom the existence of God cannot be demonstrated. Thomas* **Hobbes**, *Pierre* **Gassendi**, *Georges* **Berkeley**, *Emmanuel* **Kant**, ...

-    **Ibn al-Shatir**, [134] a *Muslim* mathematician and astronomer who fixed the hours of the prayers in the mosque of Damascus. He wasn't a typical thinker, he didn't stick to Aristotle's ideas but improved **Ptolemy**'s geocentric model. In his ingenious work, he realized numerous devices of astronomical measurements that demonstrated that all the planets could revolve around a single point, laying the groundwork for:

-    **Copernicus**, [135] a *Catholic canon*, and renowned scientist suggested timidly in 1514, the heliocentric model, where the Earth revolves around the Sun. Copernicus acknowledged that he was influenced by the writings of ancient Greeks he had studied, as well as by those of Persian and Greek astronomers. Although his ideas on the heliocentric system spread from 1533, until they reached the hands of **Pope Clement VII**, he was never worried by the Catholic Church.

But the journey to acceptance wasn't smooth. Martin **Luther**[136] and John **Calvin**,[137] early reformers but geocentrism followers, criticized heliocentrism. If

---

[134] Ibn al-**Shatir** *(Damascus, SYR 1304-1375). Having become rich in the work of the ivory he could devote himself to astronomy.*

[135] **Copernicus** Nicolaus *(Prussia: Thorn (now Poland) 1473-1543 Frauenberg, AUT). Born into a wealthy family, his maternal uncle, the bishop of Varmia, took charge of his education after his father's death. He studied law, medicine, mathematics, and astronomy at the University of Krakow. In 1496 he was appointed* canon *and left to study canon law at the University of Bologna, where he also learned philosophy and Greek. He joined the astronomer* Domenico **Maria da Novara** *(Ferrara 1454-1504 Bologna, mathematician, astronomer, physician, and a university professor,* Catholic*) who did not adhere to Ptolemy's model. He assisted him and made his first discoveries in astronomy. He began to study medicine in Padua. In 1503, having completed his studies, he returned to his diocese where he assisted his uncle and practiced medicine, a field in which he became very well known. He carried out both his ministry and his research in astronomy (where he was recognized by his peers). He published (confidentially) in 1513 "commentarius," a treatise in which he exposed his heliocentric model. Partisan of the circular orbits, he will not be able to conclude definitively. Nevertheless, from 1533* **Pope Clement VII** *was aware of his work. Bishops encouraged him to publish his work, and a few copies were circulated as early as 1540. In 1616 (73 years after his death), his work was put on the index by the Catholic Church until 1835. A crater on the Moon bears his name, as well as a chemical element: Cn112, and an asteroid: 1322. He wrote, among other things, a treatise on money and translated the work of the 7th-century Byzantine historian, Theophyled. The Catholic Encyclopedia writes* "What is most significant in **Copernicus**' character is that, while he did not shrink from demolishing a scientific system consecrated by a thousand years of universal acceptance, he set his face against the reformers of religion."

[136] **Luther** Martin *(Eisleben, Saxony 1483-1546). Theologian follower of* **St. Augustine**. *University professor, he is the initiator of* Protestantism. *Excommunicated for preferring to base his faith on the Bible - of which he made a rather polemical translation because deeply antisemitic - than on the authority of* **Pope Leo X** *and the religious hierarchy. Among other things, he condemned monastic life, and advocated the freedom of man, whose faith alone could bring him salvation, ...*

[137] **Calvin** John *(Noyon, FRA 1509-1564 Geneva, CHE). Catholic theologian, and follower of* **St. Augustine**, *he broke with the catholic religion in 1530 and became the founder of* reformed

Copernicus' dedication to understanding God's creations prevailed, the definitive printing of his work: "*On the Revolutions of the Heavenly Spheres*," was not published until shortly before his death, leaving him time to dedicate a copy to **Pope Paul III**. Copernicus would have written[138] about his thesis:

> "*To know the mighty works of God; to understand His wisdom and majesty and power; to appreciate, in degree, the wonderful working of His laws, all this must surely be a mode of worship agreeable and acceptable to the Highest, to whom ignorance cannot be more grateful than knowledge.*"

We will remember the opinion of Isaac **Newton** who, speaking of **Copernicus**, wrote later ([Ne2] T2. p.2):

> "*This great man renewed the ancient system of the Babylonians & Pythagoras, ... thus the sun was put back in the center of the world.*"

- In Meantime, in **1453** Fra **Mauro**,[139] a *Benedictine monk*, made the first planisphere of the known world. In 1459 he offered an improved copy to Alfonso V, king of Portugal, to encourage him to continue to finance explorers.

- As **Copernicus'** ideas spread, a unique dialogue unfolded at the end of the 16th century with Chinese philosophers and scientists. *Jesuits* like Matteo **Ricci**[140] brought Christianity, Western Philosophy, and advanced sciences such as Mathematics, and Astronomy to China. **Xu Guangqi**,[141] "*the servant of God*," a disciple of great value; and a renowned astronomer who *converted to Christianity*, played a crucial role in this

---

theology. *A prolific apologetic writer whose ideas were adopted by the Reformed and Lutheran Churches. There is much to say about this man who defended the protestant martyrs with all his might, but this is not the purpose of this essay.*

[138] *Cf. [Col] P. 230.*

[139] **Fra Mauro** *(Venice, ITA 1385-1460), a Benedictine monk, was the most famous cartographer of his time. The oldest known map of the world was found in Mesopotamia where it was made 2,600 years before Christ.*

[140] **Ricci** Matteo *(Macerata, ITA 1552-1610 Pekin, CHN). Italian Jesuit who served as an interface between Europe and China, giving each the other's best. He introduced our knowledge of Mathematics and Astronomy to China and the Philosophy of Confucius to Europe. He drew the maps that made the world known to the Chinese. He offered one to the emperor to whom he was the first Westerner to be so close. A copy is in Shanghai in Nandan Park where I could see it in 2016.*

[141] **Xu Guangqi** *(Shanghai 1562-1633 Beijing, CHN). Took the name of* **Paul** *in 1603 at his baptism. He held high positions at the Ming court and was appointed Minister of Rites. A great military strategist in the service of the Ming dynasty. His treatises on agriculture which he renovated were recognized. In his time, mathematics was in full decline in China. He worked with* Matteo **Ricci**, *who introduced him to Western science. Together they translated many Western texts, including the* "Elements of Euclid."

exchange. Joining the *Jesuit brotherhood* he became one of the three pillars of Chinese *Catholicism*.

We can imagine the excitement of those times, a clash of ideas, the pursuit of knowledge, and the unfolding of a new understanding of our place in the cosmos. These pioneers, from Ibn al-Shatir to Copernicus and the Jesuits, paved the way for a scientific revolution that still shapes our world today.

2.7. 17th century:
Development of experimental sciences,
Physics becomes Mathematics,
The telescope gives access to the universe => Heliocentrism*
Theory of Gravitation: Newton unifies Heaven and Earth[142]

"The most important thing in the known universe is still immediately behind the astronomer's eyes."

Holmes Rolston[143]

In the 17th century, the sciences began to take precedence over theological philosophy. Nonetheless, religious faith remained alive, fueled by the controversies between Jesuits and Jansenists. Crowds flocked to hear the sermons of the great *Catholic* orator **Bossuet**, a member of the French Academy. He did not hesitate to remind kings, including **Louis XIV**, of their duties towards the people (cf. ref. 181-p. 40).

The evolution of scientific knowledge will be the fruit of the intuitions of geniuses such as G. **Galileo**, *Catholic*, J. **Kepler**, *Lutheran*, I. **Newton**, *Anglican*, J. C. **Maxwell**, *Presbyterian*, M. **Planck**, *Protestant*, A. **Einstein**, *Spinozian deist*... who put their intuitions and mathematical theories to the test of the results of the experiment. A. **Einstein** will write (cf. [Ei1] p.133):

"... any attempt to deduce logically, from elementary experiments, the fundamental concepts and basic laws of mechanics, remains doomed to failure... The axiomatic foundation of theoretical physics is not deduced from experience, but must be established spontaneously, and freely,...

---

[142] I. **Newton** *demonstrated that the same gravitational constant G governs the fall of objects on the Earth and the movement of planets in the Universe.*

[143] **Rolston** Holmes *(Staunton VA-US 1932) Calvinist. Received a B.S. in Mathematics and Physics in 1953 from Davisson College, NC, and a Bachelor of Divinity in 1956. He was ordained a Presbyterian priest the same year. Defended a Ph.D. in Theology in 1958, and a Master in Philosophy of Science in 1968. Assistant Professor of Philosophy in 1969 and Professor of Philosophy in 1967 at Colorado State University. Winner of the prestigious Templeton Prize in 2003 and the Mendel Medal in 2005. He is a specialist in the relationship between science and religion and in ethical ecology. The epigraph was taken from ([Po], p. 28).*

The usable mathematical concepts can be suggested by experience, but never, in any case, deduced."

1609. Galileo **Galilei**,[144] a *Catholic* Astronomer, knew how to distinguish between religion and science. His name is inseparable from the birth of modern science combining experimentation and mathematical interpretation. Improving the existing glasses,[145] he made *"Galileo's telescope,"* putting the cosmos within reach of humanity. With his telescope, he observed that four small satellites revolve around Jupiter, proving that **Ptolemy**'s model of the Earth as the center of the solar system was wrong. Despite metaphysical oppositions, it was undeniable in 1609 that everything did not revolve around the Earth. But it was still necessary to be able to say it because the Aristotelian lobby was powerful and its enemies within the Roman Curia were as influential as they were jealous. Nevertheless, in **1625**, he received the support of Pierre **Gassendi**, a *Catholic*. With Galileo, we moved from purely speculative thought to the search for truth by the empirical route. His book *"Discorsi,"* published clandestinely [146] in

---

[144] **Galilei** Galileo *(ITA: Pisa 1564-1642 Arcetri). Mathematician, author of essays on the hydrostatic balance and the center of gravity of solid bodies. He was the founder of dynamics, which he distinguished from statics. Rejecting the 20-centuries-old ideas of* **Aristotle**, *he studied experimentally the fall of bodies and discovered the relationship between space and time. He gave his letters of nobility to the experimentation based on personal questioning.*

[145] **Galilei** *knew from Jacques Buvedère, a scientist at the court of Henry IV, that the Dutch optician* Hans **Lippershey** *(NLD: Wesel 1570-1619 Middelburg) had made a telescope, one of which was in Paris and the other in the Netherlands.*

[146] **Galilei** G. *A convinced Catholic, despite the attacks he suffered from the Roman Curia. His work brought him before an ecclesiastical tribunal in 1633, which found him guilty of having propagated the Copernican theory that the Earth was a satellite of the Sun. His sentence of imprisonment was commuted by the pope to house arrest (Aristotle was also exiled for not conforming his ideas with those of Athens). He was not trying to prove that the Bible was wrong, but that its use of allegories could have led the first writers to a wrong interpretation. But - with the Renaissance, then with M.* **Luther** *in 1519, followed by the Protestants and the Catholics - the theory of the four meanings (cf.* **Origen** *ref.70, p.17) had been obscured in favor of a purely literal reading. In the same way, he could not ignore that Babylonians and Pythagoreans* "regarded the Sun as the center of the universe" *(cf. [Ne]Volume 2). This error was only corrected in 1893 by* **Pope Leo XIII** *(ref.76, p.18), notably in his Encyclical* "Providentissimus Deus." *Moreover, why did they resist so much to the heliocentric model? One could not ignore at the time the works of* N. **Copernicus**, *of* J. **Kepler**, *nor that in primitive religion the Fire (representing the Sun) was sacred, which is why Abel and Cain conceived of it as such when they offered it to God respectively the firstborn of their cattle or the first products of the soil (cf. gn. 4,3). Soon altars and then prytanea were built to house the sacred fire and these prytanea were generally circular to represent the heavens with the Fire in its center. It was not the Earth that was at the center; therefore, the thesis of a misinterpretation during a translation from Latin to Greek for example seems to me to be admissible. B.* **Pullman** *cautiously suggests another reason for this condemnation: the fact that Galileo was convinced of the existence of the Atom - which was a problem for his enemies - but not for him when they spoke of transubstantiation. We must not lose sight of the fact that if the Catholic Church was wrong in this matter, it was then under attack from protestants, and its leaders lacked foresight because they were too influenced by Aristotelian philosophers who saw in his work the proof that their thinking, which*

Holland, 4 years before his death, marks the birth of modern physics. It will know a growing success and prestige in society.

According to the historian and biographer Walter **Isaacson**,[147] the *Catholic* polymath **Leonardo da Vinci** [148] associated, a century before G. **Galileo**, experimentation with observation. **Leonardo da Vinci** wrote ([Is] p.173):

> *"My method is to experiment and then, thanks to reasoning, to interpret the experiments."*

As he grew older, he closely associated experience and theory.

The last resistance to the heliocentric theory disappeared in **1687** with the work of Isaac **Newton** [Ne2]:

> *"The true system of the world having been discovered ...Ptolemy and those who wished to support this opinion of the rest of the Earth (he attributes the responsibility to Aristotle) ... were obliged to embarrass the heavens with different Epicycles, & with an innumerable quantity of circles, very difficult to conceive & to employ, for there is nothing so difficult as to put an error in the place of truth."*

1609 & 1618 Johannes **Kepler**,[149] Lutheran, was described by A. **Einstein** as *"a wonderful man,"* writing[Ei1 p.VIII]: *"clear-sighted mind ... no one before him and even since has opened new ways of thinking, research, ... genius inventor of particular guiding methods .... All these reasons make us admire him...."* Kepler

---

had prevailed for more than 20 centuries, was outdated. Moreover, Galileo did not make much effort to defend himself, focusing mainly on the phenomenon of the tides with very weak arguments. The Church began to acknowledge its wrongdoing in 1741 through **Pope Benedict XIV**, *then in 1962* **(Vatican II)**, *and 1979 and 1992 with* Pope **St. John Paul II**.

[147] **Isaacson** Walter *(New Orleans, US-LA 1952). Professor of History at Tulane, New Orleans, was CEO of the Aspen Institute, president of CNN, a publisher of Time magazine. He is the author of biographies of famous men such as Steve Jobs, A. Einstein, ... and "Leonardo Da Vinci"* [Is].

[148] **Leonardo da Vinci** *(Vinci, Tuscany, ITA 1452-1519 Le Clos Lucé, near the castle of Amboise FR).* "Universal genius," "the man of the Renaissance," *was a painter (author of, among others,* "The Last Supper" *in the Monastery of Santa Maria Delle Grazie in Milan-ITA,* "The Mona Lisa" *in the Louvre, Paris, FR ...; a sculptor (*"Il Cavallo" *equestrian statue of the Duke of Milan* **F. Sforza** *...); military and civil engineer, he designed many machines (see [Na]), architect and geometrician, a scientist specializing in dynamic mechanics, optics, hydrodynamics, anatomy.) At the age of 20, his talent as a painter was noticed. He had for patrons the princes of the time and was finally the "protégé" of* **François 1st,** *king of France.*

[149] **Kepler** Johannes *(DEU: Weil 1571-1630 Regensburg). Astronomer, a supporter of heliocentrism, wrote many articles and established what is called "Kepler's three laws" describing the motion of planets.*
1st law (1609): *The orbit of a planet is an ellipse with the sun at one of the two foci.*
2nd law (1609): *An imaginary line joining the Sun and a planet sweeps equal areas of space during equal time intervals as the planet orbits.*
3rd law (1618): *The square of a planet's orbital period is proportional to the cube of the length of the semi-major axis of its orbit.*

was the imperial mathematician of **Rudolf II** (Holy Roman Emperor) and founded dioptrics, which was to become optics, a field of Physics to which he contributed a lot. As an astronomer, he was fascinated by the study of the movement of the planets. Having the intuition that the orbit of the Earth is an ellipse with the Sun as a focus, he undertook enormous calculations (computers were not available) that allowed him to make one of the greatest discoveries in the history of science: not only, did he prove that the orbits of the planets around the Sun are elliptical, but he also established three fundamental laws describing their motion, improving the heliocentric model of N. **Copernicus** and supporting the thesis of G. **Galileo**. A. **Einstein** described his work as brilliant.

Kepler, a devoted *Protestant,* [150] wrote in *"Cosmographic Mystery"* that he felt called by God to make known to men the magnificence of the divine work. For him mathematics is uncreated, it participates in God, which did not prevent him, on the contrary, from being an astronomer supporter of heliocentrism.

In **1620**, the renowned philosopher of sciences, and statesman Lord Francis **Bacon,** [151] an Anglican, who can be considered one of the pioneers of the scientific method and empiricism published *"Novum organum scientiarum"* [Bn], his major work in which he advocates for a new method of acquiring knowledge, based on systematic observation, experimentation, and inductive reasoning, contrasting with previous deductive methods.

1637. René **Descartes,** [152] a *Catholic* scientist, had a revelation during the night of 10/11/1619 [Br]:

---

[150] **Kepler** J., *a Lutheran Protestant, entered the evangelical seminary Tübingerg Stift in 1589 where, in addition to Greek, Hebrew, and Theology, he studied Astronomy. When the Thirty Years' War broke out, he was persecuted because, refuting the cult of images and saints, he did not want to convert to Catholicism. According to* Ernst **Zinner**, *he wrote:* "...to suffer with many brothers for his religion and the glory of Christ, to endure evil and disgrace, to leave his home, his fields, his friends and his country, I would never have believed that all this was so pleasant." *Unlike* **Newton**, *he adhered to the concept of the Trinity.*

[151] Lord **Bacon** Francis *(London 1561-1626), Viscount Sant Alban, Speaker at the Parliament, was Lord Chancellor of England from 1618 to 1621. Owing to poor health he was educated at home until the age of 12 when he went up to Trinity College at the University of Cambridge for 3 years. Then he went to the University of Poitiers and Cambridge University. A philosopher of sciences, he laid the groundwork for the scientific method and encouraged a more systematic and evidence-based approach to understanding the natural world. He was knighted in 1603 by King James I of England. In 1618, he was elevated to the rank of Baron and became* "Lord Verulam."
Anglican, *he held various positions in the Church of England.*

[152] **Descartes** René *(La Haye-en-Touraine, FR 1596-1650 Stockholm, SWE). Mathematician, at the origin of analytical geometry and physicist, he established concomitantly with Willebrord* **Snell**, *the "Snell-Descartes laws" on the optical refraction and reflection of light;* Catholic *philosopher he accepted the idea that God has left to Man the responsibility of his acts. He will be followed by other philosophers like E.* **Kant**, *and G.* **Hegel**, *... Cartesianism will influence all of the West. According to him, scientific knowledge is similar to a tree whose roots are Metaphysics (God created Man, to whom he reveals the truth), the trunk being Physics (allowing us to find the rules that govern the universe), the branches (the other sciences, morals). Of fragile health, he entered only eleven years the college where the Jesuits taught him*

"... there is a fundamental agreement between the laws of nature and the laws of mathematics."

This revelation influenced the course of his life. To avoid falling victim to prejudice, he made a clean sweep of everything in which he could have the slightest doubt. The fact of denying everything he had been taught justified, according to him, that he thought, hence a truth taken from the Latin *"Cogito, ergo sum"* which appeared to him unmistakable and which he translated as:

"... I think, therefore I am." [153]

According to the *Christian* philosopher Georg **Hegel,** [154] his philosophy - at the origin of a renewed interest in the Pythagorean idea of the involvement of numbers in the formation of the universe - broke with the past. Encouraged by Cardinal **de Berulle** [155] to deepen his modern approach to Philosophy - which, although he was a *Catholic,* broke with traditional Christian philosophy - he published in 1637 the *"Discourse of the Method,"* his major philosophical work: an *intuitive* [156] and *deductive* [157] mathematical method based on the concept of a *clear idea.* [158] To illustrate this concept he writes, recognizing that man, this

---

Mathematics, Physics, and Philosophy. In 1618 he obtained a Law degree and thought about a military career, traveled all over Europe thanks to his family fortune, frequenting artists and scholars until the night of 10/11/1619 when he had a dream that led him to publish in Holland in 1628: "Rules for the direction of the mind" *(14) and in 1637 the famous* "Discourse of the method." *He also invented devices for disabled people. Jean de La Fontaine said of him:* "this mortal whom the pagan would have made a god." [Hat]

[153] See his "letter to ***" (Leiden Nov. 1640). *This concept is also called* "Descartes' cogito." S. **Hawking**'s *condensed definition of the anthropic principle three and a half centuries later in* [Ha] p. 151: "We see the universe as it is because we exist" *seems to me an updated version of this famous cogito*

[154] **Hegel** Georg W. F. (DEU: Stuttgart 1770-1831 Berlin). *Christian philosopher, a representative of German idealism, has been called the* "father of modern philosophy." *His philosophical system is interested in all the problems of society. He had a great influence on the philosophers who followed him.*

[155] de **Bérulle** Pierre *(FR: Cérilly 1575-1629 Paris) named cardinal by* **Pope Urban VIII** *in 1627. Founded in 1621 the Oratory of France, a French component of the Congregation of the Oratory (society of secular priests*, without vows) founded in Rome in 1564 by* St Philippe Néri. ** Secular priests who live among the laity, unlike regular priests (Dominicans, Franciscans, Jesuits, etc.) who live in monasteries or abbeys.*

[156] "By intuition I mean, ..., the conception of a pure and attentive mind, a conception so easy and so distinct that no doubt remains as to what we understand ... a conception which is born ... of reason ...." *Rule II pp. 43 and 44 of [Br]. He relies on the Latin meaning of the word he uses and not on the definitions adopted.*

[157] Deduction: "everything that is necessarily concluded from other things known with certainty." *Rule II p. 44 of [Br]. All intermediaries are known with certainty.*

[158] Clear idea: *those ideas which are so self-evident that, while they are held in an attentive mind, they cannot logically be doubted. Opposed to the* "distinct idea" *which is so self-evidently true that it cannot logically be dou*bted.

infinitely small being, cannot understand the design of God, who is infinitely large:

> "...I know that God is the author of all things and that these [eternal]
> truths are something, and therefore that he is the author of them. I say
> that I know it, and not that I conceive it, nor understand it; for one cannot
> know that God is infinite and all-powerful, even though our soul being
> finite cannot understand nor conceive it." [159]

His method was revolutionary at the time; it went against the flow of **Aristotle**'s method for whom the perception of things plays a primordial role in world knowledge. The *Christian* Philosopher G. **Hegel** considered Aristotle the author of the revolution of mind, marking the beginning of modern philosophy, he made him a hero. To access scientific knowledge, "*the Cartesian method*" requires, that we follow four rules. The first precept:

> "... never to receive anything as true that I did not know to be such; that
> is, to carefully avoid haste and prevention; and to understand nothing
> more in my judgments than what would present itself so clearly and
> distinctly to my mind that I would have no occasion to doubt of it."

The psychiatrist and *Christian* philosopher K. **Jasper** will state [160] two and a half centuries later:

> "There is more intelligent than exploring the unknown, for example,
> doubting the known."

R. **Descartes** applied his method successfully to the sciences and published three famous articles on Dioptric, Meteors, and Geometry (he invented analytical geometry, combining geometry and algebra). About Dioptric, every high school student knows Snell-Descartes' laws on reflection and refraction. Pierre de **Fermat**, [161] also a *catholic*, did not adhere to R. Descartes' demonstration and proposed in 1657 a more metaphysical method [Ne2] whose principle is that:

> "...the author of Nature ... made the light go from one point to another
> by the path of the shortest time,"

it is known in optics as "Fermat's Principle" and used in Mechanics as laws of reflection and refraction called: the "*Principle of least action.*"

With the success of his method in the realm of science, R. **Descartes** then applied it to the knowledge of God and His work. While waiting to arrive there,

---

[159] Letter to Mersenne. *Amsterdam, 27/05/1630. Cf. ([Br] Pp. 937-938).*
[160] *Cf. Babelio's quotations*
[161] de **Fermat** Pierre *(FR: Beaumont-de-Lomagne 1601-1665 Castres). Catholic polymath, one of the greatest mathematicians of his time, invented with B. **Pascal** the calculus of probabilities, theory of numbers 1658, ... magistrate.*

he "*formed a moral by provision (provisional) ... to obey the laws and customs of my country, ... to modify one's desires rather than the world, to overcome oneself instead of trying to dominate the elements, to act always with firmness and resolution.*"

Then, relying on his Method he:

"*... tasks to demonstrate the existence of God and the soul separated from the body.*"[162]

If he separated the soul from the body he was nevertheless convinced that the soul interacted with the body at the level of the pineal gland.[163] This spirit-matter dualism has been challenged by many philosophers, as well as by the theory of evolution. This later raises the question: when appeared the soul in man's life? Nevertheless, it has supporters among recognized scientists such as Sir John **Eccles**[164] *Deist*, one of the greatest neurophysiologists of the 20th century, co-winner in 1963 with A.L. **Hodgkin** and A. **Huxley** of the Nobel Prize in Physiology or Medicine "*For their discoveries concerning the ionic mechanisms involved in excitation and inhibition in the peripheral and central portions of the nerve cell membrane*" see p.167. J. **Eccles** claimed [Ec]:

"*The more I study the brain the less materialistic I am.*"

R. **Descartes,** a convinced *Catholic*, starting from the postulate that the effect can only draw its reality from its cause,[165] otherwise it would draw it from nothingness, deduced ([Br] - p. 294) that:

"*.... the ideas of corporeal things.... are not formally in me, since I am only a thing that thinks;..... being myself a substance it seems that they (the ideas) can be contained in me eminently. Hence there remains only the idea of God..... infinite, eternal, immutable, independent, all-knowing, all-powerful substance, and by which I and all things that are... were created and produced,*"

and gives his definition of God. According to him, there is no finality in nature but only determinisms produced by efficient causes. The free will would be only

---

[162] Letter to Mersenne*, *Leiden, March 1636. (ref. [Br]- p.958).*

[163] The pineal gland or epiphysis *is located in the brain and secretes hormones directly into the bloodstream that regulate nervous and psychological activities.*

[164] Sir **Eccles** John *(Melbourne, AUS 1903-1997 Locarno, CHE). Neurophysiologist, promoted to sir in 1958, co-***Nobel*** of Physiology or Medicine in 1963, he is the recipient of many other prizes and medals. He graduated from Melbourne in 1925 and received a scholarship to Oxford where he defended his thesis in neurophysiology in 1929 under the direction of ***Sir Charles S.*** **Sherrington** *(ENG: London 1857-1952 Eastbourne), 1932* **Nobel** *Prize of Physiology or Medicine, an* Anglican. *Sir Eccles returned to Australia in 1937, where he was a professor in New Zealand, and then at the Australian National University from 1952 to 1962. Deist at the end of his career.*

[165] Third meditation *in* **Descartes** *Œuvres et Lettres ([Des] p.289).*

illusion because if the man is conscious of his acts he does not know the causes that lead him to act as he does. Nevertheless, for him [Sc], Man is united to God by Love, a constant and eternal love conditioning his salvation.

This philosophy, called: *"modern"* for its time, postulates that reason and revelation are independent. One can distinguish three pillars (concepts) in his philosophy: God, the World, and the Self separating the matter from the spirit, thus the body from the soul. Nevertheless, he places God much higher than the other two by linking the Self and the World, because He is in both. His abundant correspondence is interesting, particularly that with the *Jesuit* Father Marin **Mersenne**,[166] *"the secretary of the scholars of Europe"* according to Cornelis de **Waard**.[167] Mersenne was a Mathematician, a pioneer in the theory of numbers. Explaining certain particular points, at a time when there was no scientific periodical, he ensured the link between scientists and philosophers of Europe. According to P. **Duhem** [[Du]], Father Mersenne submitted to him the work *"De mundi systemate, partibus et molibus ejustem, liber singularis"* (Parisiis 1643) by the *Christian* Gilles **Roberval**,[168] who long before I. **Newton** had thought of universal gravitation. But R. **Descartes** rejected this work because according to him and his followers, action at a distance was an occult quality. What would he say about Quantum Mechanics corollaries and the Principle of Indeterminacy?

I am astonished that a courageous man (although of fragile health he joined the international military school of the Prince of Orange as an observer at the age

---

[166] **Mersenne** Marin *(FR: Oizé 1588-1648 Paris). Trained by the Jesuits, he was ordained* a priest *of the Order of the Minims (mendicant hermits and penitents close to the Franciscans) in 1612. He was a musicologist (in 1627 he published a* "treatise on universal harmony" *on music), a mathematician (a convinced Pythagorean), he published numerous articles, studied cycloids, and in 1644 proposed a formula: E = 2^p -1, which should give the integers E from the known integers p. It does not give them all but it was a beginning. Not valid for p=11). Physicist (he established the 1st law of acoustics and concomitantly to G.* **Galileo** *the law for falling bodies in the vacuum). He taught Philosophy and Theology from 1614 to 1619 at the convent of the Minims in Nevers. He became friends with P.* **Gassendi** *in 1624, R.* **Descartes** *in 1626, and many other scholars. Although pro-Galilean (he published* "Les Méchaniques" *in 1633) he did not pronounce himself on the Copernican system. He created in 1635 the* "Academia Parisiensis." *Opposed to the mystical doctrines (alchemy, astrology, ...) he vigorously defended R.* **Descartes'** *philosophy, G.* **Galileo's** *ideas, ...*

[167] de **Waard** Cornelis *(Bergen-op-Zoom, NLD 1879-1963 Island of Flushing, NLZ). Historian of science, he published many original writings of 17th-century scholars.*

[168] **Roberval** "The Mathematician," *from his real name* Gilles **Personne** (FR: *Roberval (60) 1602-1675 Paris). Son of peasants, it is the chaplain of his parish, who was also the chaplain of* **Marie de Medici**, *who detected his intelligence and allowed him to have a good education in Mathematics and Physics. He frequents the greatest scholars of the time. In 1631 he was appointed to the chair of Philosophy at the Collège de Maître Gervais (University of Paris), and in 1634 he was appointed, for 3 years renewable by competition, to the chair of Mathematics at the Collège Royal. In 1654 he obtained the chair of P.* **Gassendi**. *A co-founder of the Royal Academy of Sciences in 1666. A very skilled experimenter, he invented the* "Roberval balance." *His contribution to mathematics is considerable. He popularized science by addressing workers. If he published little during his lifetime, the Academy, for example, distributed lots of his writings.*

of 22, and did not hesitate to fight a duel for the beautiful eyes of a lady), could burn his book *"The World"* when he learned the result of the trial of G. **Galileo,** whose ideas he shared, under the pretext that:

> *"... I would not want for anything in the world, that it came out of my speech, where there was the slightest word which was disapproved by the Church."* [169]

It's noteworthy that R. **Descartes** was among the first scientists to use the concept of *"imaginary number"* which is the product of a real number y by "i" (where $i^2 = -1$) giving the imaginary number "iy." The addition of the real number x gives the *"complex number"*: $x + iy$ which will be promoted a century later by L. **Euler**... G. **Leibniz** is quoted as saying of them:

> *"... imaginary numbers are a fine and wonderful resource of the divine intellect, almost an amphibian between being and non-being."*

R. **Descartes** played a central role in the 17th century, some attacked him, and others adulated him. It was the case of the prominent *scientist* Blaise **Pascal,** [170] an apologist of *Christianity* who was a precocious child. From childhood, he devoted his life to the pursuit of truth, with the help of reason. As a teenager he frequented scientific and literary salons from 1640 onwards, revealing a budding genius with an exceptional capacity for invention, attaining great fame at an early age. He had, in his own words: *"the spirit of geometry"* and a rigor of logical deduction which earned him universal recognition in the fields of:

- Physics and more particularly fluid mechanics where his measurements on the pressure of air contributed to prove the existence of the vacuum. Each of us has learned Pascal's theorem. From this came the invention of the hydraulic press, the barometer, and the syringe, ...

- Mathematics where he laid down the principles of infinitesimal calculus, probability 1 calculus, and combinatorial analysis; we can say that he opened the doors of modern science. We 1 1 owe him, among others: Pascal's triangle (representation in a triangle of the binomial coefficients,

---

[169] Lettre à Mersenne, *Amsterdam, February 1634.*
[170] **Pascal** Blaise *(FR: Clermont-Ferrand 1623-1662 Paris),* Catholic. *With a fragile health, he was a precocious child whose questions about the nature of things impressed those around him. He was a mathematical genius and an inventor (the arithmetic machine 1642). Physicist, he gave his name to many theorems. At the age of 11, he wrote the* "Treaty of Sounds," *at the age of 12 he proved that the sum of the angles of a triangle is equal to 180° (Euclid's 32nd proposition), at the age of 17 he published the* "Essay for the Conical." *Precursor of modern science, he devoted himself from 1654 to the defense of God and of the Jansenists (rigorous Catholics followers of St Augustine, opposed to the Jesuits). He opposed royal absolutism and the power of the Holy See. Charity occupied an important place in his conception of the Catholic religion:* "The only object of Scripture is charity" *(cf. [Pa] 670-270]).*

fig. 2), theorems in geometry, the resolution of the problem of the cycloid... He is the ancestor of 1 2 1 the calculating machine (1642). 1 3 3 1

1

1 1

1 2 1

1 3 3 1

1 4 6 4 1

Fig 2.

It is only in 1646, the year when he meets the Jansenists who look after his father, that he starts giving an important place to religion. He will focus on it in 1656 during what all commentators 1 4 6 4 1 have called: "*the night of fire.*" After a sermon that "*illuminated*" him, he had, during the night, a **Fig. 2** revelation of God that he recorded in "*the memorial*" as a reminder of this exceptional moment that was the origin of his renewed faith which grew to asceticism. He "*converted to Christ*" as wrote Pope Francis in the fourth century of the birth of Pascal [Fr2]. "*Marked by a fundamental awe and openness to all reality*" he did not oppose science and faith but distinguished their domains: that of reasoning on the one hand, that of authority and love on the other. When he made his famous bet[171] about God's existence, it was not because his faith was not genuine but, as a scientist addressing a rational person: he considers religion as a simple hypothesis that he wants to demonstrate through probabilities, a field in which he excelled.

From his philosophical and theologian work, devoted to God, we retain especially "*Pensées*" [Pa] collection of his writings dedicated to the defense of *Christianity* and *Jansenists* in particular; the draft of what should have been his "*Apology of the Christian religion.*" *In the last days of his life, he is reported to have said "If the physicians tell the truth, and God grants that I recover from this sickness, I am resolved to have no other work for the rest of my life except to serve the poor*" [Fr2].

Pascal, was in the line of R. **Descartes** with whom he shared the idea of separation of the eternal soul from the corruptible body, the infinitely great and small; that only God - author of all the wonders of nature - can apprehend in its globality; but in contrast to R. **Descartes**, for whom reason allows one to reach certainty; for B. Pascal the heart, which he places above the reason, seizes intuitively the truths escaping the rational demonstrations. So, in "*Submission and use of reason, in what true Christianity consists of*"[Pa] he will define the three basic qualities that one must have:

"*We must know to doubt where we need to do, to feel certain where we need to do, to submit where we need to do.*"

---

[171] Pascal's Wager [Pa]: *For B. Pascal, according to probability theory, rationalists should believe in God: if God exists, they stand to receive eternity in Heaven and many gifts; whereas if He does not exist, they will have little losses: only some pleasures....*

Pascal paved the way for the emergence of another profound thinker, Baruch **Spinoza**,[172] a *Jewish pantheist* who came into prominence a few years later. Spinoza's intellectual journey was shaped by education under the tutelage of a rabbi who followed in the footsteps of the renowned M. **Maimonides**. This period instilled into Spinoza a unique perspective: he boldly dissociated Faith from Reason. He applied the Cartesian method of the Clear Idea to elucidate the concept of monism.[173] Unlike R. **Descartes**, B. Spinoza embraced the idea of a singular Being: God, a divine entity omnipresent in everything yet possessing two distinct natures:

- The first is the divine aspect, an immutable entity with various attributes, including the infinite and unique Substance, the originating Thought of Intelligence from which souls emanate, Infinity, Eternity, Omnipotence, Goodness, Virtue, and more.

- The second is the created aspect, comprising *"modes"* resulting from God's existence. These modes exist in a hierarchical order, resembling the intermediaries found in **Plotinus'** Philosophy. According to B. **Spinoza** the entire world, encompassing objects, humans, animals, minerals, and more, represents nothing but an emanation of God and shares in His eternal nature. Each element possesses a distinct code, but humans hold the highest position. In this view, humanity is inseparable from Nature, and the soul and body are two different modes that emerge from the *"divine substance."* Spinoza rejected Cartesian dualism, which delineates the separation between the soul, capable of ideas, emotions, and desires, and the human body, governed by the laws of Physics and Biology. Instead, he posited that these are two facets of a unified reality: the *"infinite substance."* Thought and body are interdependent, with our ideas and actions being the complex result of the interplay between soul and body; both, being expressions of the same underlying substance. In

---

[172] **Spinoza** Baruch *(NLD: Amsterdam 1632-1677 The Hague). Studied the Bible and the Talmud in Hebrew. At the age of twenty, he questioned the teachings of Judaism and Christianity. For him, Good and Evil do not exist but Right and Wrong exist for Man. He was a* Jewish *philosopher with a pantheistic conception of God, therefore immanent, as opposed to the transcendental conception of God revealed by the Scriptures. In 1656, he was expelled for life from the Jewish community of Amsterdam for heresy. According to H.* **Atlan,** *the real reasons are not known. Nevertheless, he always considered himself a* Jew *and refused to convert to Christianity a religion with which he shared certain values without adhering to it. For him, Jesus was only a Patriarch. He developed in* "Ethics," *his most important work, which was published only after his death, a non-anthropomorphic vision of God that is called:* "Spinoza's pantheism."

[173] Monism: *doctrine advocated by* Adi **Shankara** *(*Tamil*, "avatar of Shiva" according to his disciples). He was born in KaladyKerala and died in Kedarnath-Uttarakhand, India). Supporting the doctrine according to which Atman = Brahman (cf. chap. 5.1.2. p.140), and that there is only one substance, the monad (from the Greek monas meaning: unity or that which is one), having various appearances according to whether it is nature, man, ... Banned by the Abrahamic religions, it is at the base of oriental religions such as Hinduism, Taoism, ...* G. **Leibniz** *developed this concept.* * Tamil*: an inhabitant of southern India and Sri Lanka.*

Spinoza's philosophical framework, freedom finds no place, as determinism reigns supreme. Our knowledge of the tangible world is considered incomplete yet not entirely erroneous. Conversely, rational knowledge is held in high esteem; it is perceived as divine and complete, leading to happiness. Through rationality, one can perceive the world according to a divine viewpoint, much like how God perceives it.

Despite his expulsion from the Amsterdam Jewish community before publishing any of his works, **Spinoza** was not an atheist. He advocated for believers to accept the dogma of God's existence as presented in religious Scriptures. He also encouraged the use of Reason to interpret all life's aspects. As a result, contemporary rabbis have sought to rehabilitate his legacy.

Furthermore, **Spinoza**'s contemplations on infinity had a profound impact, on science and theology. His ideas influenced the mathematician Georg **Cantor**,[174] a *Lutheran*, who would later develop the theory of sets.

Cantor's contributions have earned the admiration of scientists in the 21st century, highlighting the enduring relevance of Spinoza's philosophical insights.

Two centuries later, some of **Spinoza**'s ideas, such as the "*monism of body and Spirit*" find resonance in the thought of the Spinozian *pantheist* biologist Henri **Atlan**.[175] Atlan pioneered the theories of complexity and self-organization in living organisms, drawing parallels between science and biblical texts. Atlan's Judaism is distinctive, it "*does not imply a belief in God*," and it "*is not a commandment*" addressed to **Moses**. According to him "*Judaism is only half a religion*," positioned as the third monotheistic one after Catholic and Muslim.

1640. John **Wallis**,[176] after studying Theology at Cambridge was ordained a

---

[174] **Cantor** *Georg (St-Petersburg, RUS 1845-1918 Halle, German Empire). From 1860, his exceptional qualities in Mathematics were recognized, he continued his studies in this field at the University of Berlin where he defended a thesis on the theory of numbers in 1867. He was promoted to Senior Lecturer in 1872 and to Professor in 1879. Cantor's theorem induces the existence of an infinity of infinities. Although his work was not unanimously accepted, it was nonetheless important.* David **Hilbert,** *one of the greatest mathematicians of the 20th century said, talking about his work:* "No one should exclude us from the paradise that Cantor has created." *His father was Jewish and converted to Lutheran Protestantism, and his mother was Catholic, who converted to her husband's religion at the time of her marriage. He was educated according to the* Lutheran *rite, to which he remained attached throughout his life.*

[175] **Atlan** Henri *(Blida, ALG-FR 1931). Doctorate in Medicine in 1958 in Paris, and State Doctorate in Sciences in Paris VII in 1971. Biologist, writer, and philosopher, he is known for his reflections on sciences, freedom, and morals. Professor of Biophysics in Paris and Jerusalem. Member of the Ethics Committee for Life Sciences and Health for 17 years. Director of studies at the Ecole des Hautes Etudes en Sciences Sociales. Cf. his interview with A. Mercier on* "Le Judaïsme demi-religion" *Akadem Multimedi Sept. 2014 – Internet.*

[176] **Wallis** *John (Ashford 1616-1703 Oxford, ENG). Mathematician,* Presbyterian. *He studied theology, philosophy, and musicology at Cambridge then turned to mathematics where he excelled. We owe him, among other things, the symbol of the infinite, the infinitesimal calculation where he was the precursor of I.* **Newton**, *the product of Wallis, the calculation of ephemerides notably that of Venus in 1639... His work in phonetics led him to be a precursor in the education of deaf-mutes. Author of works in mathematics. He is one of the founders of the Royal Society.*

Presbyterian priest but quickly turned to mathematics, a field in which he proved to be one of the greatest geometers of the century according to G. **Leibniz**. For proof, he will be awarded the Savilian chair of geometry at Oxford.

1646. Pierre **Gassendi,**[177] *Catholic canon.* Convinced of the existence of the vacuum, by the experiments of Evangelista **Torricelli**[178] *Catholic,* and B. **Pascal** *Catholic,* Gassendi criticizes Aristotle's anti-atomism as well as his refutation of the vacuum. He works for the revival of the atomism of Epicurus of which he is a fervent defender. Mathematician and *"first true astronomer, converted to Galileo's ideas"* [Co] according to the Academician Françoise *Combes,* Professor at the Collège de France (chair *"Galaxies and Cosmology"*) he made numerous observations of eclipses of the Moon and the Sun, of the passage of Venus and Mercury in front of the Sun, of Jupiter, of the Northern Lights, ... corroborating the Copernican theory.

1651. Christian **Huygens,**[179] one of the greatest multiform scientists of all times, *Calvinist,* becomes known as a mathematician and will become a recognized specialist in geometry. In **1655** he published the first book on the calculation of probabilities. Fascinated by astronomy, observing Saturn (the 6th planet of the Earth, 2nd by its size and its mass, whose rings had been observed for the first time by **Galileo** in 1610), with a telescope carried out in the company of his brother, he discovered Titan, one of its 82 satellites. He is at the origin of other important astronomical observations. He had previously written a treatise

---

[177] **Gassendi** Pierre, **Gassend** *of his real name (FR: Champtercier-Digne (04) 1592-1655 Paris)* Catholic. "The greatest philosopher among men of letters, and the greatest man of letters among philosophers" *according to* Edward **Gibbon** *(Putney 1737-1794 London, British historian and politician). While keeping animals Gassendi became passionate about astronomy which became a vocation in 1618. In contact with the great astronomers of his time, including J. **Kepler**, he published with P. de **Fermat**. A recognized mathematician, in 1645 he was appointed professor of Mathematics at the Collège Royal (Collège de France), where he taught Epicurus' atomism until his death. Like R. **Descartes**, with whom he was for a time at odds, he corresponded with M. **Mersenne**. Philosopher; doctor in theology in 1614, he was appointed canon of Digne (04 FR) and professor of rhetoric. He led an almost monastic life.*

[178] **Torricelli** Evangelista *(ITA: Faenza 1608-1647 Florence). Educated by the Jesuits, a* Catholic *and scientist of reputation, seeing a correlation between Faith and scientific spirit, he was the pupil of Benedetto **Castelli**, Benedictine monk, professor of mathematics, and assistant of Galileo Galilei, whose disciple he became at the end of his life. After his death, he occupied the chair of Mathematics at the University of Pisa and was appointed Mathematician and Philosopher of the court of Grand Duke **Ferdinand II** of Tuscany. As a mathematician, he worked on the quadrature of the cycloid and hyperbolas. As a physicist, in 1643, he demonstrated the existence of atmospheric pressure (Torricelli's experiment) and created the first barometer. In 1644 he established the "Torricelli's law" on fluid flow.*

[179] **Huygens** Christian *(The Hague, NL 1629-1696), Calvinist, Mathematician (classical and analytical geometry, treaties on the quadrature, he worked to solve the problems that physicists posed ...), theoretical physicist (wave theory of light, calculation of centrifugal force ...), astronomer (discovered the rings of Saturn and its moon: Titan), inventor (pendulum clocks, telescopes, ...) Admired by the great contemporary scientists: I. **Newton**, G. **Leibniz**, G. de l'Hospital, J. **Bernoulli**...). Fellow of the Royal Society, In 1666, Colbert appointed him as the first director of the Royal Academy of Sciences. The asteroid 2801 bears his name.*

on dioptric, confirming the undulatory nature of light that the *Jesuit* Ignace-Gaston **Pardies**[180] had foreseen. As early as **1652** he studied **Descartes'** work and corrected his laws on elastic shocks. Working on the theory of the pendulum, he determined the period of the simple pendulum... He was at the origin of various inventions including the principle of the internal combustion engine with his assistant Denis **Papin,**[181] a *Protestant.*

The professor of philosophy and historian of science Gianfranco **Mormino** has studied [Mor] the role played by God in the work of Ch. **Huygens**. After reading the codex *"hugenorium 7A"* (a collection of Huygens' writings) the author reports that although Huygens was opposed to the interference of theological questions in physics:

"...it is not necessary, perhaps, to involve God in these questions,"

he would have admitted, on several occasions the *"necessary intervention of God"* whether in his relativistic theory of motion and rest or, in his exchanges on the atom with G. **Leibniz**, whose point of view he did not share. Speaking about the immutable grains of sand in the centuries he would have written:

"... It is that the creator once made them be born such, and in the same way for the atoms."

In *"The Cosmotheoros,"* his last work, Ch. **Huygens** does not reject the idea that God could have created on other planets *"an infinite variety of beings"* but it would be vain to believe that man can foresee their form. On the other hand, he thinks that since God imposed on nature simple universal laws, in limited numbers, it is so that the *"reasonable beings"* could find the causes.

1666. Gottfried **Leibniz,**[182] *Lutheran,* nicknamed the *"polymath"* (universal

---

[180] **Pardies** Ignace-Gaston *(FRA: Pau 1636-1674 Paris)* Catholic Priest *and scientist, he entered the Society of Jesus in 1652. After his ordination, he taught philosophy and mathematics at the "college de Clermont" in Paris. In 1662 he criticized Galileo's theory, published an instrument for building various sundials, and the* "Traité complet d'Optique" *quoted by Ch.* **Huygens** *and other scientific treatises.*

[181] **Papin** Denis *(FR: Blois 1647-1713). Doctor in medicine in 1669, but passionate about physics, is appointed two years later assistant of Ch.* **Huygens**. *He is at the origin of various inventions: the steam engine, the vacuum machine, the compressed air gun, the digester (precursor of the autoclave), the pneumatic water pump, a submarine .... Member of the Royal Society of London, and of the Academy of Berlin.*
Protestant, *foreseeing the revocation of the Edict of Nantes (Louis XIV in 1685) he leaves, from 1675, to join the* Anglican R. Boyle in London.

[182] **Leibniz** Gottfried W. *(DEU: Leipzig 1646-1716 Hanover). Virtually self-taught, he learned Greek and Latin at a very young age. He was initiated to Theology, Philosophy, Logic, read ancient authors ... before discovering in 1661: G.* Galileo, R. **Descartes** *to whose rationalism he adhered, B.* **Pascal**, *... . He studied Mathematics in Jena from 1661 and defended in 1663 a thesis in Philosophy: "On the Principle of the Individual" which will be an introduction to his work on monads, he will defend another one in Law in 1666... hence his various nicknames: "The living dictionary" given by the writer and scientific popularizer Bernard le Bouyer de Fontenelle, better known under the name of* **Fontenelle** *(Rouen 1657-1757 Paris), "A thinking*

mind), spanned the intellectual landscape, being at the same time a philosopher (his treatise of logic De Arte Combinatoria published at 20 years old put in light his thesis and himself), jurist, diplomat (ambassador in Paris from 1672 to 1676 he tried to convince Louis XIV to conquer Egypt). In **1672**, under the direction of Christian **Huygens**, a *Calvinist*, he devoted himself to mathematics and physics. Inventor, he created the calculating machine then, in 1683 he proposed a project for the desalination of water. In **1675** he laid the foundations of infinitesimal calculus (differential and integral) that he developed, independently of I. **Newton**, but which would be the subject of a quarrel between them. If we know today that I. **Newton**'s work predated that of G. **Leibniz**, we also know that Leibniz, who did not know Newton's work, published it before him.

His scientific research did not prevent him from devoting much of his time to Theology; on the contrary, they were complementary. Whereas R. **Descartes** - relying on the concept of the Clear Idea and the immutability (which cannot change) of God - had established the law of conservation of the quantity of motion (which he considered as an algebraic quantity, which is false), he established, by scientific reasoning, the laws of conservation of energy (which he called living force, an algebraic quantity) and the quantity of motion (product of the mass by the velocity vector, thus a vector quantity). This distinction between energy, an algebraic quantity, and momentum, a vector quantity, led him to write:

> "*It appears more and more, although all the particular phenomena of nature can be explained mathematically or mechanically by those who hear them, that nevertheless the general principles of bodily nature and of mechanics itself are rather metaphysical than geometrical and belong rather to some indivisible forms or natures as causes of appearances than to body mass or extent.*"

- In **1676** he will have fruitful exchanges with B. **Spinoza,** he will go further than the latter in the idea of coherence and perfection of the world. Trying to reconcile reasoning and mathematics (algebra), he postulates the "*Principle of sufficient reason*":

---

machine" by Denis **Diderot** *(FR: Langres 1713-1784 Paris) a leading figure of the Enlightenment in France, a contributor to the Encyclopedia.* Lutheran, *Irenist (supporter of the rapprochement between all the confessions and particularly with the Catholics)* **Leibniz** *tried in vain to reunify Catholics and Protestants. He was a member of the French Academy and Permanent Secretary of the Royal Academy of Sciences. His work is prolific. In "Theodicy," 1710, he exposed his ideas on divine justice. It is only in 1714 that he stated the question of the preamble (p.1) in: "Principes de la Nature et de la Grâce fondés en raison" (7ème art.). His conception of preestablished harmony - arising from the fact that, according to him, although there is no communication between monads, each being a world in itself, they agree with each other - will be questioned by P.* **Teilhard de Chardin** *(ref.505-p.155), a Jesuit priest, a scientist who was a follower of evolution.*

"Nothing exists without sufficient reason,"

His authorship was contested by some who opposed him to the principle of causality of R. **Descartes** but this seems unjustified.

- In **1684**, he published: *"Reflections on Knowledge, Truths, and Ideas"* establishing a direct relationship between the ideas of God and man.

- It was not until **1695** that he revisited the Pythagorean notion of *"monads"* [Lz2]:

a) *"The monad is nothing other than a simple substance, which enters into the compounds; simple, that is, without parts....... These monads are the true atoms* of nature and, in a word, the elements of things...they [the monads] can only begin by creation and end by annihilation instead of what is composed, beginning or ending with parts. ....."*

* his atom differed from the material atom of **Democritus**, whom he fought vigorously, by the fact that his monads have energy or force of their own and also because he could not admit that a *"material being (the atom of Democritus) could be both indivisible and endowed with a true unity"* ([Lz2]-2443). Let's note, that **Bohr**'s atom (ref.298, p.90), as we know it today, is divisible, contrary to **Leucippus**' definition (ref.46, p.13), and has electronic and nuclear energies too.

b) *"... monads must have some qualities, otherwise, they would not even be beings.... they would be indistinguishable from each other......[that] each monad be different, from each other. ... Monads have no windows through which anything can enter or leave them......Monads cannot be altered or transformed...the natural changes of monads come from an internal principle...they are endowed with perception and appetite [internal principle making the passage from one perception to another]."*

This idea will be taken up by the *Jesuit* **Teilhard de Chardin** and the *atheist* J. **Haldane** (ref.672-p.172).

c) *"I agree that the general name of monad and entelechy suffices for simple substances because they have in them a certain perfection, there is a sufficiency which makes them sources of their internal actions and, as it were, incorporeal automata.... and that only those whose perception is more distinct and accompanied by memory are called souls.*

G. **Leibniz** postulates that: the monads, in infinite number, are staggered one above the other, without empty space (for purely theological reasons he always denied the void, which was a serious error) between them; from the least perfects: elements of matter to the most perfect that of God, which he calls Soul, dominating all the other monads. Between them, he places the monads of the beasts, which have clear Ideas, and those of the finite spirits (men), which have clear, sometimes confused, but distinct Ideas, endowing them with "*intelligent or reasonable souls*" which alone know their actions, and which not only do not perish naturally but even always keep the basis of knowledge of what they are, allowing Man to access the knowledge of necessary and eternal truths. He will state:

> "... *the study of the preformation of plants and animals confirms my system of the pre-established harmony between soul and body, where the body is inclined by its original constitution to execute, with the help of external things, all that it does according to the will of the soul*" because "*the principle of action is in the souls.*"

Lutheran *Irenist*, G. **Leibniz** maintained an important correspondence with the great French *Catholic theologian* **Bossuet**[183] but, his spiritualist and optimistic rationalism will be one of the causes of the nonadhesion of this one. Some of his theological arguments were taken up and developed in the 20th century by the great *Deist* mathematician Kurt **Gödel**[184] in "*Gödel's Ontological Proof*," a work published after his death, in which he asserted his faith in God.

In **1669** the distinguished Isaac **Newton**[185] - affectionately referred to as the

---

[183] **Bossuet** Jacques-Bégnine *(FR: Dijon 1627-1704 Paris)* "the Eagle of Meaux." *Follower of scholastic, orator, and polemicist, he was a great* Catholic theologian. *Some of his sermons and Oraisons are famous. Although accepting some Thomistic and Cartesian ideas, his faith is based on the existence of God as presented in the Scriptures.*

[184] **Gödel** Kurt *(Brno, AUT 1906-1978 Princeton, NJ-US). Brilliant student, he won all the prizes in Mathematics, Languages, Philosophy, and Religion, ... In 1924, he joined the University of Vienna, where he had lectures in Theoretical Physics, Mathematics, and Philosophy. Then, he joined the* "Vienna Circle," *a group of scholars and philosophers. He obtained a Ph.D. in 1930 and published his first incompleteness theorem (ref.613-p.212) followed by other results having a big impact on* "Mathematical Logic." *After the Anschluss, he immigrated to Princeton where he befriended A.* **Einstein** *and revisited the equations of the theory of relativity by finding solutions.* Protestant, *emulator of G.* **Leibniz***, Gödel continued his theological work and undertook to logically demonstrate the existence of God:* "Gödel's ontological proof" *which was not published until after his death. He is a holder of the* "Albert Einstein Award," *the* "National Medal of Science," ...,

[185] Sir Isaac **Newton**(ENG: Woolsthorpe 25/12/1642 or 01/1643-1727 Kensington) was the first scientist to be knighted for his contribution to the Physical and Mathematical Sciences (he created the "differential calculus" at the same time as G. **Leibniz**). Astronomer, Newton worked in the field of optics where he established the basic laws, invented the telescope that bears his name, and polished himself the glasses of the optical instruments he needed; Alchemist, Theologian (ref.204-p.44). He is, with G. **Galileo**, G. **Leibniz**, M. **Planck**, and A. **Einstein** one of the greatest scientists of our time. In 1687 he made a leap in science by establishing the "Theory of Gravitation" or "Law of Universal Attraction": two objects of masses M and m, distant by d, attract each other with *forces F of the same intensity but of opposite direction. The modulus of F is proportional to the product M by m and inversely proportional to d^2. This relation is comparable to Coulomb's law expressing the electrostatic force acting between 2 charges Q and q distant of d. He introduced for that the* "constant of Gravitation G" *which is one of the 7 universal constants (G = 6,672 59. 10^{-11} $m^3.kg^{-1}.s^2$). This fundamental theory laid the foundations of classical mechanics, known as* "Newtonian mechanics," *and made it possible*

*"Great **Mr. Newton**,"* by the eminent *Deist*[186] **Voltaire**, [187] a leading figure of the Enlightenment era - succeeded Isaac **Barrow**,[188] his professor at Trinity College, Cambridge. Barrow, the first occupant of the **Lucasian**[189] Chair of Mathematics, initially Anglican had converted to *Catholicism* and left a lasting legacy. During Newton's time at Trinity College, he forged a friendship with the *Anglican* John

---

to establish the equation for the "motion of free falling bodies," called "Newton's second law of motion" *and to demonstrate* **Kepler's** *three laws. Like* **Galileo** *and* **Pascal**, *he bequeathed to us a method of work free of any metaphysical prejudice, based on observation and validation by experimentation. Entered Trinity College, Cambridge in 1661, Bachelor of Arts in 1665. He studied the ancient masters (**Aristotle, Euclid, Epicurus**...), as well as* J. **Kepler**, G. **Galileo**, R. **Descartes**, P. **Gassendi**, R. **Boyle**, R. **Hooke**, ... *The period 1664-65, was the most fertile for his work on light, hence the name "Annus Mirabilis" (Wonderful). It marked the break with the Aristotelian theses on the colors of the light. In 1696 he joined the Mint (issuing department of metallic coins) and became its director in 1700. President of the Royal Society in 1703.*

[186] Deist, *believes in God, creator of the universe, generally, but not always, appealing to pure Reason (cf. [Ka2] p.553) but does not adhere to any religion. Contrary to the theist, he thinks that God does not intervene in the affairs of Man. This idea of a "Supreme Being," "Archea" or "Principle" as some call it, does not fundamentally go against the trinitarian theories of Hinduism. We find there are some of the absolute characteristics that we attribute to God. On the other hand, for* B. **Pascal** *(Pensées VIII, 556):* ... *"Deism, [is] properly as far from the Christian religion as atheism."*

[187] **Voltaire** [Vo], *whose real name was* François Marie **Arouet** *(Paris 1694-1778). The "multiform" according to* **d'Alembert**, *so much he shone in all the literary genres that he approached. Author of numerous tragedies of historical character: Œdipe, Zaïre, La Henriade, the Death of Caesar, Mahomet, ... of philosophical tales: Zadig, Micromégas, Candide. The publication of the Philosophical Letters (he was strongly influenced by the parliamentary monarchy that he discovered during his exile in England from 1726 to 1729), obliged him to find refuge in Cirey, a province of Lorraine not belonging to France at this time. He was at the same time a courtier of the great ones of Europe, including* **François II of Prussia**, *but also a rebel, which earned him enmity and led him, in 1764, to buy* "La Ferney," *an estate straddling France and Switzerland. He had a church built there dedicated to God, marking his deity, which is confirmed in his will:* "I die adoring God, loving my friends, not hating my enemies and hating superstition." *A leading figure of the* "Age of Enlightenment," *he had many enemies including* **J.-J. Rousseau**. *He devoted his time and money to the defense of the innocent (Calas and Sirven trials, etc.) but was not entitled to a religious funeral.*

[188] **Barrow** Isaac *(London, ENG: 1630-1677). Mathematician, he is the precursor of infinitesimal calculus, and his work on Geometry is recognized. He was also interested in Philosophy and wrote about the Greek philosophers. Having studied the works of* J. **Kepler**, R. **Descartes,** *and Thomas* **Hobbes,** *his lectures on optics contributed to the improvement of knowledge in this field. Member of the Royal Society since 1662, the first holder of the Lucasian chair of Mathematics at Cambridge in 1663 after having been a professor of Greek at Cambridge.*
He resigned his chair in 1669 in favor of I. **Newton**, *his student, and devoted to Theology.* Anglican, *ordained in 1660, his sermons and writings on religion were very appreciated. In 1670 he was made a Doctor of Divinity (ranked ahead of all doctors at Cambridge) and* chaplain *of* **Charles II** *(1630-1685; the king of England, Scotland, and Ireland from 1660). In 1672 he was appointed principal of Trinity College, where he had been a student.*

[189] *In 1663 the Anglican Reverend Henry* **Lucas** *(ENG 1610-1663) donated to the University of Cambridge to create a chair of Applied Mathematics, where the greatest scientists followed one another:* G. **Stokes**, P. **Dirac**, S. **Hawking**, ...

Ray,[190] who shared his passion for natural history.

John **Ray** became one of the first British parson-naturalists: priests in charge of a parish for whom natural science study is connected with their religious work. For him, like later for another eminent parson, the naturalist, and *Christian apologist* William **Paley,**[191] the world's complexity was evidence of the existence of a Creator. In 1659, he was influenced by the future *Anglican bishop* of Chester (Province of York), John **Wilkins**[192] a natural philosopher, and co-founder of the Royal Society of London.

1685 Jacques **Bernoulli,**[193] a *Protestant,* published his first important papers in Mathematics: on probability, logic, and algebra. Known for the Bernoulli differential equation that he solved, his discovery of the sequence whose limit is the value of the mathematical constant "e"...

1687 Isaac **Newton** publishes: *"Opticks"* [Ne1] in Latin, a treatise on light (completing a first edition published in 1675) qualified by the *Christian* scientific abbot Alexis de **Rochon:**[194]

---

[190] **Ray** John *(Black Notley, Essex, ENG: 1627-1705).* Anglican *naturalists contributed to modern taxonomy (the scientific process of classifying living things). He attended Trinity College and Cambridge University. He was a lecturer in Greek in 1651, in Mathematics two years later, in Humanity. He was ordained* a priest *of the Church of England in 1660.*

[191] **Paley** William *(ENG: Peterborough 1743–1805 Lincoln).* Anglican, *philosopher,* theologian, *and anti-slavery. Senior Wrangler (top mathematics undergraduate) in 1763 at Christ's College, Cambridge; appointed as a tutor in 1768. In 1672 he was appointed principal of Trinity College, where he had been a student. He lectured on Moral Philosophy. A stall in his honor was erected in St Paul's, London.* Vicar *of Dalston (Cumberland, N-W of England) in 1780, the* Archdeacon *of Carlisle (West Cumberland) in 1782. Author of theological books relying on anatomy and natural science subject to controversy.*

[192] **Wilkins** John *(ENG: Fawsley, West Northamptonshire 1614–1672 London).* Anglican bishop, *a man of great culture and empathy. Educated in Oxford he obtained his BA degree in 1631, and 1634, a Master's degree in Art. Student of the* Puritan *(affiliated to Calvinists) astronomer John* **Bainbridge** *(ENG: Ashby-de-la-Zouch 1582–1643 Oxford). Warden (college's principal) of Wadham College in Oxford. President of Oxford and Cambridge universities he was a co-founder of the Royal Society in which he was elected fellow, then secretary.* Ordained priest *of the Church of England in 1638,* chaplain *of different Lords before being the one of Prince* **Charles Louis,** *nephew of* **King Charles I.** *Bishop of Chester in 1668. Living in a period of great religious and political controversy he reached out to men of all Religious and political stripes.*

[193] **Bernoulli** Jacques *(Basel, CHE 1654-1705) Swiss mathematician and physicist, exchanged with G.* **Leibniz** *(cf. [Th] p. 527). Against the wishes of his apothecary father, who intended for him to study Theology, he turned to the sciences (mathematics, physics, and astronomy). As his father financed him, he could meet the great European scientists of the time. Do not confuse him with his brother Jean (1667-1748, mathematician and physicist recognized who was with Jacques one of the first mathematicians to apply the infinitesimal calculus) or his nephews Daniel (1700-1782, scientist author of Bernoulli's Theorem and the theory of gases) and Nicolas (1695-1726, polymath, who occupied the chair of physics in Padua).*

[194] de **Rochon** Alexis Marie *(FR: Brest 1741-1817 Paris). Physicist, and astronomer of the Navy since 1766, he entered the Academy of Sciences in 1767. He traveled a lot and invented various instruments including the* "lunette de Rochon." *Although a Catholic priest, he was passionate about science.*

"... of immortal work, one of the most beautiful monuments of the genius of Newton...."

He was aware of the works of *Canon* Pierre **Gassendi**, of the *Jesuit* Francesco **Grimaldi**,[195] of the Irish scholar Robert **Boyle**,[196] *Anglican*, an ardent defender of the Christian faith, whose name remained attached to that of *Father* Edme **Mariotte**,[197] physicist and botanist, for their evidence of a law of perfect gases, the law of BoyleMariotte,[198] and of R. **Descartes**, whom he quoted. In the conclusion of Opticks, which he presents in the form of questions, I. **Newton** deals in the last one (the XXXIst) with some experiments that challenge him and starts talking about God, the Creator. He is a believer *(Anglican)*. We have reported in particular these passages:

> *"All this considered, it seems to me very probable that God formed matter at the beginning of solid, heavy, hard, impenetrable, mobile particles, of such sizes, figures, and other properties ... so hard that they never wear out or break, nothing being able (according to the ordinary course of Nature) to divide what was created primitively by God."*

And this one occulted by Marat in his French translation:

> *"...there is no doubt that, if the worship of false gods had not blinded the Pagans, their philosophical morality would have gone beyond the stage*

---

[195] **Grimaldi** Francesco, *M. (Bologna, ITA 1618-1663).* Jesuit, *Physicist (his work on the diffraction of light - which he named - was known by* I. **Newton** *who quoted him), Astronomer (he studied the moon, one of whose craters bears his name). From 1638 he taught Philosophy, then Physics and Mathematics from 1648.*

[196] **Boyle** Robert *(Lismore, IRL 1627-1691 London, ENG),* Anglican. *Son of a wealthy Irish earl, he could travel around Europe after graduating from Eton and make progress in natural sciences. Influenced by the works of* G. **Galileo**, *and* P. **Gassendi**, *... he successfully launched himself into experiments in Physics and Chemistry. In his famous work* "The Sceptical Chemist" [Boy] *he rejects by interposed characters the notion of the 4 elements of* **Empedocles** *adopted by the Aristotelians and defines the simple and primitive bodies, that he will not name however atoms, but this did not prevent* B. **Pullman** [Pu] *from considering him as one of the Christian precursors of atomism. He also defined compound bodies. Co-responsible for the creation of the Royal Society, which he chaired, he contributed financially to the propagation of faith by creating a foundation where Christian scientists would give lectures. An ardent* Christian theologian, *he wrote many books in defense of Christianity.*

[197] **Mariotte** Edme *(Dijon 1620-1684 Paris), Chemist, Physicist, Botanist, and member of the Academy of Sciences in 1666. Author of numerous scientific works on sight, the percussion of bodies, and the vegetation of plants,... His work is important, one can consult* "Œuvres de M. Mariotte" *in 2 volumes at Pierre Vander - MDCCXVII.* Mariotte invented "the Newton's cradle," an apparatus demonstrating the conservation of momentum and energy, and consequently Newton's first law. He was a Catholic priest.

[198] The Boyle-Mariotte law *discovered in 1662 by* **Boyle**, *and independently by* **Mariotte**, *indicates that in the case of a perfect gas, the product of its pressure P by the volume V that it occupies is constant at a given temperature.*

of the 4 Cardinal Virtues; [199] *and instead of teaching the Transmigration of Souls, and the worship of the Sun and Moon, and the departed Heroes, they would have taught us to worship our true Creator and Benefactor, as their ancestors did under the Government of Noah and his Sons before they were corrupted."*

This passage gives us a glimpse of his obsession: the return to the original religion, taught by Christ and his apostles.

The same year, based on the work of G. **Galileo** and J. **Kepler**, Newton published his famous treatise: *"Philosophiae Naturalis Principia Mathematica"* [Ne2], a monumental work in which he established the *"Theory of Gravitation"* or *"Law of Universal Attraction,"* confirmed by the **Cavendish**'s experiment. He demonstrates that the kinematics of the planets resides in the forces exerted on them. His expression of the gravitational force (in $1/d^2$, where "d" is the distance between the object and the center of Earth) could be verified experimentally, from astronomical dimensions up to 1/10mm; but, below nobody succeeded, leaving an interesting field of investigation to cosmologists. Newton laid the foundations of classical mechanics called *"Newtonian mechanics,"* [200] and demonstrated the three laws of J. **Kepler**, applying to all material bodies, including planets. The Theory of Relativity will correct the slight discrepancies observed between Newtonian theory previsions and certain observations.

The concept of the *"material point"* that he introduced allowed all physical phenomena, including light, to be interpreted by Mechanics. We also owe him the introduction of differential equations, a subject of conflict with G. **Leibniz** who claimed authorship.

Newton formulated rigorously the *Principle of Inertia (*called *"Newton's first law"*), that Galileo had partially formulated.

Newton's laws teach us that there is no absolute rest, [201] which has considerable implications in Physics because we can no longer know if two events that occurred at different times took place, in the same place, or not, in space. On the other hand, he believed in absolute time, a notion that seems obvious but is not when it comes to bodies moving at speeds close to the speed of light, which is what the theory of relativity postulates. If some of the bases of the Newtonian theory, turned out to be erroneous, we should not forget the leap that it made to science, and the successes that it obtained in Mechanics as well as in Astronomy, which justifies that it is still taught and that I. **Newton** deserves the glory that he knew and that only A. **Einstein** has equaled to this day. Nevertheless, some rationalists wanted to see in Newtonian mechanics a proof of the non-existence of God, because it

---

[199] *Cf. [Bi] New Testament Matthew 22:37.*
[200] *In* Newtonian mechanics, *the motions of planets, objects, ... are reduced to a simple relation: the product of their mass by their acceleration is equal to the force acting on them.*
[201] *This was refuted by the Aristotelians for whom the Earth and the stars, for example, were immobile,*

was deterministic. I. **Newton** did not, fortunately, subscribe to this idea, we shall see that Quantum Mechanics proves that in some cases uncertainty prevails.

In Book 3 (volume 2 of [Ne2] pp. 174-175) he develops in the *"General Scholia"* some of the religious ideas inspired by his work, leaving no doubt about his faith:

> *"... the arrangement of the Sun, the planets & the comets can only be the work of an almighty & intelligent Being."*

> *"... we see that the one who arranged this universe, put the fixed stars at an immense distance from each other, lest these globes fall on each other by the force of their gravity."*

He ends this work by leaving the door open to new experiments that could explain the forces of gravity:

> *"I have not yet been able to deduce from phenomena the reason for these properties of gravity, and I cannot imagine any hypotheses. For everything that cannot be deduced from phenomena is a hypothesis & hypotheses, whether metaphysical, physical, mechanical or those of occult qualities, must not be accepted in experimental philosophy."*

It is well known that he refuted R. **Descartes**' conception, which he considered too materialistic, preferring a spiritual cause that he did not want to put forward without strong scientific arguments. This did not prevent his friend Nicolas Fatio de **Duillier**,[202] a *Protestant*, mathematician, and astrophysicist, from supporting the research on the causes of gravitation undertaken by the physicist Georges-Louis **Le Sage**[203] for whom *"ultramontane corpuscles"* were responsible for the forces of gravitation. This concept based on mechanistic concepts was quickly abandoned; but, the string theory will introduce (ref.345, p.87) the "graviton," a hypothetical elementary particle, a vector of the gravitational force.

According to specialists, Sir Isaac **Newton**'s theological work of[204] [Ba], is, as important as his scientific work, if not more so. Newton was particularly

---

[202] **Fatio de Duillier** Nicolas *(Basel, CH 1664-1753 Worcester, ENG).* Protestant, *mathematician who adhered to the theses of expatriate prophets from Cevennes*

[203] **Le Sage** Georges-Louis *(Geneva, CHE 1724-1803). Physicist, Member of the Royal Society, and correspondent of the Academy of Sciences.*

[204] Sir Isaac **Newton**, *the theologian (ref.185-p.56). Little known for his writings on religion, to which he devoted more time than to his scientific work (he was an exceptional man). His interpretations of the prophecies were preceded by their decoding. His rational study of religion and especially of its evolution over the centuries and his rewriting of the Gospel were only known, until his death, by a limited circle of close friends (scientists, theologians, etc.). Their publication was in 1727.*
Classified as a Unitarian* *by some - because he believed in God as the only the same spirit (Arian doctrine, see below ref. 206), thus opposed to the dogma of the Trinity (identity of nature*

interested in the interpretation of prophecies (the book of Daniel, the Apocalypse according to St. John) of which he thought he had found the codes, and in the history of the Church. A follower of a *"religious rationalism,"* he was a fierce supporter of the *"true religion,"* which he summarized in the two great commandments, [205] namely, the love of God and of one's neighbor with its corollaries. It was, according to him the religion of **Noah, Abraham, Moses,** and **Christ**, which had to be constantly put back on track over the centuries, because it was quickly corrupted by men, hence these cycles: *"reformation - corruption"* corruption that he denounced, just like the dogma of the Trinity[206] defended by **Athanasius of Alexandria**[207] *Church Father*, of Greek culture, to which he was opposed without being as some detractors said, an *"**Arianist**."*[208] He was also very

---

between the three forms of God; the Father, the Son, and the Holy Spirit) adopted after the Council of Nicaea (325), shared by the majority of Christians - deist *and even* millenarian *by others, he nevertheless remained an* Anglican, *a heretic for the* orthodox Christians. *He desired to restore the primitive religion that led him to fight against the "Trinitarians" and especially against the* Catholics *who, since the 16th century, in the wake of* Luther *(in 1519) and of the protestants, had put aside the theory of the four meanings of the reading of the Holy Scriptures to retain only the literal meaning.*

* *The taxonomy of intellectuals can be controversial.*

[205] *Cf. [Bi] New Testament Mat 22:37.*

[206] *According to the* dogma of the Trinity *(a notion that is rather difficult to grasp and is part of the Mysteries) adopted by most* Christians *(Catholics, Orthodox, Protestants, Evangelicals) God is unique in three persons or hypostases, consubstantial (the Father, the Son, and the Holy Spirit, equal and of the same essence). According to the Scriptures, God revealed his existence and uniqueness to Moses, cf. Old Testament. He also announced the coming of the incarnate* Word. *Thus, in the New Testament, the divinity of Jesus Christ, son of the Eternal Father, and that of the Holy Spirit, the breath of God, are affirmed. To be distinguished from tritheism (three gods) some strands of Hinduism, for example, focus on specific deities such as Brahma, Vishnu, and Shiva, considering them separate and independent gods.*

[207] **Athanasius of Alexandria** (Damanhour, EGY 296/298-373). He passed through the ranks of the clergy. Participated in the synod of Alexandria of 321 which deposed **Arius** (cf. below) and his followers. Appointed bishop of Alexandria in 328, he went into *exile from 335 to 337 following the tensions within the Church. He was accused of crimes, including embezzlement, ... by the* **Arians** *who had returned to power.* **Constantine II** *gave him back his cathedra in 337. His supporters being in the majority, he moved to Alexandria while* **Eusebius of Nicomedia** *(280-341), an* Arianist, *was bishop of Constantinople. However, the followers of* **Arius**, *supported by the emperor* **Constantine II**, *attacked him, leading him to exile in Rome in 339. Although* **Pope Julius I** *(brother of* **Constantine II***) had ordered an end to the persecution of the followers of* **Athanasius** *in 344, he did not return to Alexandria until 21 October 346. He received a triumphal welcome from his supporters and sat until the Council of Milan (355) during which he was dismissed. Exfiltrated from the church of St Thomas to avoid arrest, he remained hidden while* **Constantine II** *ferociously persecuted his followers in Alexandria. At the end of 361* **Constantine II** *died, the emperor* **Julian**, *who called himself a pagan, succeeded him and promulgated, in 362, an Edict authorizing the return of all the banished bishops.* **Athanasius** *defended monasticism all his life (he had the real life of a monk).*

[208] Arianists, *followers of* **Arius** *(Cyrenaica, LBY 256-336 Constantinople, TUR). Ordained deacon in 308 and priest in 311,* Arius *began in 312 to profess a doctrine that had many followers but was at the origin of many inter-religious conflicts, involving the emperors of the West and the East during the fourth century. According to him: the Logos (Christ) is not God but an*

circumspect about the religion he called *"circumstantial"*: that of rites and ceremonies, which he accused of having separated itself from the *"Religion of the People of God."* He was interested, among other things, in the religious writings of his assistant and successor at the Lucasian pulpit, the *Presbyterian theologian* and mathematician William **Whiston.**[209] He exchanged an important correspondence with R. **Boyle** on these subjects.

2.8. 18th Century: The Age of Enlightenment championed by rationalist thinkers

"Faith consists in believing when it is beyond the power of reason to believe.
It is not enough that a thing be possible for it to be believed."
Voltaire

The 18th century saw the emergence of atheism, notably represented by figures such as Sigmund **Freud,** the pioneer of psychoanalysis. This era also witnessed intellectual clashes, including some between Freud and Leonhard **Euler,**[210] a prominent *Protestant* physicist and revered mathematician of the time, whom Pierre S. de **Laplace** referred to as *"our Master to all."* Euler did not shy away from engaging in vigorous debates. In *"Letters to a German Princess"* he stated:

---

intermediary between God and the world, before the world but not eternal, begotten by God, and subordinate to Him. Jesus helped God to create the Holy Spirit who is of separate substance and entity from both.

[209] **Whiston** William *(ENG: Norton-juxta-Twycross 1667-1752 Rutland). Mathematician, a specialist in the study of comets with* Edmond **Halley,** *maintained that the comet of 1682, later called* "Halley's comet," *was the one that caused the Flood, which he dated to 18/11/2349 BC.* Presbyterian pastor, Arian *theologian, and historian, he translated the works of* **Flavius Josephus.** *He resigned from his position as a lecturer due to health problems. His book* "A New Theory of the Earth, From its Original, to the Consummation of All Things" *tending to prove that the creation of the earth in 6 days is possible, obtained the congratulations of* I. **Newton.** *In 1701 he resigned from his curatorship and became* I. **Newton'**s *assistant at Cambridge, succeeding him in 1702.*

[210] **Euler** Leonhard *(Basel, CHE 1707-1783 St Petersburg, RUS). The most famous mathematician and physicist of the 18th century. During 12 years he was the first prize winner of the French Académie des Sciences. His father Paul, a* Calvinist *pastor, taught him the first rudiments of Mathematics but, noticing his precociousness, entrusted him to his friend* Jean **Bernoulli** *who, being too busy, served only as his mentor. Leonhard entered high school at the age of 9, went to university at the age of 13, and obtained a Master's degree in Philosophy at the age of 16. He finished his studies in Mathematics at the age of 19 and looked for a university position which led him to St Petersburg where he was offered a chair in Physiology. In 1731 he obtained the chair of Physics and in 1733 the chair of Mathematics. In 1741 he joined Berlin. His scientific production (Mathematics, in all fields, and in Physics: Optics, Astronomy, Mechanics, ...) will be considerable. He returned to Russia in 1766 because he enjoyed great prestige there. He will remain there until his death. Holder of many prizes. A* Calvinist, *ardent defender of Christianity, he believed that the Bible had been inspired. He opposed the monadology of* G. **Leibniz** a Protestant.

"Since we are creatures of God, we ought to conform ourselves to His will, and to acknowledge His supreme dominion. This is the first and most essential of our duties. But how can we fulfill this duty, if we are ignorant of God, and of His relations to us? It is, therefore, necessary, to please Him, that we should know Him, and that we should know ourselves as His creatures. This is the foundation of all religion, natural and revealed."

1728. James **Bradley**, [211] *an Anglican* (vicar of Bristow) was the first astronomer to prove experimentally that the Earth revolves around the Sun. This, in 1741, led **Benedict XIV**[212] the *"Pope of the Enlightenment"* to authorize the publication of the complete works of G. Galileo, implicitly revising the sentences against him. He authorized the publication of works on heliocentrism in **1757**. Let's note that Vatican II (1962-1965) made amends for the trial of Galileo and that, in **1979**, Pope St. **John Paul II** requested a review of the trial of Galileo, whose conclusions published in 1992 explicitly recognize the wrongs of the Church.

The end of the century will see the birth of the *"Prince of Mathematics"* namely Johann C. **Gauss**,[213] a *Christian,* author of a considerable body of work, most of which was not published until after his death. According to Wikipedia, A. **Einstein** said of him:

"Without Gaussian geometry, the theory of relativity would not exist."

I have not been able to verify its truth, but it is probable when we know that to define what is commonly called *"the Riemannian geometry"* (Gottingen 1854, subject of his habilitation dissertation) on which A. **Einstein** relied, the *Lutheran*

---

[211] **Bradley** *James (ENG: Sherborne 1693-1762 Chalford). Astronomer, he discovered the nutation (periodic oscillation of the Earth's axis around the average direction of its axis (N-S)), improved the knowledge of the speed of light. Succeeded* Edmund **Halley** *as director (3rd) of the Greenwich Observatory. He was sponsored by him to join the Royal Society in 1718.* Anglican vicar.

[212] **Lambertini** *Prospero L. (ITA: Bologna 1675-1758 Rome) was elected Pope in 1741 under the name of* **Benedict XIV**. *Passionate about literature, undertook studies in Law and Theology. Archbishop of Bologna in 1731 he was interested in Medicine and Physics.*

[213] **Gauss** *Johann C (DEU: Brunswick 1777-1885 Göttingen). Coming from a poor family he was spotted at the age of 8 by his teacher who helped him and introduced him in 1791 to the* **Duke of Brunswick** *who granted him a scholarship. Between 1792 and 1795, I remember among many other works, his formulation of the method of least squares. At the age of 22, he obtained a Ph.D. in Mathematics. Elected to the Royal Academy in 1804. Because of his scientific contribution, he was appointed professor of Astronomy and director of the Göttingen observatory in 1807. He was awarded the Copley Medal and many others. From 1856 commemorative medals with his effigies were engraved by the king of Hanover. His work is considerable but most of it was not published until after his death. A very devout Christian.*

Bernhard **Riemann** [214] drew inspiration from the concept of *"Gaussian Curvature"* (see Google scholar, for confirmed mathematicians). On his tombstone (Bizangolo, IT) [215] one can read: *"Here rests in God"* Georg F. B **Riemann**... followed by the 8:28 verse of **St Paul Apostle's** epistle to the Romans:

> *"We know that all things work together for good to them that love God, to them who are called according to his purpose."*

2.9. 19th-century: The Atom Electromagnetic waves - Maxwell's equations

"God made the integers, everything else is the work of man."
L. Kronecker [216]

For the *Christian* mathematician, L. **Kronecker** who worked on number theory and algebra, only natural numbers (positive integers) were real and concrete, while other numbers: fractions, irrational numbers (real numbers that cannot be represented by the ratio of two integers, as $\pi = 3,14159...$, e $= 2,718...$, etc.), complex numbers, etc) were artificial and abstract creation of human minds. He was a critic of Georg **Cantor**'s work on set theory (ref.174-p.51). The above epigraph was reported in 1893, in a scientific article by Heinrich Weber, a student of Leopold Kronecker, who once had said it to him.

In **1801** John **Dalton**, [217] *Quaker*, mathematician, meteorologist, and chemist

---

[214] **Riemann** (Georg Friedrich) Bernhard *(Breselenz, Kingdom of Hanover 1826–1866 Selasca, Kingdom of Italia). Famous mathematician. At a very young age, he became interested in reading the Bible. Although he was very gifted in Mathematics, he wanted to become a pastor like his father, so at the age of 19 he began studying Philology and* Christian Theology. *In 1846 he entered the University of Gottingen where he studied Mathematics under the direction of F.* **Gauss**, *who became his thesis director, which led him to abandon his studies in Theology. In 1847 he studied at the University of Berlin where he met the great mathematicians of the time, including* Carl Gustav J. **Jacobi** *(Kingdom of Prussia: Potsdam 1804-1851 Berlin). Ashkenazi* Jew *who converted to* Christianity *in 1825. Head of Gottingen Mathematics department. His contribution to Mathematics (geometry, analysis, number theory) is numerous. His writings are of great value.* Lutheran.

[215] *Cf. Wikipedia B. Riemann biography.*

[216] **Kronecker** Leopold (Legnica, POL. 1823-1891 Berlin, DEU). Jewish mathematician who converted to Christianity. Specialist in number theory, algebra, and logic. He left his name to various theorems and mathematical concepts. After his secondary *education at the Gymnasium in Legnica, he entered the University of Berlin in 1841. After completing his studies, he managed the family business and from 1851 onwards he undertook self-financed research in Mathematics. Elected in 1861 to the Academy of Berlin he refused the chair of Mathematics of Göttingen until 1883 when he accepted that of Berlin.*

[217] **Dalton** John *(ENG: Eaglesfield 1766-1844 Manchester). He adopted and practiced during all his life the religion of his parents who were* Quakers*. Appointed professor of Mathematics and Natural Philosophy in 1793 in Manchester.*

stated what is called: "*Dalton's Law.*"[218] As early as 1794 he published some of his results obtained in the field of meteorology but especially an article on dyschromatopsia (he had noticed that he could not distinguish certain colors), hence the term "*color blindness*" given to people suffering from this retinal malformation. In **1805** he published the first volume of "*A New System of Chemical Philosophy,*" a work in which he developed the atomic theory, attributing to the different elements, and masses relative to that of the hydrogen atom of mass 1.

In **1811** Amedeo **Avogadro,**[219] *Catholic* stated Avogadro's law that some call Avogadro - Ampère because the latter published the same result 3 years later: "*equal volumes of perfect gases, in the same conditions of temperature and pressure, contain the same number of molecules.*"

N= $6.022\ 140\ 857\ 10^{23}$ mole^{-1} (N, called Avogadro's number is one of the 7 universal constants; it is equal to the number of molecules contained in a mole of any chemical compound.

In **1814** Pierre **Laplace,**[220] published: "*Essai philosophique sur les probabilités*" *(philosophical essay on probabilities),* a work in which he gives his interpretation of "*determinism.*" According to Laplace universe is determined by a set of laws that, if we know the initial conditions, allow us to predict everything that will

---

*The Quakers *are dissident* Anglicans, *pacifists, philanthropists, austere, partisans of a very simple way of life... who founded in England, in the 17th century, a* Protestant church *which spread mainly in Holland and the USA*

[218] Dalton's law: *At a given temperature T, the product PV (P = pressure and V volume) of an ideal mixture of perfect gases is equal to the sum of the products P_iV_i of its constituents considered separately at the same temperature. Where P_i and V_i are the pressures and volumes of each component i of the mixture. When each gas is considered separately as occupying the entire volume V, it takes a simplified form: P = sum of P_i, which is called the* "law of partial pressures." *Appointed a member of the Royal Society in 1822, he received the Royal Medal in 1826. A world-renowned scientist, he is one of the eight foreign members associated with the French Académie des Sciences.*

[219] **Avogadro** Amedeo *(Turin, ITA 1776-1856).* Catholic *physicist and chemist. He began his studies as a jurist but abandoned them in 1806. He became famous in 1809 with a communication to the Royal Academy of Turin and was given a professorship at the Royal College of Turin. A chair of physics was created for him at the University of Turin in 1820. His law was very controversial at the time. It was confirmed and accepted only later.*

[220] **Laplace** Pierre-Simon, *Marquis (FR: Beaumont-en-Auge 1749-1827 Paris)* Catholic. *Astronomer, mathematician, and physicist, his work is considerable. He left his name to the* "Laplace equation," *to the* "Laplacian operator," *and to the* "forces and laws of Laplace" *governing the interactions between electric currents and magnetic fields. His work* "Celestial Mechanics" *is a synthesis of the previous works of* I. **Newton**, E. **Halley**, ... *Was one of the first, after J. **Michell**, to propose the existence of stars emitting no light (Black holes ref.279-p.84) based on the theory of gravitation of* I. **Newton**. *He reformulated* **Bayes'** *theorem* which he discovered in another way. Joined the Senate in 1979 and became vice president in 1803. Appointed Count of the Empire by* **Napoleon** *in 1806, he rallied in 1814 to King* Louis XVIII *who made him Marquis and Peer of France.*

***Baye** Robert *(London 1702-1761 Tunbridge Wells, Kent). Mathematician, Member of the Royal Society, a* Presbyterian pastor.

follow. This is verified in the case of planet movement, for example, which we can predict perfectly. But he also applied this principle to human behavior, which alienated the Church because it went against the idea that God made Man free of his choices. It was not until the beginning of the 20th century that this principle was questioned. Raised by his uncle, a Priest, he was *Catholic* according to his friend the *Catholic* chemist JeanBaptiste **Dumas**.[221]

In **1822** Augustin **Fresnel**,[222] a *Catholic* engineer and physicist, patented a groundbreaking innovation: "*the Fresnel lens.*" Using multiple prisms arranged concentrically to capture and concentrate light from a single source (no more than 250W) he revolutionized lighthouse optics and significantly improved the efficiency and range of light signals emitted from lighthouses, improving their effectiveness in guiding ships and warning them of dangerous coastlines. During the 19th and 20th centuries, they greatly improved maritime safety and navigation. As a leading physicist, his most significant contribution to science lies in the field of wave theory. His wave theory revolutionized the understanding of light, explaining various optical phenomena such as reflection, refraction, diffraction, and interference. Fresnel's work laid the foundation for modern physics and the development of numerous optical devices and technologies.

In the **1840**s the mathematician and physicist Sir William Rowan **Hamilton**,[223] an *Anglican*, contributed to the development of optics, dynamics,

---

[221] **Dumas** Jean-Baptiste *(FR: Alès 1800-1884 Cannes) fervent* Catholic. *Initially a pharmacist, then an assistant at the Faculty of Sciences in Paris to Baron* Louis J. **Thénard**, *an eminent French chemist, and professor at the Collège de France. Dumas succeeded him in 1835 as the chair of Chemistry of Polytechnique. From 1836 he held the chair of organic chemistry at the Faculty of Medicine (his student was* L. **Pasteur***). Member of the French Académie des Sciences in 1832, and of the Royal Society in 1840, he took an important part, in the development of organic chemistry, isolating numerous organic compounds, inventing methods of analysis and determination of atomic masses... A politician since 1842, he became a deputy, city councilor, minister, and irremovable senator. President of Paris' Council when* Baron Georges **Haussmann** *developed the French capital.*
[222] **Fresnel** Augustin *(FR. Broglie 1788-1827 Ville d'Avray).* Catholic *experimental and theoretical physicist, he entered the* "Ecole Polytechnique," *Paris in 1804, and he joined the* "Ecole Nationale des Ponts et Chaussées," *Paris in 1809 where he began his career building roads and lighthouses. His research in optics began in 1814 was quickly recognized by the international community, and received scant public recognition during his lifetime and until now. He was the first to produce polarized light. In 1823 he became a member of the French Academy of Sciences and of the London Royal Academy.*
[223] Sir William R. **Hamilton** *(Dublin 1805-1865). Mathematician, physicist, and astronomer left his mark on the 19th century in all three disciplines to whose development he contributed. A precocious child, from the age of 3 he showed great aptitude for reading and arithmetic. His uncle, a highly literate Anglican priest, took charge of his primary education. At 5, he knew Greek and Latin, at 13, he spoke 13 languages, including Arabic, Sanskrit, Persian, French, and Italian. At 15 he discovered Newton's Principe Mathematica. At 17, he amended the Laplace's Mécanique Céleste. At 18, he was admitted to Trinity College Dublin, where he collected top marks. Awarded the Cunningham Medal (twice in 1834 & 1848), the Royal Academy Medal, and the Royal Medal of the Royal Society in 1835. The same year he was knighted... He remained a*

and algebra. His work on the theory of quaternions intended initially for use in classical mechanics will find applications later in quantum mechanics and quantum field theory.

In **1843** James P. **Joule,**[224] Called "*The Great Experimenter who was Guided by God*" by the novelist Ann Lamont,[225] Joule derived from his experiments on mechanical energy (W) conversion into heat Q, the conversion factor J = W/Q. But it will be only in 1847 - after a new presentation of his experiments attended by scientists among whom the *Calvinist Presbyterian* Michael **Faraday,** (ref. 235, p.71) and the *Anglican Evangelical* Sir George G. **Stokes**[226] *(Church of Ireland), who* supported his demonstration - that his work was retained. The young Professor of Physics William **Thomson** (Lord **Kelvin**)[227] who attended was

---

faithful Anglican *all his life, despite a rapprochement in 1833 with the Oxford Movement*, from which he broke away when some of its members converted to Catholicism.*
* This movement was created in Oxford when the state tried, through the Reform Bill 1833), to take control of the Anglican Church.

[224] **Joule** James P. *(ENG: Salford, near Manchester1818-1889 Sale, Cheshire)* Anglican. *Suffering from a spinal disorder he studied at home until the 15th. Then, he shared his time working in the family brewery and following his education with private tutors. The physicist and chemist well-known J.* **Dalton** *(ref. 217-p66) a Quaker tutored him and his brother in Mathematics, Physics, and Chemistry for 3 years. In 1839, his father became ill, so he set up a laboratory in the familial home and began a series of experiments. In 1840, Joule's first paper to the Royal Society, on the conversion of energy-heat, was ignored until 1847. Recognized, he was called: the* "law of conservation of energy," *becoming* "the first law of thermodynamics." *In recognition of his work, the unit of energy in physics was named* "the Joule (J)"

[225] **Lamont** Ann *(San Francisco, US-CA 1954) Novelist having covered various articles on Christianity.*

[226] Sir George **Stokes** G. (IE, County Sligo 1819-1903 Cambridge, ENG), Made 1st baronet in 1889. He was graduated from Pembroke College, Cambridge with honors in 1841. Mathematician and physicist, at the origin of geodesy (modern geography), a *specialist in hydraulics (Navier-Stokes equations on the flow of viscous fluids) - his name is the symbol (St) of the kinematic viscosity unit - Waves, light: polarization, fluorescence,... chemical analysis, ...Appointed in 1849 Lucasian Professor of Mathematics at Cambridge,* President of the Royal Society from 1885 to 1890, *Member of various Academies, and recipient of numerous awards.* In *1893, he received the Royal Society's Copley Medal (the most important scientific recognition of this epoch), the Helmholtz in 1900,... He published a hundred scientific papers.* Anglican Evangelical.
Member of the Church of Ireland, a *leading authority on natural Theology, he was the most influential* Anglican *of the Victorian era, for him, the Bible if correctly understood, and supernatural revelation were the foundations of his faith. He thought that man had been created by God with an innate sense of right and wrong. Assimilating God into a merciful and loving father he was against the doctrine of eternal punishment for beings. President of Victorian Institute.*

[227] **Lord Kelvin** William *(Belfast, IRL 1824-1907), son of* James **Thomson***. *William Thomson went to Peterhouse College in Cambridge where he won many prizes. Mathematician, physicist, and inventor (70 patents), he taught Physics at the University of Glasgow from 22 to 72 years old. Named 1st Baron of Kelvin in 1892 for his work in the fields of Thermodynamics (his name was given to the absolute temperature Kelvin (K): K = °C + 273.15), heat, electricity (discoverer of the Thomson effect), magnetism, ... the asteroid 8003 bears his name. Has 625 publications to his credit. Recipient of numerous prizes including the First Smith's Prize in 1845, the Royal*

initially skeptical. Nevertheless, in 1852, the two started a cooperation, which led to the discovery of the "*JouleThomson effect*." This effect justifies the temperature variation observed when gases or liquids expand, it is notably used to create cold in refrigerators.

J. **Joule** was a *sincere Christian*, I will say: a Long Life Learning gifted experimenter, but also an intuitive scientist. According to A. Lamont, concerning the relation Energy-Heat, he wrote [Lam]:

> "... *the phenomena of nature, whether mechanical, chemical, or vital, consist almost entirely of a continual conversion... into one another. This is why order is maintained in the universe, nothing is disturbed, nothing ever lost, but the whole machinery, however complicated, works smoothly and harmoniously...the whole being governed by the sovereign will of God.*"

It's difficult to separate the two Irish protestants Sir George **Stokes**, a *passionate Evangelical*, and Lord **Kelvin** *a devoted Christian*, very close friends for 60 years despite their different personalities. They were among the most prominent scientists of the 19th century and played leading roles in Britain's scientific institutions. They made pioneering work in hydrodynamics, optics, and electricity. "*The Correspondence Between Sir George G. Stokes and Sir William Thomson, Baron of Largs*" has been edited (two volumes, about 650 letters) by Cambridge University Press, NY, 1990. Kelvin played central roles in theory and in technology, he contributed among others to the telegraph link between the US and Europe.

In **1846** Urbain **Le Verrier**, [228] *Catholic*, was famous thanks to the observation in Berlin, by the astronomer Johann **Galle**, [229] a *Catholic* too, of

---

Medal in 1836, the Copley Medal in 1883, ... President of the Royal Society of London, member of most of the scientific academies of Europe including that of Paris. A practicing Christian, he attended mass every morning.

* James **Thomson** *(Ballynahinch, IRL 1786 - 1849 Glasgow) Presbyterian minister, member of the Church of Scotland. Professor of Mathematics at the University of Glasgow (at the origin of the creation of the Glasgow School of Thermodynamics) at the Royal Academical Institution. His eldest son* **James** *(Belfast, IRL 1822-1892 Glasgow) also studied brilliantly, and became a renowned physicist and engineer; less so than his brother but was nevertheless a Fellow of the Royal Society of Edinburgh and President of the Engineers and Shipbuilders of Scotland.*

[228] **Le Verrier** *Urbain (FR: Saint-Lô 1811-1877 Paris) graduate of Polytechnique, Paris, Fr. Astronomer, mathematician, specialist in celestial mechanics, he presented a memorandum to the French Académie des Sciences in 1839 in which he mathematically justified the existence of Neptune, whose trajectory he had calculated to justify, within the framework of Newtonian theory, the anomalies detected in the displacement of Uranus. He founded modern meteorology in France. Member of the French Académie des Sciences in 1846, and of most of the European scientific academies... Copley Medal (1846), Gold Medal of the Royal Astronomical Society (1876). Catholic.*

[229] **Galle** *Johann (Radis, DEU 1812-1910), Catholic. He observed Neptune according to the data sent by U. Le Verrier. Appointed director of the Breslau Observatory in 1850 and Professor of*

Neptune a new planet whose trajectory he had just communicated to him the same day. John C. **Adams**,[230] an *Anglican*, had also predicted the existence of this planet but had not published his work.

In **1847** Maria **Mitchell**[231] discovers a new comet which will be named "*Miss Mitchell's Comet*" and calculates its orbit. What, for some, may seem an anecdotal discovery, interests me because her career is remarkable. She was educated by her father, a schoolmaster but, above all, passionate about Mathematics and Astronomy, passions that he transmitted to his daughter by making her his assistant from the age of 11 or 12 in his small observatory. Although she did not go to university, she became a professor of astronomy (the first in this discipline), and then the director of Vassar College. In 1869 she was elected to the American Philosophical Society. Educated by her father in the *Quaker religion* she left the Quaker faith in the 1840s, followed Unitarian principles, and integrated the *Unitarian Church* in the 1860s.

In **1859** Charles **Darwin**,[232] initially *Anglican, Unitarian* then *Agnostic* (but never materialistic (cf. [Da] p.6), a great observer of nature, an admirer of the eminent *Anglican* scientist William **Whewell**,[233] published: "*The Origin of*

Astronomy in 1856. Discovered 1984 planets and published numerous works. Two craters, one on the Moon and the other on Mars bear his name.

[230] **Adams** John C. *(ENG: Laneast 1819-1892 Cambridge).* Anglican *mathematician and astronomer. Gold medal of the Royal Astronomical Society in 1866. Although he came from a modest family, he studied at St John's College in Cambridge where he was distinguished for his results in Greek and Mathematics. In 1845 he communicated to* Georges Airy, *astronomer royal at Greenwich, the position of Neptune so that he could observe it. But the latter did not give the follow-up he had hoped for, leaving the field open to U. **Le Verrier**. Adams contributed a lot to the development of celestial mechanics. Elected to the Royal Society ...*

[231] **Mitchell** Maria *(US-MA: Nantucket 1818 -1889 Lynn).* Unitarian *Astronomer. Mainly educated by his father a* Quaker *from 1831 to 1834 she attended the unitarian school of her town. From 1836 to 1856 she worked, during the day, as a librarian, and in the evening as an astronomer with her father. With the discovery of the planet eponym, she became a celebrity, obtained a computing and research position from the US administration, joined the American Association for the Advancement of Science, and visited the most prestigious European observatories. She published many articles, received various awards, and was granted honorary degrees from different universities.*

[232] **Darwin** Charles *(ENG: Shrewsbury 1809-1882 Down). He was a naturalist, botanist, and geologist who participated in many expeditions in South America, Australia, Galapagos, from which he brought back important collections of animal and plant fossils. Apostle of natural selection, initially an* Anglican Unitarian, *he brilliantly obtained his degree in Theology in 1831 and declared himself* Agnostic *only in 1851 after the death of his daughter Anne (10 years old); a death that was very difficult to bear and justify. Especially since Mary, his first daughter, died at one-month-old. Until then his faith in Christianity was present and had not been altered by his discoveries. He did not accept the idea of eternal hell, for example. He is buried in Westminster's Abbey, near Sir I. **Newton**, an* Anglican.

[233] **Whewell** William *(ENG: Lancaster 1794-1886 Cambridge),* Anglican Theologian, *polymath: Mathematics, Physics, Geology, History of sciences, Philosophy, Economics, ... After studying at Trinity College in Cambridge, he taught Mineralogy and Philosophy from 1821, the year he joined the Royal Society. He was awarded numerous distinctions, including the Royal Medal in 1837.*

Species" which will make the effect of a tsunami even if it was not its goal; he wrote in fact:

"I see no good reason why the views given in this volume should shock the religious feelings of anyone. ..... A celebrated author and divine has written to me that:"

He has gradually learnt to see that it is just as noble a conception of the Deity to believe in that He created, a few original forms capable of self-development into other needful forms, as to believe that He required a fresh act of creation to supply the voids caused by the action of His laws [Dar].

This, brings us back to the question: did He (God) feel obliged to create Man? But it goes into the direction of deism or even theism as some, F. **Collins** [Col2] for example, wrote. Let's note, however, that Darwin's work, in the continuity of Jean-Baptiste **Lamarck**'s work, was mostly condemned by *Protestants* and *Anglicans*. The *Catholics,* who could not ignore the theses of **St. Augustine** (4th century, cf. [Ag1] pp. 26, 132 & 148), did not attack this theory, provided that *"the presence of God was not denied at the beginning"* which was not **Darwin**'s idea. They did right, the great majority of Catholic scientists today adhere to the theory of evolution if we don't attribute the small variations that occur as hereditary but accidental, contrary to what Darwin thought.

In chapter 5.3. we will discuss the perception, by some contemporary believing scientists, of this theory.

In **1865**, the distinguished physicist James Clerk **Maxwell**,[234] an *Evangelist Presbyterian*, forged a crucial connection between previous experiments in electricity and magnetism, particularly those by the *Calvinist Presbyterian* Michael **Faraday**.[235] Maxwell deduced that the electric and magnetic forces are essentially one force, which he coined the *"electromagnetic force."* Photons were identified as the *"messengers"* or *"vectors"* of this force. Consequently, he reduced the number of fundamental forces to 4[236] and introduced the concept of

---

[234] Clerk **Maxwell** James *(ENG: Edinburgh 1831-1879 Cambridge). Brilliant student, he published his first article:* "On the Description of Oval Curves" *at the age of 14. His work on the kinetics of gases contributed to the development of* **Boltzmann***'s equation. But we especially remembered:* "Maxwell's laws," *the* "electromagnetic force" *and the notion of* "electromagnetic wave." *An* Evangelical Presbyterian, *he became an* Elder of the Church of England *at the end of his life.*

[235] **Faraday** Michael *(ENG: Newington 1791-1867 Hampton Court). Physicist, he is experimentally interested in the fields of the electric and magnetic forces in space and time, the starting point of Maxwell for the mathematical form. He played an essential role in the experimental discovery of the magnetic properties of electric currents, including induction phenomena (inventor of the first electromagnetic motor, he created the first direct current generator, etc.). Chemist, and founder of Electrochemistry, he stated the quantitative laws of electrolysis (Faraday's laws), discovered the importance of dielectrics, liquefied various gases, and ... He was a member of* the Church of Scotland.

[236] The three other fundamental forces are*: the* gravitational force, *responsible for the universal attraction, whose messengers (the gravitons) have not yet been discovered, the* strong nuclear

"*electromagnetic wave*" (E.M.W) traveling at the speed of light in a vacuum. He formalized this understanding through a set of fundamental equations known as: "*Maxwell's equations.*"

Later, A. **Einstein** mentioning Maxwell, wrote [Ei1]: "*Before him, I conceived physical reality ... as a set of material points. ..... After him, I conceive the physical reality as represented by continuous fields represented by partial differential equations.*" During the centenary of Maxwell's birth (1931) A. **Einstein**, speaking of Maxwell's work, will say: "*The most profound and fruitful that physics has known since the time of Newton.*" In the meantime, in **1877** the "*father of statistical physics,*" Ludwig **Boltzmann**[237] formulated the entropy formula, also known as "*the Boltzmann-Planck equation*" linking the entropy S (a thermodynamic quantity introduced in 1865 by Rudolph **Clausius** [238] *a Protestant*) of an ideal gas to its statistical probability W of being in a given macroscopic state "i": $S = k.\log W$, where $k = 1.38.10^{-23}$ J.K^{-1} represents the **Boltzmann** constant. As W depends on the number N_i of particles in each possible state "i," it follows that higher entropy corresponds to greater disorder. Therefore, entropy is a measure of order or disorder in an isolated system, a crucial concept for understanding our universe's state in Planck's time (ref.13-p.4). The dilemma hinges on determining whether the universe was disordered (high entropy) or ordered (low entropy); a quandary that we will address in Chapter 3, by showcasing the role of the attractive gravitational force with its extensive reach, adhering to known laws of Physics, in the creation of our universe.

Although L. Boltzmann was hailed as a 19th-century "*hero of scientific materialism*" by some Marxists, including Vladimir **Lenin**, I resonate more with Ranjbar's perspective. **Ranjbar** [239] encapsulates [Ra] **Boltzmann**'s version of materialism, asserting that it doesn't negate the existence of a deity. Instead, he emphasizes the presence of an external, scientifically knowable world:

---

force, *acting on the quarks that make up the nucleons (protons and neutrons), whose messengers are the gluons, and the* weak nuclear force, *acting on a subatomic scale and responsible for radioactive decay, among other things, with the bosons as vectors.*

[237] **Boltzmann** Ludwig *(Vienna, AUT 1844-1906 Duno, ITA). Austrian physicist and philosopher, the father of Statistical Physics, improved 1877 his paper of 1872 on the relationship between entropy S and the disorder of atoms in a gas. His work was crucial in understanding the fundamental principles of thermodynamics; particularly the relationship between macroscopic properties of a system and the behavior of its microscopic constituents. He was raised as a Roman Catholic but the probabilistic nature of physical processes influenced his philosophical outlook and became more inclined towards a materialistic and atheistic worldview.* Deist materialistic.

[238] **Clausius** Rudolf *(Koszalin, POL. 1822-1888 Bonn, DEU)* Protestant. *In 1865 he introduced the notion of entropy S, a thermodynamic quantity, in the differential form $dS = dQ/dT$ where dQ is the variation of heat communicated to a system at temperature T during a reversible transformation.*

[239] **Ranjbar** *Vahid Houston says he is a Theoretical Physicist working in Brookhaven Laboratory. I cannot confirm this fact, he appears on Intern as a contemporary medium writing books about connections between science and religion. Cf.* **Cercignani** *(ref.767-p.287).*

"His version of materialism did not entail the rejection of deity but rather the claim, that an external world does exist and is knowable by science"
[Ra].

1873, is marked by the reception of Louis **Pasteur,**[240] often hailed as the *"benefactor of humanity,"* and rightfully so, at the French Academy of Sciences. Placing him within the landscape of the 19th century becomes a challenge, considering the sheer magnitude of his scientific contribution.

In the 1840s, Pasteur embarked on an intellectual journey that seamlessly intertwined light, chemistry, and crystallography. His groundbreaking work illuminated the effect of specific crystals on light, laying the sturdy foundation of stereochemistry. This early phase showcased his brilliance and set the stage for future advancements in synthetic chemistry.

Transitioning into the crossroads of the 1850s and 1860s, Pasteur delved into the intricate mechanics of fermentation, a venture that would steer him toward the realm of biology. His advocacy for asepsis, coupled with the revelation of *"aerobic"* and *"anaerobic"* microorganisms, not only explained the mysterious formation of ferments but, also revolutionized industries like vinegar and beer production. His agricultural impact was no less remarkable, introducing innovative vine care methods.

Among his numerous inventions, pasteurization stands out as a game changer, transforming the food safety landscape. As the 1870s and 1880s unfolded, Pasteur's focus shifted towards researching disease-causing microorganisms. During this decade he pioneered several breakthroughs, leaving an indelible mark on medical science.

Beyond his brilliance, Pasteur's unwavering faith as a *Catholic* added a unique dimension to his persona. His humanity commitment, coupled with a profound understanding of the interplay between science and spirituality, solidified his legacy as a figure of immense significance in the tapestry of 19th-century discoveries. In this mosaic of scientific progress, Pasteur's contributions shine as a beacon of innovation and compassion, transcending the boundaries of his era and leaving an enduring impact on the trajectory of human advancement.

---

[240] **Pasteur** Louis *(FR: Dôle 1822-1895 Villeneuve - l'Etang). Son of a tanner, he integrated the Ecole Normale Supérieure (ENA) in 1842, which he directed later. Chemist, and biologist, he is considered as the* "Father of Microbiology." *His contributions to science, medicine, and technology are nearly without precedent, earning him the title of* "Benefactor of Humanity." *In 1847 he was appointed laboratory assistant at the Ecole Nationale Supérieure (ENS). Professor of chemistry at the University of Strasbourg in 1848. In 1857 he was appointed ENA's Director of Studies. He was the director of the Pasteur Institute from its creation in 1888 until his death. He did not hesitate to question himself. Even in the face of personal tragedy, such as the loss of his daughter Jeanne in 1859. Pasteur's unwavering faith (he was a* Catholic) *found solace in the belief that she* "has just gone to Heaven to pray for us," *extracted from a letter to a friend. He was a Member of the French Academy of Sciences.*

In **1880**, Pierre **Curie** [241] and his brother Jacques, demonstrated experimentally, with an electrometer that they had realized, that quartz, tourmaline, and other crystals are piezoelectric. [242] This effect predicted by the renowned mineralogist and *Catholic priest* Rene J. **Haüy** [243] is based on the geometric properties of these crystals. One year later, in **1881**, the Franco-Luxembourgish Gabriel **Lippmann**, [244] a *Jewish* physicist, known for his numerous works in Electricity, Optics, Thermodynamics, and Photochemistry… predicted the converse piezoelectric: mechanical strain is generated in piezoelectric materials when an electric field is applied. In 1894, he published the theory of reproducing color, the result of 7 years of experimental research. He will be awarded in 1908, the **Nobel** Prize in Physics *"for his method of reproducing colors photographically based on the phenomenon of interference."*

While piezoelectric applications were limited to weight measurements, pedestrian and vehicle traffic counting, and speed measurements for many years, they are currently experiencing significant development in the medical sector. Among the pioneers, we can mention the teams of D. **Vasilescu,** [245] who

---

[241] **Curie** *Pierre (Paris 1859-1906). Physicist known for his work on magnetism (Curie's Law and Curie's Point), piezoelectricity, and radioactivity. In 1877, he brilliantly obtained a degree in physical sciences and then joined the Physics Laboratory of the Faculty of Sciences in Paris, where he defended his thesis in 1895. Lecturer in 1900, then full professor in 1904. He was shared in 1903 with his wife Marie half of the* **Nobel** *Prize, obtained in 1904 the creation of a laboratory, and joined the French Académie des Sciences in 1905. He died in a street accident.*

[242] Piezoelectricity (1880): *property that some materials have to accumulate an electrical charge under the action of mechanical stress. The converse piezoelectric effect (a mechanical deformation is produced in some crystals when an electric field is applied) was highlighted in 1881 by* Gabriel **Lippmann.**

[243] **Haüy** René Just *(FR: St Just-en-Chaussée 1743-1822 Paris).* Although ordained a Priest *in 1770, he devoted himself to the sciences: Botany and Mineralogy. He is one of the founders of geometrical crystallography. He was elected to the French Académie des Sciences in 1783. His demonstrations were followed with interest by the great scientists of the time, like* Simon **de Laplace,** *and: the* Catholic Antoine de **Lavoisier** *(Paris, Fr: 1743–1794) philosopher, economist, and prolific chemist, the father of modern chemistry, associating experimentation and mathematics. Financial he worked for the royal and revolutionary administrations that he will defend, but falsely convicted he was guillotined...*

[244] **Lippman** Gabriel *(Bonnevoie, LUX 1845-1921 died in the Atlantic Ocean aboard the steamer: "France"). He joined France in 1848 with his father, a French Jewish, and attended ENS in Paris in 1868. He was sent in 1972 to Heidelberg University, specializing in electricity under G.* **Kirchhoff***'s direction, and was awarded a doctorate in 1874. Returning to Paris he presented a Ph.D. at the Sorbonne in 1875. He was successively a physicist, photographer, and inventor. In 1883 he was appointed at the University of Sorbonne, Paris as a Professor of Mathematical Physics, then Director of the Research Laboratory in Physics..... Titular of numerous awards, and a member of various institutions such as the French Academy of Sciences of which he was President, the Royal Society of London...President of the French Société Astronomique. He was laureate of the 1908 Nobel Prize in Physics.*

[245] **Vasilescu** Dan, Cornillon Raphaël, *and* Mallet Georges [Va2] *"the evidence of piezoelectricity in 19 amino acids playing a preponderant role in the constitution of proteins, 7 of which have a vital role for man."* Nature Vol. 225, P. 635,14/02/1970.

highlighted piezoelectricity in amino acids in 1970, followed by **Colin**'s team[246] who highlighted it in biological polymers.

Since piezoelectricity operates in a range of medical implications:

- Biomechanics and Tissue Engineering in the field of orthopedics and regenerative medicine,

- Sensors and diagnostic tools, exploiting piezoelectric properties of proteins by AI allow the detection of early diseases, and the monitoring of biological processes.

In **1887** A. **Michelson,** [247] one of the giants of the late 19th century, collaborated with chemist Edward **Morley** [248] on one of the most important experiments of the 19th century. To reconcile the theories of I. **Newton** and J. C. **Maxwell** postulated the existence of a colorless, odorless, and imperceptible gas, named "*ether*," to act as a medium for EMWs. Despite their sophisticated and precise measurements, they observed neither the expected ether wind nor a variation in the speed of light. Initially deemed a failure, Albert **Einstein**, a humble patent office employee in Bern, would vindicate their experiments through the theory of relativity, asserting the invariance of light speed in a medium. His explanation left many physicists of the time stunned.

A. **Michelson** will earn the 1907 **Nobel** Prize in Physics "*for his optical precision instruments and the spectroscopic and metrological investigations carried out with their aid.*"

In **1888**, Heinrich **Hertz,** [249] an engineering physicist, affirmed Maxwell's theory by successfully generating EMWs, solidifying its validity.

---

[246] A. **Colin**, J. Dufourcq, J.F. Faucon and R. Marchessault "*Piezoelectric Behaviour of Biologicals Polymers*" Nature, Vol. 289, N° 5799, Pp. 596-598, 19/02/1981.

[247] **Michelson** Albert A. *(Strzelno, POL. 1852-1931 Pasadena, CA-US). He entered the US Naval Academy as a Cadet in 1867 and was promoted to Officer in 1873, he was an instructor in Physics and Chemistry from 1875 to 1879. He resigned in 1880 to pursue his studies in Europe: Berlin, Heidelberg, and France. Returning to the US, he obtained a teaching position in Cleveland-OH and then in Chicago, IL. In 1881 he invented an interferometer with which he measured the speed of light very precisely. From 1881 he tried vainly to measure the* "ether wind" *that physicists were waiting for; first alone, then, associated until 1887 with* Edward W. Morley. *Their failure proved positive in 1905 when Albert* **Einstein**, *proved that this ether is not indispensable if one accepts the idea of relative time. The same year,* A. Michelson *was awarded the prestigious Copley Medal by the Royal Society in London, and the 1907* **Nobel** *Prize in Physics. He had previously received other distinctions. He participated in the 14-18 war as a naval officer.*

[248] **Morley** Edward W. *(USA: Newark, NJ 1838-1923 West Hartford, CT). Professor of Chemistry at Western Reserve College in Cleveland, US-OH from 1869 to 1906. After his experiment with* A. **Michelson** *made other measurements on ether with* Dayton **Miller**. *Davy Medal in 1907, Willard Gibbs Award in 1917...*

[249] **Hertz** Heinrich *(Hamburg, German Confederation 1857-1894 Berlin, DEU) Engineering Physicist. Student of Gustave Kirchhoff and Hermann von Helmholtz, he was appointed lecturer*

In **1895**, the *Catholic* turned *Anglican*, Guglielmo **Marconi**,[250] improving on **Hertz**'s instrumentation, made in Salvan (Swiss Alps) the first telegraph link between his house and a receptor located at 2km behind the "Celestini" hill. This experiment earned him the **Nobel** Prize for Physics in 1909, along with Karl Ferdinand **Braun**, "*in recognition of their contributions to the development of wireless telegraphy.*" Let's note that G. **Marconi** underestimated the importance of the work done by John A. **Fleming**,[251] an expert in power engineering, a *devout Christian*, whom he had hired as a consultant. J.A. Fleming was knighted in 1929 and co-founded 1932 the "*Evolution Protest Movement,*" a British creationist organization.

In **1898** Ernest **Rutherford**,[252] the "*father of nuclear physics,*" highlighted the presence of particles α and β during the disintegration of uranium. An *Anglican*

---

in Kiel in 1883 and professor in Karlsruhe in 1885. He won the Rumford Medal, awarded to researchers who have achieved outstanding results while working in Europe.

[250] **Marconi** Guglielmo *(ITA: Bologna 1874-1937 Rome). He was not a good student, but his family fortune allowed him to conduct experiments on radio waves in 1895, which brought him fame. Not being supported by the Italian state, he left for the UK where he obtained numerous patents, some of which were contested. He was Laureate in 1932, of the* **Nobel** *Prize in Physics. His declared support for Mussolini tainted his nomination to the head of the Italian Royal Academy. Catholic becomes* Anglican.

[251] Sir John A. **Fleming** *(ENG: Lancaster 1849–1945 Sidmouth). Physicist and Electrical Engineer invented the first valve (vacuum tube) and, as a consultant for Marconi Wireless company, the first radio emitter allowing G.* **Marconi** *to realize transatlantic radio communications. Teenager he builds various engines in his workshop. Learning Mathematics and Physics he graduated with a BSc degree at the University College London. Then he studied Chemistry at the Royal College in London. He published scientific papers.*
From 1877 he followed **Maxwell**'*s courses and obtained a D.Sc. In 1882 he took the post of an electrician to the Edison Company,*
In 1884 he was elected to the chair of Electrical Engineering at the University College London, followed in 1897 by a better-paid Pender chair. He is the author of a great number of publications and books. He was awarded different medals and was elected "Fellow of the Royal Society" *in 1935. Member of the* Congregational Church, Christian *devoted to the indigent, he bequeathed an important part of his estate to Christian charities.*

[252] Lord Ernest **Rutherford** *(Brightwater, NZL 1871-1937 Cambridge, ENG). 1908* **Nobel** *Prize in Chemistry. Gifted student, he obtained a BSc in 1894 and continued his studies in Cambridge in the Cavendish laboratory directed by J.J.* **Thomson**, **Nobel** *Prize for Physics. In 1898, he was appointed Chair of Physics at McGill University (Canada) where he worked on radioactivity. In 1901 he invited F.* **Soddy** *to join him. In 1902 he invented a radio wave detector, which led to his participation in the sonar development during the First World War. In 1907 he left NZL for the University of Manchester, where he had* Hans **Geiger** *as an Assistant. 1919 Professor Cavendish at Cambridge. This giant of physics directed and received many future* **Nobel** *prizes:* Niels **Bohr** *(1922) who improved his atomic model,* James **Chadwick** *(1935),* Patrick **Blackett** *(1948),* John **Cockcroft** *(sir in 1947), and* Ernest **Walton** *(1951), ... Winner of prestigious medals, President of the Royal Society from 1925 to 1930 (Cf [Cam]). In 1931, he was raised to the peerage as Baron Rutherford of Nelson. Little is known about his religion if he ever had one. Some classify him as a partisan of* Young Earth Creationists *he married in an* Anglican *church and was buried in Westminster Abbey (Royal Church, independent economically of all institutions) near I.* **Newton**.

near YEC (ref. 653–p.235) he was awarded the 1908 **Nobel** Prize for Physics *"for his investigations into the disintegration of the elements, and the chemistry of radioactive substances."*

2.10. 20th century Theories of quantum and Relativity, Einstein unifies space and time, mass and energy
Expansion of the universe
Telescopes operating: in I-R, Radio waves, and X-rays, allow us to see a new facet of the universe that is no longer limited to our galaxy and invisible to our eyes.

"It is in the smallest things that God is the greatest."
Jean-Henri Fabre[253]

We hope that the reader, who has persevered up to this point, will understand how stimulating the dialogue - between the experimentation that controls, and the mathematical concepts that advance the ideas - is for both parties. These exchanges became more difficult with the advent of scientific cosmology whose predictions require a very sharp technology, but it continues.

In **1900,** Karl **Landsteiner**[254] biologist, physician, and immunologist *"the father of transfusion,"* identified three blood groups which would become 4, following the 1910 **Descatello** et al. work. We had to wait **8** years before this discovery became the object of extensive work. In **1909** he discovered the polio virus. This discovery will be of very great importance in many fields of medicine. K. Landsteiner was awarded the **Nobel** Prize in Physiology *"for his discovery of human blood groups"* in **1930**. Born into a *Jewish* family but converted to Christianity in 1890 as a *Roman Catholic.*

---

[253] **Fabre** Jean-Henri *(FR: Saint-Léons 1823-1915 Sérignan-du-Comtat). Humanist, poet, and polymath universally recognized, he is one of the precursors of ethology (study of the behavior of animal species in their natural environment) and entomology (part of zoology interested in insects). His "Entomological Memories" and books for children have been translated into 15 languages.*
Christian, sympathizer deist, *influenced by religion, refuted the theory of evolution, which did not prevent Darwin from writing (cf. [Dn]) that he was an* "incomparable observer." *He was awarded numerous distinctions*
[254] **Landsteiner** Karl *(Baden bei-Wien, AUT 1868-1943 US-NY),* Catholic. *Biologist, physician (1891), and immunologist. He studied medicine at the University of Vienna, then chemistry in München, DE, and Zürich until 1893. In 1896 he became the Assistant of the bacteriologist* Max von **Gruber** *at the Hygienic Institute of Vienna where he studied the mechanisms of immunity and the nature of antibodies. From 1897 to 1908 he was the assistant of the pathologist and bacteriologist Anton* **Weichslbaum** *in Vienna. In 1919 he moved with his family to the Netherlands where he was appointed as a medical examiner in a Catholic hospital in the Hague. In 1923 he integrated the Rockefeller Institute. He wrote a lot of scientific papers and made numerous contributions to pathology, immunology, and histology.* He was Laureate of the 1930 **Nobel** Prize in Physiology or Medicine.

In **1901** Max **Planck**[255] *Lutheran,* a prominent theoretical physicist opened the door to the first quantum revolution, that would last some thirty years. It will concern all scientific domains and give rise to the transistor, the laser, computers, smartphones, and GPS (Global Positioning Systems),... In 1933, Louis **de Broglie**, *Catholic,* 1929 **Nobel** Prize in Physics will say, during his reception at the Academy of Sciences, speaking of Mr. **Planck**:

- *"... the work he accomplished is one of those which give its author an immortal glory..."*

Physicists of the centuries to come will always speak of Planck's constant *"h"* and will not cease to repeat with admiration the name of the one who revealed to men the existence of the quantum Theory.

After fruitful research on thermodynamics and particularly on black radiation law, M. **Planck** published in Annalen der Physics, two articles including: *"On the Elementary quanta in Matter and Electricity,"* [256] laying the foundations of Quantum Physics, introducing, what we call *"Planck's postulate"* according to which the emission and absorption of electromagnetic energy (including light) by the matter are discontinuous and are done by discrete quantum (quantum of elementary action or quantum energy) whose energy E of the photon is proportional to its frequency v: $E = hv$. The *"quantum energy"*[257] hv is the smallest quantity of electromagnetic energy carried by a photon having a v frequency. In **1918**, he was awarded the **Nobel** Prize for Physics *"in recognition of the services he rendered to the advancement of Physics by his discovery of energy quanta."* All his colleagues revered him for his work and his humanity.

In **1903,** E. **Rutherford** published with Frederick **Soddy**[258] that radioactive elements spontaneously transmute from one chemical element to another of lower

---

[255] **Planck** Max, Karl, *E., L. (DEU: Kiel 1858-1947 Göttingen). 1918 Nobel Prize for Physics.* Lutheran *(ref. 771-p.202). After his doctorate, he obtained 1885 a position as an assistant professor at the University of Kiel and then in 1892 as a full professor in Berlin where he continued his work in thermodynamics, electromagnetism, and statistical physics. Reformulating the second principle of thermodynamics, he became interested in black body radiation and established in 1900 - thanks to the experimental work of Heinrich **Rubens** (DEU: Wiesbaden 1865 - 1922 Berlin) - the law that bears his name. He was awarded numerous medals including the Lorentz in 1927, the Copley, and the Max-Planck in 1929... He could also have been a great genius in music; but, he had to choose at the beginning of his career.*
[256] M. **Planck** *"Ueber die Elementarquanta der materie und der Elektricität,"* Annalen der Physics, Vol. 309 Issue 3, pp. 564-566.
[257] Quantum Postulate: *The relationship* $E = hv$ *(h = 6.626 068 74.10^{-34} J.s.) is essential to interpret the photoelectric effect, an interpretation that we owe to A.* **Einstein.** *This relation which seems simple is not obvious because it associates two different quantities: the energy E (dynamic quantity to which A.* **Einstein** *will associate the mass m which varies with the speed) and the frequency (kinematic quantity, which is associated with the time).*
[258] **Soddy** Frederick *(ENG: Eastbourne 1877-1956 Brighton). Nobel Prize in Chemistry in 1921 he graduated in Chemistry from Merton College in Oxford in 1898, and he remained there as a researcher until 1901 when E.* **Rutherford** *called him to work on radioactivity. In 1904 he was appointed lecturer in Glasgow, in 1914 he was awarded the chair of Chemistry at the University*

mass. In 1908, he received the Nobel Prize in Chemistry *"for his research into the disintegration of elements and the chemistry of radioactive substances."* At the same time, he supervised the scattered experiments of alpha particles by thin gold foils made by Hans **Geiger**[259] and Ernest **Marsden**[260] (Sir in 1948) which allowed him, thanks to the invention of the Geiger-Muller counter, to establish in 1911 the nuclear nature of the atom.[261]

The same year, Henri **Becquerel**[262] was awarded half of the **Nobel** Prize in Physics: *"In recognition of the extraordinary services he has rendered by his discovery of spontaneous radioactivity,"* the other half going to a couple of French scientists Marie **Curie**,[263] of Polish origin, and her husband Pierre **Curie**

---

of Aberdeen, and in 1919 the chair at Merton College in Oxford. At the same time, he advocated a reform of monetary relations.

[259] **Geiger** Hans *(DEU: Neustadt an der Weinstraße 1882-1945 Potsdam), physicist. E.* **Rutherford***'s assistant from 1907 to 1911. Moved to Berlin to work in the Metrology Department and set up a radioactivity laboratory. Professor of Physics at the University of Berlin in 1924. Hughes Medal in 1928 was co-inventor with his assistant* Walther **Müller** *(Hanover 1905-1979 Walnut Creek, CA) of the Geiger-Müller counter. Nazi, he tried in vain to develop the German atomic bomb.*

[260] Sir **Marsden** Ernest *(ENG: Manchester 1889-1970 Wellington). As a student in Manchester, he met E.* **Rutherford***, whose collaborator he will be for a time. Royal Engineer in France during the First World War, he founded the Department of Scientific and Industrial Research in New Zealand. In 1926 he oriented his research in the field of agriculture. During the Second World War, he worked on the development of radar. Fellow of the Royal Society of London in 1946, President of the Royal Society of New Zealand in 1947.* Among various distinctions, he was appointed Knight Bachelor in 1948 for services to science.

[261] Rutherford's atomic model: *the positive charges (equal in absolute value to that of the electron) and the majority of the mass of the atom are concentrated in the center of the atom, in a sphere (which will be named: the* "nucleus"*) very small, of a diameter of the order of 10^{-14}m for the hydrogen. The electrons move further away on circular orbits of a diameter 10,000 times larger.*

[262] **Becquerel** Henri *(FRA: Paris 1852-1908 Le Crosic). Son and grandson of physicists (Edmond and Antoine). Polytechnique from 1872 to 1874 then Ponts et Chaussées. Engineer in 1877, he became interested in optics and defended his thesis on the absorption of light in 1888. Member of the Académie des Sciences in 1889. Professor at Polytechnique in 1895. Discovered the radiation emitted by uranium that* Marie **Curie** *called "Radioactivity," which earned him the Rumford Medal in 1900, awarded by the Royal Society - thanks to a donation from* Benjamin **Thompson** *(Woburn, MA 1753-1814 Auteuil, American* Anglican*) - then the* **Nobel** *Prize for Physics in 1903 with* **Marie** *and* **Pierre Curie***. Member of the Royal Society in 1908. The Unit of radioactivity initially called the Curie was modified and bears his name.*

[263] **Curie** Marie *(Warsaw, POL 1867 - 1934 Passy, FR) born* **Sklodowska***. The daughter of a professor of mathematics and physics and a schoolteacher was an excellent student and won a gold medal in 1883. Higher education being forbidden to girls (Russian Empire) she worked as a governess to earn a little money which allowed her to enter a laboratory in Warsaw in 1886. She enrolled in the Faculty of Science in Paris in 1891 and obtained a degree in Physics in 1893 and Mathematics in 1894. She joined* Gabriel **Lippmann's** *laboratory of general physics and collaborated with* Pierre **Curie** *who worked on magnetism and piezoelectricity. They got married in 1895. 1st in the Aggregation in Mathematics in 1896, she became interested in the radiation produced by uranium and obtained numerous prizes. In 1898 she discovered the radioactivity of Polonium while* **Pierre Curie** *was working with her. At the end of the year, she*

(ref.241, p.74) *"In recognition of the extraordinary services they have rendered by their joint researches on the radiation phenomena discovered by Professor* Henri **Becquerel**." In 1911, Marie **Curie** received the **Nobel** Prize in Chemistry *"in recognition of her services to the advancement of chemistry by the discovery of the elements radium and polonium, by the isolation of radium and the study of the nature and compounds of this remarkable element."* During World War I (the Great War), she participated in the creation of mobile radiobiological units to radiograph the wounded. Her daughter **Irène** accompanied her to the front in 1916. In 1918 Marie took over the head of the Radium Institute and demonstrated the therapeutic virtues of radium, in the treatment of breast cancer. In 1920, she realized that the illnesses she suffered were due to excessive radiation exposure. She died in a sanatorium.[264]

Marie Curie was the first scientist to have received twice the Nobel Prize in Science. She will be followed by four other scientists: the engineer John **Bardeen**[265] (1956 and 1972), the biochemist Frederick **Sanger** (in 1958 and 1980), cf. (ref. 438, p.130), and Karl B. **Sharpless**[266] (2001 and 2022).

In **1904 Lord Rayleigh**[267] (3rd Baron) born John William **Strutt**, a *devout Anglican,* mathematician, received the **Nobel** Prize in Physics *"for his*

---

[264] *announced the radioactivity of Radium. Appointed* "chargé de conférences" *in 1900. In 1903 she defended her doctoral thesis* "Recherches sur les substances radioactives," *was co-recipient with Pierre of half of the* **Nobel** *Prize in Physics (see* **Becquerel**), *and received the Davy Medal awarded by the Royal Society. Was the first woman to receive the* **Nobel** *Prize. In 1906, after Pierre's death, she became the first woman director of a university laboratory and was appointed professor at the Sorbonne.*

[264] Marie **Curie***'s work had other developments than radiotherapy. It also allowed for the dating of objects and in particular, the measurement of the age of the oldest elements of the Earth (4 billion years), as well as of a hundred meteorites and a few stones brought back from the Moon, its satellite (4.5 billion years). Her daughter* Irène **Joliot-Curie** *(Paris 1897-1956), as well as his son-in-law* Frédéric **Joliot** *(Paris 1900-1950), were awarded the* **Nobel** *Prize in Chemistry in 1935* "in recognition of their synthesis of new radioactive elements." *The other daughter* **Eve** *(Paris 1904-2007 New York) led, with the encouragement of her mother Marie, a literary career and participated in the war effort from 1941 to 1944, which earned her the congratulations of General de Gaulle.*

[265] **Bardeen** John *(USA: Madison, WI 1908–1991 Boston, MA). Physicist and engineer, he is the only person awarded twice, the* **Nobel** *Prize in Physics. In 1956 with W.* **Shockley** *and W.* **Brattain** "for their researches on semiconductors and their discovery of the transistor effect." *In 1972 with L.N.* **Cooper** *and J.R.* **Schrieffer** "for their jointly developed theory of superconductivity, usually called BCS-theory." *Bardeen was introduced to quantum theory by J. H.* **Van Vleck***, laureate of the 1977* **Nobel** *of Physics.*

[266] **Sharpless** Karl B. *(Philadelphia, 1941)* **Nobel** *Prize in Chemistry in* **2001** "for his work on chirally catalyzed oxidation reactions" *and 1/3 in* **2022** *with* Carolyn R. **Bertozzi** *and* Morten P. **Meldal** "for the development of click chemistry and bioorthogonal chemistry." *Was awarded a Ph.D. from Stanford University in 1968 and began working in 1990 as W. M.* **Keck** *Professor of Chemistry at the Scripps Research Institute in La Jolla, US-Ca.* Quaker.

[267] **Lord Rayleigh** *was born John W.* **Strutt** *(Essex, ENG: Langford Grove 1842–1919 Witham). In 1873, he inherited the Barony of Rayleigh upon the death of his father. Mathematician, he worked on fluid dynamics, optics, and also on black body radiation; work which was appreciated*

investigations of the densities of the most important gases and his discovery of argon in connection with these studies." He made extensive contributions to science, studying the light scattering by particles smallest that the light's wavelength used, he described what we call "*Rayleigh scattering.*" On this occasion, he quoted Psalm 111:2:

> "*The Works of the Lord are great, sought out of all them that have pleasure therein,*"

This law justifies the blue color of the sky; a question that one day everyone has asked or will ask a scientist.

1905, a year that would forever alter the scientific landscape, Albert **Einstein,** [268] a profound thinker aligned with *Pantheist Spinozian* beliefs (he didn't believe in a personal God who interferes with human affairs, but rather in a cosmic spirit that manifests itself in the harmony of natural laws), made an astonishing entrance into the scientific community. His contemplations on field theory would shape the unveiling of the principle of invariance of the speed of light and pave the way for the inception of the **Theory** of **Special Relativity** [269]

---

by M. **Planck**. *His name is implicated in a lot of laws or equations. Teenager, he attended Eton College, then he entered 1861 Trinity College, Cambridge, and graduated in 1865. Elected Fellowship of Trinity. From 1879 to 1884 was the second (after J.* **Clerk Maxwell***) Cavendish Professor of Physics at Cambridge. Professor of Natural Philosophy at the Royal Institution. 1904* **Nobel** *Prize in Physics.*

[268] **Einstein** Albert *(Ulm, DEU 1879-1955 Princeton, USA).* Jewish Pantheist *Spinozian* (cf. [Ei1] – p.186). *Eminent physicist of the 20th century naturalized Swiss in 1900 then American in 1940. In 1905, while he was an engineer at the Office of Technical Inventions in Bern, he published 3 crucial articles for physics: the statistical theory of Brownian motion, the interpretation of the Photoelectric Effect, and the theory of Special Relativity postulating the Energy-Mass equivalence: $E = mc^2$. The works that followed: the Theory of Gravitation, the Theory of General Relativity, and the introduction of* "4-dimensional space-time," *curved and finite, produced a tsunami in the scientific community, comparable to* **Newton**'s *theory of gravitation that he revisited. An Israelite, he left for the United States in 1933 when the nazis arrived and taught at Princeton. He signed the letter warning* President Franklin D. **Roosevelt** *that the nazis were on the verge of acquiring nuclear weapons*. Roosevelt then initiated, in association with the United Kingdom and Canada, the* "Manhattan Project," *the objective of which was to create the first atomic bomb. According to [De] p. 9, A.* **Einstein** *wrote to H.* **Goldstein**: "I believe in Spinoza's God," a *quote from* "Conceptions" *N° 114 - p.25. 1921 Nobel Prize in Physics.*

* *Following the works of* Enrico **Fermi** *– (Rome, ITA 1901-1954 Chicago, IL-US) an experimental and Theoretical physicist,* **Nobel** *Prize for Physics in 1938 - and those of* Otto **Hahn** *(ref.345, p.104),* Lise **Meitner** *(ref.347, p.104) and* Fritz **Strassmann** *(ref.346, p.104) who, in 1936 achieved nuclear fission of uranium, there was no doubt in anyone's mind that the Germans were on the verge of making the atomic bomb.*

[269] *This theory is qualified as restricted because it is incompatible with* **Newton**'s *theory of gravitation.* Stephen **Hawking** [Ha1] *explains the problem very simply. The theory of gravitation allows us to describe perfectly the movement of the Earth around the Sun. Let us assume that the Sun disappears instantaneously; considering the speed of light and the distance T-S it would take us about 8 minutes to be in darkness. On the other hand, our orbit would be modified*

(T.S.R.). This theory necessitated a paradigm shift in our fundamental understanding, particularly challenging established notions of space and time.[270]

Einstein, challenged bravely the traditional concepts of ether and absolute time, justifying **Michelson** and **Morley's** experimental results, and proposing the abandonment of these concepts. He boldly redefined the essence of relativity (under the name of *"Principle of relativity"* [271]), amalgamating Newton's fundamental principle of physics – that the laws of the universe remain consistent for all observers in motion - with **Maxwell**'s revolutionary laws of electromagnetism. This amalgamation explained that the speed of light, an Electromagnetic Wave (EMW), solely depended on the medium it traversed, eradicating the need for ether and unshackling it from the constraints of a specific system. The repercussions were monumental:

- Time and distances were now relative in systems in motion relative to each other,[272] necessitating a new understanding of simultaneity and dissolving the concept of absolute time. **Lorentz**'s transformation[273] became indispensable in comprehending these complex space-time relationships. This re-evaluation allowed for a more nuanced

---

instantaneously because the attraction exerted on Earth by the Sun, would disappear instantaneously. This means that the gravitational effect would have moved with an infinite speed, which is incompatible with the speed limit c imposed by the theory of Special Relativity. Hence the interest of the General Relativity Theory.

[270] *On the other hand, the common language, although referring to more vague definitions (from a mathematical point of view), will always be useful to explain discoveries, because it is in direct contact with reality whereas, the scientific concepts used by the mathematical formalism applying to a restricted field, remain too abstract.*

[271] Principle of relativity: *the laws of physics must be the same for all observers in motion, whatever the speed at which they move* [Ha1].

[272] *Let's consider an observer O,' stationary in the middle of a train of length 2x, moving at a finite speed v in the direction of travel (from left to right), and an observer O stationary on the platform. Two alarms S_1 and S_2 are placed on the platform, at an equal distance x from O, (S_1 on his right and S_2 on his left). S_1 sounds when the head of the train reaches him and S_2 when the tail reaches him, i.e. at the same time. O will perceive both sounds simultaneously after a time t. On the other hand, O' (who is in front of O when the 2 signals are triggered) will hear the sound emitted by S_1 after a time t_1 smaller than t and the signal of S_2 after t_2 larger than t. There will be no simultaneity, therefore no absolute time, it is relative, and its vision can be very different from one individual to another. Explanation: O' moving towards S_1 will receive the signal emitted by S_1 after a time $t_1 < t$ because the distance traveled by the signal emitted by S_1 will be smaller than x whereas the signal emitted by S_2 will have to travel a distance greater than x hence needing a time $t_2 > t$. We could do the same applies to the length which is also relative.*

[273] **Lorentz** Hendrik *(NLD: Arnhem, 1863-1928 Haarlem). Physicist, he shared, in 1902, the* **Nobel** *Prize of Physics with* Pieter **Zeeman** *(NLD: Zonnemaire, 1865 - 1943 Amsterdam), a freethinker: "in testimony of the extraordinary service rendered by their researches on the influence of magnetism on radiation phenomena." To interpret the experiments of* **Michelson** *and* **Morley,** *those of* **Fizeau,** *... he proposed in 1904 a mathematical transformation that became the subject of* **Poincaré'**s *work.*

interpretation of biblical events, embracing a less dogmatic view of their dating.

- the iconic relation $E = mc^2$ emerged, unveiling the intimate connection between energy E and mass m*. It illuminated that an *ordinary body*[274] of mass m (variable with speed) could never surpass the speed of light in vacuum: the celerity c.

The equivalence between mass and energy is known since the works on the radioactivity of Becquerel, Curie, and Rutherford. It is easily demonstrated that one gram of uranium has an energy of $9.10.^{13}$ J,

- This transformative theory, not only challenged Newtonian mechanics but also offered a refined perspective on gravitation. Unlike **Newton**'s theory, the gravitational constant could be repulsive, enabling an interpretation of the expansive Universe (cf. § 4.1.1.) marking a paradigm shift in cosmology.

Yet, even with these revolutionary ideas, Einstein acknowledged the enduring validity of Newton's mechanics within the realm of low velocities compared to the celerity c. The conservation laws, once distinct for mass and energy, unified into the singular *"law of conservation of mass or energy."* In the face of skepticism, Einstein responded with profound insight:

"The Lord is subtle, but He is not cruel."

In the same year, A. **Einstein** applied **Planck**'s theory to elucidate the *photoelectric effect*[275] discovered by H. **Hertz** in 1888, affirming the quantum nature of light, a radiation made up of photons,[276] and furthering the quantum theory's foundations.

Einstein would be awarded the 1921 **Nobel** Prize in Physics *"for his contributions to theoretical physics especially for his discovery of the law of the photoelectric effect."*

Einstein's genius didn't stop there, in **1915,** uniting inertia and gravity, and

---

[274] Non-ordinary body: the Tachyon, *a hypothetical particle [to be taken with care] that can only be imaginary if we want it to satisfy Einstein's equivalence principle, which nobody has been able to disprove.*

[275] Photoelectric effect. *Let's apply a continuous potential difference between two metal plates placed in a vacuum. When a beam of light strikes the plate at the lowest potential (the photocathode), an electric current flows between the two plates, provided that the frequency of the light is higher than a certain frequency characteristic of the photocathode... It can be seen that the energy of the emitted electrons is independent of the intensity of the beam. The interpretation of this effect and its equation by A. Einstein was not only further proof of the validity of Planck's theory but also boosted the quantum theory.*

[276] Photon: *A quantum of light that behaves as a particle or a wave, depending on the experimental conditions.*

introducing the *Principle of Equivalence*, [277] he formulated the theory of the gravitational field, also known as the Theory of General Relativity (TGR), revealing that gravitation emerged from the curvature of *space-time*[278] due to the presence of mass and energy.

In **1916, Einstein** predicted gravitational waves resulting from matter displacement, anticipating their confirmation a century later. In effect, the 15/09/2015, we will learn their evidence by LIGO (ref.455-p.135). He also envisaged the existence of regions with such immense space-time curvature that not even light could escape, birthing the concept of "***black hole***," [279] a name proposed in 1969 by the *Unitarian* theoretical physicist John **Wheeler**, [280] a specialist in the Theory of General Relativity, and one of the pioneers of quantum

---

[277] Principle of Equivalence: *Extends the Relativity Principle to observers who, under the influence of a field (e.g. gravitational), are not moving freely*:, "in a limited area of space (cf. Einstein's elevator experiment), it is impossible to know if one is at rest in a gravitational field or if one is moving with a constant acceleration in empty space" [(Ha1)] *One of the consequences of the Equivalence Principle is that **time is not the same at all points in** space (Ha, p. 58), it is a function of the gravitational field and is slower the greater the gravitational force is.*

[278] Space-time. *It does not decompose into space and time in an absolute way. It is nevertheless a four-dimensional space integrating the time dimension ([Ei1 p.152]). In this concept, time intervenes in the expressions of the spatial coordinates. Contrary to the 3 dimensions of space, time flows irreversibly. As we have seen when an observer O' moves about another O, two events that appear simultaneous for O are not for O,' but the position and time nevertheless retain a clear meaning for each. In the same way, the notion of curvature involving an abstract quantity has nothing to do with the notion that we commonly use. We should not try to represent this new concept but consider it an element of a mathematical whole essential to the development of this new theory. As with any concept, it may be abandoned one day in favor of another more appropriate.*

[279] Black hole (BH): *A star (488-151) whose gravitational field is so intense that it traps all bodies and radiation passing between its outer surface and an imaginary sphere that surrounds it, called the* "event horizon." *It can be immobile or in rotation on itself. BH are classified according to their mass M_{BH} compared to M_S (mass of the Sun):* **1.** supermassive BH: $M_{BH} > 10^6$ M_S, **2.** stellar BH: $M_{BH} < 100 \, M_S$. See R. Penrose *for its creation.* **3.** Intermediate BH: *between* **1** *&* **2.** *In 1971.* Stephen **Hawking** *suggested that, given the very high densities of matter that existed in early space, black holes with masses of up to one billion solar masses formed. In the same way, BH of $M_{BH} < M_S$, could then have been formed taking into account the high pressure and temperature prevailing at the beginning. Let's point out that the space-time distortion due to black holes, was detected on 15/09/2015 (see p. 93). In 2005 S. Hawking wrote [Ha1, p. 95]* "Today, thanks to the Hubble Space Telescope and other telescopes that detect X-rays and gamma rays, ...we know that black holes are commonplace objects...." *On 10/01/2019 the journal Science published a study by MIT researchers, having highlighted the absorption of a star by a supermassive Black Hole, belonging to a galaxy located 300 million light years from Earth. This absorption, called* "tidal disruption," *is manifested by the emission of X-rays detected from the Earth.*

[280] **Wheeler** John A. *(USA: Jacksonville, FL 1911-2008 Hightstown, NJ). Doctor of the John-Hopkins University in 1933. Professor at Princeton, he directed the thesis of R. Feynman, Nobel Prize in Physics in 1965. Franklin Medal in 1969 for his work in nuclear physics, he participated in the Manhattan Project and led one of the teams that developed the first hydrogen bomb. Recipient of numerous awards and medals: the A. Einstein Award in 1965, and the Enrico Fermi Prize (1968), ... He was a* Unitarian.

gravitation. However, as pointed out by S. **Hawking** [Ha1-p.94], as early as **1783** the astronomer and physicist John **Michell,** [281] an English *Unitarian Clergyman,* based himself on the light's corpuscular nature (according to Newton), and the fact that the speed of light is finite, as measured in 1675 by Ole **Römer,** [282] asserted that: if a star were sufficiently massive, it could exert a gravitational force sufficiently great for no particle to escape. Assuming that light is corpuscular, these stars would not emit any radiation, remaining invisible. He named such a star: *"**Black Star**,"* which later became: *"**Black Hole**."* In **2020** (cf. p.139) R. **Penrose**, R. **Genzel**, and A. **Ghez** will share the Nobel Prize. R. **Penrose** for having *"mathematically proved that Black Holes are a direct consequence of A. Einstein's general theory of relativity;"* R. **Genzel** and A. **Ghez** *"for finding an extremely heavy, invisible object (Sagittarius A*) that pulls on the jumble stars, causing them to rush around at dizzying speeds"* (Cf. Nobel lecture 2020).

Einstein tested his theory by calculating with accuracy the *"advance of the perihelion of Mercury,"* [283] an effect known for 6 decades but which the Newtonian theory does not correct exactly.

Illustration of the space-time curvature. Taking again the example proposed by B. Greene [Gr] we can imagine that space-time would be perfectly flat in the absence of any mass or energy, like a bare floor. On the other hand, under the effect of stars, of energies of all kinds, it deforms like our parquet floor on which we have placed objects of varying weight. If the source of energy (star, light) is mobile, the variation of curvature also moves at the speed of light. Newton's space and time are no longer static: space-time is dynamic. The motions of the planets are no longer managed by the force of gravitation as it is in Newtonian theory, but by the fact that the planets must follow geodesics [284] minimizing the path traveled.

---

[281] **Michell** John *(ENG: Nottinghamshire 1724-1793 Thornhill, Yorkshire). Physicist, one of the* "fathers of seismology," *published a treatise on magnets, invented the torsion balance - allowing one to measure weak forces - and proposed to measure with it the force of gravitation, but he died before.* (Henry **Cavendish**, agnostic, *inherited his balance and improved it). Michell was the first astronomer to propose a realistic distance between the Earth and the stars and to suggest the existence of* "binary stars" *(close stars orbiting around their center of gravity). Geologist, and co-founder of modern seismology, he was elected to the Royal Society in 1760, explained the origins of the Lisbon earthquake (1755), and taught geology from 1762 at Cambridge. Without having been aware of Michell's work,* Pierre S. de **Laplace** (ref.220–p.66) *published in 1796 the idea of invisible stars. Thornhill Church's Pastor from 1767 to his death.*

[282] **Römer** Ole Ch. *(DNK: Aarhus 1644-1710 Copenhagen). Astronomer, based on the work of* R. **Descartes** *measured the variations of the eclipses of the satellites of Jupiter and deduced that the speed of light was 298,800 km/s which was correct for the time.*

[283] Advance of the perihelion of Mercury. *By applying Kepler's laws, Newtonian mechanics allows us to calculate the elliptical trajectory of the planets around the Sun (fixed focus); but allows only an approximation of the calculation of the speed of the perihelion (closest point to the Sun) which remains slightly higher than the theoretical speed. The major axis of the ellipse covered by Mercury (the planet closest to the Sun, therefore the most affected) turns by one degree every 10,000 years!*

[284] Geodesic. *The shortest line between two points on a surface. In our 3-dimensional space, it is the curved trajectory followed by an airplane, for example, going from Nice to New York thus*

On this basis, A. Einstein formulated the Theory of General Relativity (T.G.R.). An important consequence: Euclidean geometry does not allow the description of solid bodies' motion laws, but Riemann geometry does.

"Space is no longer rigid, it can participate in physical events" [Ei1].

In **1917**, a pivotal year, **Einstein**'s enhancements to TGR paved the way for a burgeoning community of *"cosmologists,"* physicists, and astronomers. Their mathematical models soared beyond the confines of everyday Euclidean space (bounded by three dimensions), venturing into n-dimensional realms to unravel the mysteries of the universe beyond our solar system.

Their exploration extended far beyond our solar abode, delving into the vast expanse of the *"Milky Way,"* a mere fraction of the universe, as elucidated in Chap. III).

Einstein, in the early stages, grasped that resolving TGR's equations required a universe either expanding or contracting toward the grand spectacle of a **Big Crunch**, a cataclysm collapse. However, Einstein, a visionary of his era, envisaged a perpetual universe, steadfast and stationary,[285] resisting the notion of a beginning or an end. This perspective stood at odds with some contemporaneous scientists, notably Alexander **Friedmann**.[286] Consequently, to counterbalance the attractive effect of the gravitational constant **G**,[287] beyond the Solar System, **Einstein** reshaped its expression. **Newton**'s formulation relied merely on masses and distances separating them, but Einstein introduced the *energy* from atoms and the *pressure*[288] exerted, intertwined with their temperature.

In certain cosmic domains, the repulsive pressure of the atoms could overpower, rendering the gravitational constant negative, propelling an expansion (as detailed § 4.1.1.). To bolster this notion, **Einstein** introduced the *"cosmological constant,"*[289] energy (qualified by G. **Lemaître**: *"exotic energy"*) filling all the Universe, akin to the long discarded *"ether"* after **Maxwell**'s work. In the realm of Science, experimentation holds the trump card, and we awaited the revelations of 1929 and 1979 for intriguing turns. Another great success of

---

reducing the distance by about 150km compared to the straight one. In space-time, the trajectory of the planets as well as that of light rays are also curved in the vicinity of the Sun, under the effect of its gravitational field (cf. Eddington-Dyson Mission 1919, p. 72).

[285] *He was probably influenced by his religious upbringing.*

[286] **Friedmann** Alexander *(RUS: St Petersburg 1888-1925 Leningrad). Physicist and mathematician. In 1922 he became interested in the theory of relativity and its equations which made him predict that it would allow the study of the universe. Introducing the idea of an expanding universe, he opposed A. **Einstein**, a supporter of a static universe.*

[287] *The effect of this repulsive force should not affect the theories of* 1. **Newton** *and Relativity, its effect could intervene only at a very great distance (at the level of the Cosmos).*

[288] Brownian motion. *Under the effect of heat, the atoms acquire a thermal agitation energy, and +-pressure, depending on the temperature.*

[289] Cosmological constant. *According to* S. **Hawking**, *it is an* "antigravitational force" *linked to the structure of space-time which, under its effect, can expand, which cannot be explained by* **Newton**'s *law of gravitation.*

TGR emerged: a foresight that masses in gravitational fields would induce a contraction of length and a dilation of time.

Among the earliest adherents to TGR was Sir Arthur **Eddington,** [290] a *Quaker*, and then Secretary of the Royal Astronomical Society of England. He had the distinction of being the first to peruse **Einstein**'s submission for publication. Collaborating with the Royal Astronomer Sir Frank **Dyson,** [291] a *Baptist,* Eddington spearheaded a mission to the island of Principe (Gulf of Guinea). Here, during the momentous solar eclipse of May 29, 1919, they confirmed Einstein's calculated angular deviation of light rays near the sun. Sir A. Eddington went on to popularize this theory, encapsulating his insights in the seminal "*Mathematical Theory of Relativity*" lectures, a work lauded by Einstein as "*the best presentation of the subject in any language.*"

Meantime, in **19**10 Frederic **Soddy** highlighted the existence of isotopes and was awarded the 1921 **Nobel** Prize of Chemistry "*for his contributions to our knowledge of the chemistry of radioactive substances, and his research on the origin and nature of isotopes.*"

Note. The $^{A}X_Z$ nuclei are represented by their symbol X, the atomic number Z which is equal to the number of protons (p), and their atomic mass A, equal to the sum of protons and neutrons (n). Two nuclei with the same Z (e.g., hydrogen $^{1}H_1$ (1p), deuterium $^{1}H_2$ (1p +1n), and tritium $^{1}H_3$ (1p + 2n)) are called isotopes.

While, in 1912 the *Catholic* Alexis **Carrel,** [292] surgeon, and biologist, had been awarded the **Nobel** Prize in Physiology or Medicine "*in recognition of the work on vascular suture and the transplantation of blood vessels and organs.*" In 1903 he left France for Canada because, having claimed in 1902 that he witnessed during a pilgrim in Lourdes (FR), a miraculous cure, he had made a lot of enmities

---

[290] Sir Arthur **Eddington** *(ENG: Kendal 1882-1944 Cambridge) was one of the most famous astrophysicists of his generation. In 1902, as a gifted student, he was awarded a scholarship to Trinity College, Cambridge. In 1904 he was Senior Wrangler (best student in Mathematics). In 1906 he was appointed Chief Assistant to the Astronomer Royal at Greenwich Observatory and was promoted to Professor of Astronomy in 1913, and to Director of the Cambridge Observatory in 1914. Was the first, in 1920, to interpret the origin of the radiation of the stars.*
Philosopher *he defends the existence of harmony between scientific research and religious mysticism in many books for the general public including* "Why I believe in God, science, and religion: as a scientist sees it" *Haldeman & Julius Publications 1925. Named Knight Bachelor (sir) in 1930, Order of Merit in 1938, ...* Quaker.
[291] **Dyson** Frank *(Measham, ENG. 1868-1939 at sea),* Baptist. *Astronomer Royal of Scotland from 1905 to 1910 then of the Royal Observatory of Greenwich from 1910 to 1933. Holder of the Royal Medal, ...*
[292] **Carrel** Alexis *(FR: Sainte-Foye-lès-Lyon 1873–1944 Paris),* Catholic. *He was a pioneer in transplantology and in thoracic surgery. In 1906 he joined the Rockefeller Institute of Medical Research in New York. His contributions to science are numerous, let's quote: vascular structure for which he developed different technics, during World War I he co-developed the* **Carrel-Dakin** method *for wound antisepsis and organ transplants: he co-invented with* Charles A **Lindbergh***, his closest friend, the perfusion pump allowing organ transplantations. He was awarded the French Legion d'Honneur, and the lunar crater Carrel was named after him in 1979.*

among his university's colleagues, and he could not obtain a hospital appointment. In this epoch, the trend in the French university was strongly anti-clericalist. In 1942, he will bravely repeat:[293]

> "*I believe in the existence of God, in the immortality of the soul, in Revelation, and in all the Catholic Church teachers.*"

The same year (1912), the *Catholic* physicist Max **von Laüe**[294] published the mathematical formulation of the passage of waves of light through a periodic crystalline arrangement of particles, establishing the fact that Xrays are electromagnetic waves, and are diffracted by crystals, a theory confirmed by the P. Knipping and W. Friedrich's experiments. His scientific work extended over a wide field: General Relativity, superconductivity, and entropy, ... He exerted a great influence, on the direction and development of German scientific work. He received the 1914 **Nobel** Prize in Physics "*for his discovery of the diffraction of X-rays by crystals.*" Before his death, von Laüe requested that his tom bear the inscription: "*He had died trusting firmly in God's mercy.*" His work opened the way to the later work of two eminent, Anglican Physicists: Sir William **Henry Bragg**[295] and his son Sir William **Lawrence Bragg**,[296] were awarded jointly the 1915 **Nobel** Prize in Physics "*for their services in the analysis of crystal structure using X-rays.*" In 1969, during an interview he gave to David Edge, Sir William **Lawrence Bragg**, speaking of his father, said:

---

[293] According to Wikipedia biography. *He also commands* **A. Presse** *to administrate the Catholic sacraments on his deathbed.*

[294] Von **Laue** Max *(DEU: Pfaffendorf 1879-1960 Berlin). After his military service, he went in 1898 to the Strasbourg University, FR to study Mathematics, Physics, and Chemistry. He moved to Göttingen, Munich, and in 1902 to the University of Berlin to work under Professor Max **Planck** where he obtained his doctorate in 1903. He became Planck's assistant in 1905 and Professor of Physics in 1912 at Zurich University. He was then appointed to diverse universities. Anti nazi he worked in England from 1944 to 1946. He was awarded many honors and distinctions, held Honorary Doctorates from numerous universities, and was a member of various Academies, ...* Catholic.

[295] Sir William **Henry Bragg** (ENG: Westward,1862-1942 London); Elected minor scholar of Trinity College, Cambridge 1881, he studied mathematics in 1884 and Physics in 1885 with remarkable results. In 1986 he was elected to the Professorship of Mathematics and Physics at the University of Adelaide, S-Australia, then Cavendish Professor of Physics at Leeds, Quain Professor *of Physics at London, and Fullerian Professor of Chemistry in the Royal Institution. He embraced with success many topics of research and was knighted in 1920. Author of many books and many medals and awards were bestowed upon him.* Anglican.

[296] Sir William **Lawrence Bragg** *(Adelaide, S-Australia 1890-1971 Waldringfield, ENG). Went to the University of Adelaide in 1906 where he studied Mathematics, Physics, and Chemistry, graduating in 1908. He entered Trinity College, Cambridge in 1909. He excelled in Mathematics and Physics and graduated with first-class honors in 1911. He wrote in 1912 the simple Bragg equation relating X-ray wavelengths and the distance d between two crystalline plans, with the angles of incidence reflection. His father built an apparatus that allowed them to confirm the equation. During the war, he was seconded to Royal Engineers to develop a method to localize enemy artillery from the sound emitted.* Anglican.

"In Adelaide, he was a churchwarden, and went to church constantly but I don't think he ever talked to us about religion."

In **1913**, Niels **Bohr**[297] - delving into quantum physics, rather than into the Electromagnetic theory from J.C. Maxwell, crafted *"Bohr's atomic model"*[298] by combining classical mechanics with **Planck's** quantization. This model, justified in the great lines, atomic radiation observed, and surpassed **Rutherford**'s earlier attempts. with whom he had worked, did not. To realize this, he had to postulate two revolutionary concepts: stationary states and the absence of electrodynamic radiation (cf. J.C. **Maxwell**) by the electron orbiting around the nucleus. The most controversial point, by the greatest of the time: E. **Schrödinger**, W. **Pauli**, ... was that of quantum jumps of the electron producing the radiations observed. N. **Bohr** was aware of this but, even if it was impossible to represent this behavior; although incomplete, his model had an experimental justification. Theoretical physicist Arnold **Sommerfeld**[299] improved it, hence the name *"Bohr-Sommerfeld Atomic Model,"* allowing the interpretation of the fine structure of the line spectrum of the hydrogen atom. This model is taught only at the college level or cited as an example of scientific intuition.

In **1919**, E. **Rutherford**, and two of his doctoral students, demonstrated experimentally the existence of the proton, and in 1920 that of the neutron which was highlighted in 1932 by Sir James **Chadwick**[300] honored with the 1935 **Nobel** Prize of Chemistry *"For the discovery of the neutron."*

In **1922**, A. **Friedmann**, relying on the initial equations of General Relativity demonstrated that the universe is expanding without convincing A. **Einstein**.

---

[297] **Bohr** Niels *(Copenhagen, DNK 1885-1962). Studies at Trinity College, Cambridge. Doctorate from the University of Copenhagen in 1911, joins E. **Rutherford** in Manchester. Leading theoretical physicist, Nobel Prize in Physics in 1922* "For his contributions to research on the structure of atoms and the radiation they emit." *He contributed to the development of Quantum Mechanics and received numerous medals. He was a member of the most prestigious scientific academies. Raised in the Catholic tradition, he declared himself an Atheist. In 1975* **Aage**, *one of his sons was also awarded the* **Nobel** *Prize in Physics.*

[298] Bohr's atomic model. *As in* Rutherford's *Atomic model, the nucleons (protons p and neutrons n) are confined in a nucleus with a diameter of about 10^{-14} m in the case of the hydrogen atom; the electrons, equal in number to protons, orbit on circumferences ("stationary orbits") 10^4 times larger without emitting electromagnetic radiation, contrary to the laws of the electromagnetic theory. But, they jump from one orbit to another by emitting or absorbing radiation of well-defined energy corresponding to the energies of the observed radiations. However, this theory could not justify everything, hence the need to improve it, which is what* Arnold **Sommerfeld** *did.*

[299] **Sommerfeld** Arnold *(Königsberg, Prussia 1868-1951 Munich, DEU), theoretical physicist. Four of his Ph.D. students (W. **Pauli**, W. **Heisenberg**, P. **Debye**, and H. **Bethe**), 3 of his post-docs (L. **Pauling**, I. **Rabi** & M. **von Laüe**) obtained the* **Nobel** *Prize in Physics while he, a prominent figure of 20th-century physics, nominated 84 times, did not get it!*

[300] Sir James **Chadwick** *(GB Bollington 1891-1974 Cambridge). 1932 Nobel Prize for Physics.*

What the latter will regret in **1929** when Edwin **Hubble**[301] proved that the universe is in expansion.

In **1924**, E. **Hubble** observed that our galaxy (the Milky Way)[302] is not unique: there are currently more than 100 billion galaxies, each containing billions of stars, not counting the dwarf galaxies …

The same year Louis de **Broglie**,[303] *Catholic*, supported a thesis: *"Research on the theory of Quantum"* at the origin of the principle of the wave-particle duality[304] of corpuscles. According to which:

> *"… the theory of relativity leads to associate the uniform motion of any independent particle of matter or radiation to the propagation of a wave."*[305]

He was able to establish the relation $\lambda = h/p$ between the wavelength λ of the wave associated with the particle (that we call *"matter-wave"*) and its momentum

---

[301] **Hubble** Edwin *P. (USA: Marshfield, MO 1889-1953 San Marino, CA). Astronomer, he classified the extra-galactic nebulae and observed in 1924 that the universe is populated by galaxies. In 1929, he demonstrated experimentally the universe expansion, and published the law of spectral displacements, called 1928* "**Hubble**'s law."

[302] The Milky Way, *our galaxy, containing about 200 billion stars of varying sizes and about 100 billion planets, is one of several billion galaxies of significant masses that make up the universe. To which are added a few tens of billions of* "dwarf galaxies" *(galaxies containing between 10^5 and 10^9 stars). Each galaxy also contains black holes, dark matter, gas, and dust. We know today that the Milky Way can be assimilated to a disk with a spherical light bulb in its center. Its diameter D is at least 15 billion km. It is part of a barred spiral galaxy, which is constituted by a luminous central bar, from the ends of which the spirals (consisting of stars), seem to start winding around the center forming spirals. It moves at a speed of 600km/s*

[303] **de Broglie** Louis Victor *(FR: Dieppe 1892-1987 Louveciennes), duke, then prince at the death of his elder brother* **Maurice** *(FR: Paris 1875-1960 Neuilly sur Seine). Bachelor of Arts in 1910; under Maurice's guidance, obtained a Bachelor of Science in 1912. Mobilized from 1913 to 1918, he defended his thesis in Theoretical Physics in 1924. In 1933, he was elected to the chair of physical theories at the Henri-Poincaré Institute and the French Académie des Sciences. In 1944, his brother* **Maurice** *welcomed him at the Académie Française. Max-Planck Medal (gold medal in the effigy of Max Planck awarded by the Deutsche Physikalisch Gesellschaft for outstanding work in theoretical physics) in 1938 and the Gold Medal of the CNRS in 1955 … Is one of the founding members of CERN.* Catholic.

[304] By duality *we mean that in certain situations we must consider the corpuscles as waves and, in others the waves as corpuscles, the two aspects being complementary and not contradictory. De Broglie assimilates the electron (matter) as the sum of different waves forming what we call: a* "wave packet" *propagating with velocities greater than the light velocity! New questions arise: is matter sustainable, can it spread into a certain area of space? In 1952, N.* **Bohr** *receiving W.* **Heisenberg** *and* W. **Pauli** *in Copenhagen said during a conversation:* "But whoever understands quantum theory will not talk about dualism here; but, will conceive the theory as a unified description of atomic phenomena, a description that is susceptible of presenting diverse aspects only when it is translated into natural language given its application to experiments." *The same could be said of allegorical language.*

[305] The wave, called "matter wave," *guides the displacement of the particle in such a way that the wave intensity, represents at each point P and at each time t, the probability that the associated particle is present.*

p = mv (v is the velocity and m the mass). His application to atomic electrons, associating standing waves to the stationary states of **Bohr's** atom, such that the circumference of the orbit contains an integer number of wavelengths, lifted the mystery surrounding the quantization of atomic states and electronic transitions. As early as 1927, his thesis was validated experimentally by Clinton J. **Davisson** and Lester **Germer,**[306] who demonstrated the diffraction of electrons by crystals, a phenomenon observed until then only on waves. Also, in 1929 (he was 37 years old!), the Nobel Prize in Physics was awarded to him: "*For his discovery of the wave nature of the electron.*" Other diffraction experiments followed, demonstrating that other atomic particles could behave like waves. From then on the case was solved, and in **1937** Clinton **Davisson** & Sir George P. **Thomson**[307] an *Anglican*, received the Nobel Prize in Physics "*For their experimental discovery of the diffraction of electrons by crystals.*" Note that G. P. Thomson is the son of the discoverer of the electron, the apological *Anglican* Sir Joseph-John **Thomson,**[308] **Nobel** Prize in Physics in 1906: "*In recognition of the great merits of his theoretical and experimental investigations of the conductivity of electricity by gases.*"

Louis **de Broglie**'s thesis was the starting point of Wave Mechanics and Erwin **Schrödinger**[309] equation. His groundbreaking work was the foundation for the development of quantum mechanics. Erwin, a *Catholic* born of a *Catholic*

---

[306] **Davisson** *Clinton J. (USA: Bloomington, IL 1881-1958 Charlottesville, VA)* & Lester **Germer** *(USA: Chicago, IL 1896 - 1971 Gardner, NY). Physicists at Bells Laboratories in New Jersey. 1937* **Nobel** *Prize in Physics.*

[307] Sir George Paget **Thomson** *(sir in 1943) [Cambridge, ENG 1892-1975]*, Anglican. *Sir Joseph J.* **Thomson's** *son (below) studied Mathematics and Physics at Trinity College. In 1914, after 1 year of research under the direction of his father, he was enlisted in the infantry. He then worked in Cambridge and Aberdeen. In 1930 he was appointed Professor at Imperial College in London and became interested in nuclear physics and its military applications. In 1952 he was elected Director of Christi College in Cambridge. Fellow of the Royal Society, in addition to the* **Nobel** *Prize in Physics (1937), he received the most prestigious medals: Hughes, Royal, Faraday, ...*

[308] Sir Joseph-John **Thomson** *(sir in 1908) (ENG: Manchester, 1856-1940 Cambridge)* Anglican apologist. *Joined Trinity College, Cambridge, as a student in 1876. Associate Professor in 1883, Cavendish Chair in 1884. Author of numerous scientific works and recipient of many international scientific medals among which the 1906 Nobel Prize in Physics. Fellow of Trinity College in 1880, of the Royal Society in 1884, and President of it (1916-1920). Over awarded by various countries in the world.*

[309] **Schrödinger** Erwin *(Vienna, AUT 1887-1961). Physicist, he defended his thesis in theoretical physics in Vienna in 1910. Assistant to* Max **Wien,** *he was appointed professor in Breslau in 1921. Five years later he published an article presenting what will be called "Schrödinger's equation» and joined* M. **Planck** *in Berlin. He was awarded the 1933* **Nobel** *Prize in Physics. In 1936 he returned to Gratz, received the Max Planck Medal in 1937, and went into exile in 1938 because his position on Nazism was considered unreliable. He holds positions at Oxford and Ghent. In 1940 he agreed to participate in the creation of an Institute in Dublin where he would remain until 1957, becoming Irish.*
Philosopher, Catholic, *impregnated with* Hindu metaphysics*, and very familiar with the Upanishads.*

father and a *Lutheran* mother, was sympathetic to *Hindu* philosophies [Sc] (cf. chapter 5) dealing with the ways of salvation for the soul. Speaking of particles represented by wave functions he stated:

"The unity and continuity of Vedanta are reflected in the unity and continuity of wave mechanics." Reasoning by the absurd, he wrote (cf. [Sc]-p.27):

> *"If from the outset we decide to leave aside, as too naive and childish, the idea of a soul that would inhabit the body as if it were its home, which would leave it at the moment of death and continue to exist outside of it, then we could ... arrive at illogical conclusions such as the fact that the disappearance of the Ego would lead to the disappearance of the World."*

The years **1925** to **1927** stand as a pivotal period, witnessing the birth of quantum mechanics and the emergence of foundational principles that continue to shape our understanding of our universe. This period saw the convergence of revolutionary ideas and formulations. The intertwining of science, philosophy, and spirituality during this time laid the groundwork for a new era in physics.

a) The "*Exclusion Principle*" and Metaphysical Explorations. In 1925, the eminent physicist Wolfgang **Pauli**,[310] a devout *Catholic* introduced the Exclusion Principle, a fundamental concept stating that 2 electrons in an atom cannot be in the same quantum state.[311] This principle, a cornerstone of quantum mechanics, is also called "*Pauli's Principle*."

W. Pauli is also known for his work on the theory of relativity and atomic theory with W. **Heisenberg**, N. **Bohr**, and A. **Einstein** .... We also owe him the prediction in 1930 of the existence of the neutrino to explain the disintegration β (cf. Rutherford experiment in 1898). The neutrino, which plays an exceptional role in nuclear physics, and serves as a messenger of the Big Bang, was only discovered in 1956.

Pauli's deep exploration into the metaphysical implications of Quantum Theory showcased his dedication to scientific understanding and the philosophical and theological dimensions of existence.

b) Matrix Formulation and Statistical Interpretation. The Matrix formulation of quantum mechanics is a mathematical framework that allows a rigorous description of quantum phenomena using matrices and linear operators.

---

[310] **Pauli** Wolfgang *(Vienna, AUT 1900-1958 Zurich, CHE)* Catholic. *Theoretical physicist, student of* A. **Sommerfeld** *in Physics in 1919 in Munich, wrote in 1921 an article on relativity which made* A. **Einstein** *say:* "Pauli is an amazing guy... he can be proud of his article...." *Defended his thesis on the hydrogen atom in 1921 and then worked with* N. **Bohr**. *Professor in theoretical physics in Zurich, he left in 1935 for the US, became an American in 1946 then a Swiss in 1949. Mains medals: Lorentz in 1931, Franklin in 1952 & Max Planck in 1958. He received the* **Nobel** *Prize in Physics in 1945.*

[311] Quantum states in atoms. *There are four:* n, *the principal (can take the values 1,2,3, ...;* l, *the orbital angular momentum (0,1, ..., n-1);* m_l, *the magnetic (-l, -l +1, ..., l-1, l), and* m_s, *the electron spin (+/- ½) characterizing the electron spinning around his axis.*

It was created by two *Lutheran* physicists, Werner **Heisenberg**[312] and Max **Born**,[313] as well as Pascual **Jordan**[314] who had defended his thesis in 1924 in Göttingen under the direction of M. **Born**. According to this formulation, the physical properties of particles are represented by matrices that can evolve in time according to certain rules; their eigenvalues correspond to the possible values of observables, such as energy, position, and momentum. The matrix elements are calculated using the *"principle of correspondence"* which relates the classical variables (energy, position..) to quantum operators.[315]

Following closely, in 1926, W. **Heisenberg** formulated the *"Uncertainty Principle"* that S. **Hawking** prefers to call the *"Principle of Indeterminacy"*[316] because it *"refers to the absence of a determination of a physical quantity."* A consequence of this principle is that at the atomic scale, the result of any measurement is not unique, it is affected by a certain probability, so it cannot be predicted with absolute certainty. This indeterminacy results from the fact that in the atomic domain, any measurement involves a disturbance of the quantity to be

---

[312] **Heisenberg** Werner Karl *(DEU: Würzburg-German Empire 1901-1976 Munich)* Lutheran. *Theoretical physicist, he defended his thesis in 1923 under the direction of A.* **Sommerfeld**. *From 1924 to 1925 he worked with* Niels **Bohr** *at the University of Göttingen. He was appointed professor of theoretical physics in 1927 (at the age of 26!) at the University of Leipzig, where he remained until 1941 when he joined the Kaiser Wilhelm Institute of Physics. Appointed Professor of Physics in Munich in 1958. Max Planck Gold Medal in 1935. He received many prestigious medals and prizes from different countries.*

[313] **Born** Max *(Breslau-German Empire 1882-1970)* Lutheran of Jewish *ancestry. Attends the Universities of Breslau, Heidelberg, Zurich, Cambridge, and Göttingen. He submitted his thesis in 1906 in Göttingen (where he was appointed professor of theoretical physics from 1921 to 1933). In the meantime, he worked with* Hermann **Minkowski***, M. **Planck**, *and* A. **Einstein**. He was appointed professor Of Natural Philosophy in Edinburg from 1936 to his retirement in 1951. *½ 1954 Nobel Prize in Physics. He received a lot of prestigious medals and prizes.*

* **Minkowski** Hermann *(Alexotas, RUS 1864-1909 Göttingen, DEU), mathematician, combined space and time to give the* "Minkowski space," *an idea that was taken up by A.* **Einstein** *(whose professor he was in Zurich) when he developed the theory of special relativity.*

[314] **Jordan** Pascual *(NLD: Hanover 1902-1980 Hamburg). After studies in Mathematics, Physics, and Zoology he defended his thesis in 1924 in Göttingen under the direction of M.* **Born** *with whom he collaborated, as well as with* W. **Heisenberg**. *His adhesion in 1933 to the NDSAP (Nazi party) meant that he was not awarded the Nobel Prize, despite the support of A.* **Einstein** *on two occasions. Einstein was aware of the importance of Jordan's work in matrix quantum mechanics, among other things!*

[315] *Quantum operators are specific tools, used by mathematicians and physicists. Nevertheless, when you multiply, add numbers, derive, or integrate functions, you use operators too.*

[316] Heisenberg's Uncertainty principle or Principle of indeterminacy. *Some pairs of observables (called* "conjugate quantities," *like the momentum p (= mv) and the position (x,y,z)) cannot be measured simultaneously. Simplified version: One can never be completely sure, at a given time, of the position (x) and the velocity according to the x-axis (v$_x$) of a quantum particle. The product of their uncertainties Δx on x by Δv_x on the v$_x$ component of v satisfies the following relation: $\Delta x. \ \Delta v_x = h/2\pi$. But more importantly, as we shall see in Chapter 4, it applies to fields for which the intensity and the rate of change cannot be accurately determined simultaneously.*

measured. This principle challenged classical determinism by highlighting the inherent unpredictability at the atomic scale (cf. below).

c) Schrödinger's Wave Mechanics and the Schrodinger Equation. In 1926, Erwin Schrödinger, building on Louis **de Broglie**'s wave-corpuscle duality, introduced the Schrodinger equation, a fundamental differential equation describing the evolution of a wave associated with a subatomic particle. This formulation known as Wave Mechanics, united wave-like behavior with the behavior of particles.

These two formulations of quantum mechanics are equivalent but, the matrix formulation has some advantages over the wave formulation, such as being more suitable for discrete systems.

In **1927** Paul **Dirac**[317] a *Deist* at the end of his life, with a fascinating philosophical journey from *Atheism to Agnosticism* completed the Quantum Theory of electromagnetic fields. His formulation of *"the Dirac equation"* profoundly linked quantum theory with special relativity, ultimately predicting the existence of antiparticles[318]!

It's noteworthy that the whole of these works not only consolidated quantum mechanics but also garnered recognition from physicists like A. **Einstein**, and Louis de **Broglie**,.... Besides:

-    Werner **Heisenberg** received the 1932 **Nobel** Prize in Physics for *"the creation of quantum mechanics, the application of which led, among other things, to the discovery of the allotropic varieties of hydrogen;"*[319] while,

-    P. **Dirac** will share the 1933 **Nobel** Prize in Physics with E. **Schrödinger** *"for the discovery of new and useful forms of atomic theory."*

---

[317] **Dirac** Paul, A., M. *(Bristol, ENG 1902-1984 Tallahassee, FL-US) is one of the fathers of Quantum Mechanics. Within 6 months of entering Cambridge in 1923, he wrote his first scientific papers. He collaborated with N. **Bohr** and then W. **Heisenberg**. Worked to establish a link between the quantum theories of W. Heisenberg and E. **Schrödinger**. Invited to Solvay in 1927 (25 years old!). He was not initially a follower of any religion but his ideas on the existence of God, creator of the universe, became more and more accepted over the years, going from negation in 1927 to justification in 1971. Copley Medal.... 1933 **Nobel** Prize in Physics.*

[318] Antiparticle. *A particle having the same mass and spin (intrinsic kinetic moment) as a known particle, but whose quantum numbers and charge are opposite. For example, the positron, the antiparticle of the electron has the same mass as the latter but its charge (identical in absolute value) is of a different sign (it is positive = $1.6.10^{-19}$ Coulomb while that of the electron is negative). Another peculiarity, the antiparticles have very low lifetimes. When an antiparticle meets its corresponding particle they annihilate, giving rise to two photons of equal energy emitted in diametrically opposite directions.*

[319] Allotropic varieties of hydrogen: *orthohydrogen, whose two protons of the molecule have parallel spins in the same direction, and parahydrogen for which the spins are antiparallel (their axes are parallel but in opposite directions).*

- W. **Pauli** will be awarded the 1945 **Nobel** Prize in Physics "*for the discovery of the principle of exclusion,*"

- Max **Born** will be awarded half of the 1954 **Nobel** Prize in Physics "*for his fundamental research in quantum mechanics, especially for his statistical interpretation of the wave function,*" the other half going to Walther **Bothe,**[320] *Lutheran too "for the coincidence method and for his discoveries made therewith.*"

The year **1927** marked a pivotal moment when the profound philosophical debates around determinism, indeterminacy, and the nature of reality were epitomized by the renowned Solvay Congress. **Einstein**, a key figure in these discussions, famously asserted:

"*God does not play dice*" [He],

emphasizing his belief in a deterministic universe.

N. **Bohr**, an *atheist,* on the other hand, retorted, advocating for complementarity and acknowledging the intrinsic indeterminacy at the heart of quantum mechanics:

"*Who are you, Einstein, to tell God what to do?.*"

Schrödinger, like **Einstein**, opposed the probabilistic interpretation of quantum physics. This shows that even the greatest can be wrong. Nevertheless, for *Catholics* for example this would not be a problem, because it supports the hypothesis that God has left us free will, which was not the case with deterministic theories like Newtonian and relativity.

These debates highlighted the ongoing dialogue between scientific discovery and philosophical or religious beliefs. Eminent physicists like W. **Pauli** and S. **Hawking** emphasized the fundamental role of indeterminacy in our universe, acknowledging its deep connection to the nature of reality.

According to W. **Pauli** [He], *Catholic,* the Principle of Indeterminacy rejecting that of causality, the "*narrow*" perception of the physicists of an objective world,

---

[320] **Bothe** Walter *(DEU: Oranienburg 1891-1957 Heidelberg) Lutheran. Studied Physics, Mathematics, and Chemistry at the University of Berlin. He defended his Ph.D. in 1914 under the direction of* M. **Planck**. *Taken prisoner of war by the Russians, he devoted himself to mathematics. In nuclear physics, he developed the coincidence method. He discovered a radiation that penetrates all matter except water, which* J. **Chadwick** *named* "neutron" *in 1932. Professor at the University of Giessen in 1930, then director in 1932 of the Institute of Physics and Radiology in Heidelberg. He was dismissed from his position by the nationalist movement* "Arian Physics" *but regained it in 1946. In the meantime, he was appointed director of the Institute of Physics at the Heidelberg Medical Research Center where he built the first German cyclotron (invented by* Ernest O. **Lawrence** *in 1931, who received the Nobel Prize in Physics for this invention in 1939). Holder of the Max Planck Medal, Bothe was awarded the Nobel Prize in Physics in 1954.*

subject of the divergences of view with the religion, which is subjective, should lead the physicists to:

"... *respect a middle ground, perhaps that of Bohr's complementarity.*[321] *A science that has adopted this way of thinking will not only be more tolerant of the various forms of religion; it will perhaps, thanks to a better overview, be able to contribute to the world of values.*"

Pauli's opinion is shared by another great *Catholic* scientist: V.F. **Hess** (ref.332-p.99), 1936 **Nobel** Prize in Physics.

For the *Agnostic* Stephen **Hawking** [Ha2]:

"*The principle of indeterminacy is indeed one of the fundamental properties of the Universe in which we live.*"

In "*Une Belle Histoire du Temps*" [Ha1], Hawkins states (p.107):

"*... one could postulate the existence of a series of laws determining events which would be accessible to a supernatural being capable, unlike us, of observing the present state of the universe without disturbing it.*"[322]

Notes. **1.** While A. **Einstein** and I. **Newton's** universes are local: an interaction between two objects A and B can only take place if someone or something connects them, the Quantum Mechanics universe is not necessarily local: it reveals quantum entanglements (interactions) between binomials located at very great distances (light-years) from each other. The pre-requisite, which is not easy to achieve experimentally, is that these two particles A et B have been created in the same quantum state in the same place at the same moment. In these conditions, an action produced here on A must have an immediate effect on its binomial[323] B located at infinity. We are plunged into black magic and yet some experiments conducted in recent years tend to prove that it is feasible.

Between 1980 and 1982, Alain **Aspect**[324] et al., using specific photons made pairs of entangled particles and demonstrated in 1982 at Orsay - with 2 detectors

---

[321] Bohr's principle of complementarity: *a single phenomenon can be grasped by two different modes of interpretation. These 2 modes must be mutually exclusive and complementary.*" *This new notion in physics was not new in philosophy nor in some Asian religions for whom God is not a personal god.*

[322] *Let's recall that S.* **Hawking** *was at least, Agnostic or even Atheist.*

[323] Binomial particles and entanglement for a non-scientific. *Let us suppose that A and B are two true twins. One goes into the Arctic while the other goes into the Antarctic. You punch A, and B instantly feels the effect of this action!*

[324] **Aspect** Alain (Agen Fr 1947). *Graduated from the Ecole Supérieure de Paris, with a Ph.D. degree in 1971, and a habilitation thesis in 1983, made fundamental experiments in atomic physics and quantum optics having an international impact. Besides the 2022 Nobel Prize in Physics, he has been awarded numerous prestigious prizes since 1983 and a lot of Honours:*

spaced 12m apart - that this Quantum Mechanics prediction is right. In 1988 a Swiss demonstrated it for particles 30km apart, followed in 2017 by **Juan Yin** et al. who sent entangled photons from the QUESS (QUantum Experiments at Space Scale) satellite, orbiting 500km from the Earth to terrestrial satellites spaced at 1203km. The effect is instantaneous, without an exchange of information between the two. In these conditions the principle of causality is not affected, nor the celerity c. Consequently, the present and the past become indistinguishable because when we measure an observable in the system, its value is the same as if the two particles were still together or very close! His work marks the 2nd quantum revolution. He paved the way for quantum computing and extra-secure communications,... Consequently, Alain **Aspect**, John **Clauzer,** and Anton **Zellinger** shared the 2022 **Nobel** Prize in Physics *"for experiments with entangled photons, establishing the violation of Bell inequalities and pioneering quantum information science."* The Nobel Jury had finally realized that their work had triggered the 2nd quantum revolution.

Based on such work, on the indeterminism highlighted by E. **Heisenberg** in the case of sub-atomic particles (emphasis added), ... *"pseudo-scientists"* directed their research several decades ago towards a mystical interpretation of Quantum Mechanics. For Amit **Goswami,**[325] for example, it is necessary to introduce into science a new paradigm: that of consciousness primacy over matter, making God and the world one thing. This would be a justification for pantheism, in my opinion.

As we read this essay, we discover that quantum mechanics has led renowned scientists, even Nobel Prizes in Physics, to ask philosophical questions. Richard **Feynman,** 1965 **Nobel** Prize in Physics (ref.361-p.108), describing himself as a *freethinker*, demonstrates that the present no longer derives from a single past. It is the combination of several possible pasts. His explanation is based on **Davisson** 1937 **Nobel** Prize in Physics and **Germer's** experiment (ref.306-p.91). To simplify we will say that the image of the electron striking the two slits of the Young spectrometer is the result of the combination of the two possible probability waves corresponding to the passages through each slit. Sir R. **Penrose** (ref.368, p.109), 2020 Nobel in Physics, as a handful of physicists believes that consciousness follows a quantum process. Chapter 3, Pp. 164-170, we will speak of Consciousness with notorious scientists.

Commander of the Legion d'Honneur, of the Palmes Académiques, ... Co-founder of the start-up Pasqal which develops quantum processors.
[325] **Goswami** Amit *(India 1936)* Hinduist. *Professor in the physics department of the University of Oregon from 1968 to 1997. Ph.D. in Physics in 1964 from the University of Calcutta, calling himself a* "quantum activist," *at the beginning of the 1970s, he turned his research towards quantum cosmology and mysticism, what he calls:* "Science within consciousness." *A prolific writer, he appeared in movies too. For him* "consciousness is the ground of all being."

The same year (**1927**), Professor Arthur H. **Compton**[326] a *Presbyterian* won half part of the **Nobel** Prize in Physics *"for his discovery of the effect named after him."* He demonstrated that, when the matter is submitted to X-rays, there is an emission of secondary X-radiations of energies depending on the direction of emission. This effect is known as *"Compton scattering."*[327] Let's notice that, following the discovery of X-rays, by Wilhelm **Rontgen**[328] - who received 1901 the first **Nobel** Prize of Physics *"in recognition of the extraordinary services he has rendered by the discovery of the remarkable rays subsequently named after him"* - the *Methodist* Charles G. **Barkla**[329] received the 1917 Nobel Prize of Physics *"for the discovery of the characteristic Rontgen radiation of the elements;"* he put in evidence the fact that scattered X-rays are polarized to a certain degree and have some properties characteristics of the diffracting element. Barkla considered his work to be:

"part of the quest for God, the creator."[330]

Carl D. **Anderson**[331] published the discovery of the first antiparticle: *"the anti-electron,"* (which is named: *"positron"*) which earned him in 1936 half of the

---

[326] **Compton** Arthur H. *(Wooster, OH-US 1892–1962 Berkeley, CA-US). 1912, Ph.D. in physics from Princeton University. During World War I he developed instrumentation for aircraft. In 1919 he joined the Cavendish Laboratory of the University of Cambridge where he studied the absorption and diffusion of γ-rays. During World War II, Compton was under the direction of* Robert **Oppenheimer** *one of the leaders of the Manhattan Project whose goal was, to quickly develop a nuclear bomb and to counter the Nazis. From 1945 to 1953 he was Chancellor of Washington University in St Louis, IL-US. Besides the 1927 Nobel Prize in Physics, he received various prestigious medals: Franklin, Hughes, ...He was a* Deacon of a Baptist Church.

[327] *This effect is disturbing when we must interpret the γ-ray spectrum emitted during the disintegration of radioelements. Project leader for the design of Nice Sophia Antipolis University's van de Graaff proton beam accelerator, I was confronted at the end of the 1960s with this effect. Also, I patented an "anti-Compton spectrometer."*

[328] **Rontgen** Wilhelm *(DE: Lennep 1845–1923 Munich). Learning in mechanical engineering, in 1869 he graduated with a Ph.D. from the University of Zurich. Lecturer in 1974 at the University of Strasbourg, he was successively appointed as Professor at Hohenheim-Württemberg, Strasbourg, Giessen, Würzburg, and from 1900 at the University of Munich. He donated the amount of the Nobel Prize to the University of Würzburg and rejected the nobiliary particle "von." Some say that he was a* Christian, *but I cannot confirm.*

[329] **Barkla** Charles G. *(Widnes, ENG 1877–1944 Edinburg, Scotland). Attending Liverpool University, he began with mathematics before specializing in physics. In 1899 he was admitted to the Cavendish Laboratory of Trinity College, Cambridge under the direction of* **J.J. Thomson**. *In 1913 he was appointed Professor of Natural Philosophy at the University of Edinburg. The lunar crater Barkla was named in his honor. 1901 Nobel Prize in Physics,* Methodist, *he was a religious man.*

[330] *McTutor Index Barkla or mathshistory.st-andrews.ac.uk/Biographies/Barkla/*

[331] **Anderson** Carl D. *(USA: New York 1905-1991 San Marino, CA). Made all his career at Caltech (Bache Anderson CA). He worked under the direction of* R. **Millikan** *with whom he studied cosmic rays in which he discovered traces of the positron as early as 1932. He then conducted the first experiment to demonstrate this antiparticle, which earned him half of the 1936 Nobel Prize in Physics.*

Nobel Prize in Physics "*For his discovery of the positron.*" The other half was awarded to the *Catholic* Victor Franz **Hess**[332] "*for his discovery of the cosmic ray.*" In conclusion of "My Faith,"[333] he says:

> "*What is certain is that, when Faith comes, there follows a great serenity of soul and a deep peace in the human heart.*"

In **1928** Sir Alexander **Fleming** [334] a *Presbyterian* physician, and microbiologist returning from a holiday noticed in his Petri dishes an unusual "*mold juice*" secreting something, which proved to be able to inhibit the growth of bacteria. Although he published his discovery in 1929, it was not until 1939 that the team of the pharmacologist and pathologist Sir Howard **Florey,**[335] *an Agnostic,* began the work of extracting, purifying, and transforming penicillin, into a potent antibiotic, and in 1940 the experimentation. They contributed to the saving of tens of thousands of Allied troops. In the meantime, the chemist Ernst **Chain**[336] *devout Jew,* who had been recruited in 1935 by Florey, and helped by some crystallographers had succeeded in identifying the structure of penicillin by X-ray crystallography. The introduction of penicillin in the 1940s heralded the era

---

[332] **Hess** Victor Franz *(Peggau, AUT 1883-1964 Mount Vernon, US-NY)* Catholic, *1936 Nobel in Physics. Under graduated from the University of Graz (AUT) he received his Ph.D. in 1910 at the same university. Assistant of* Stefan **Meyer** – *a Jewish physicist involved in the research of radioactivity - until 1920 when he moved to the United States Radium Corporation in Orange (NJ-US) until 1923. Then he went back to Graz where he was appointed Professor of Physics in 1925. Professor and Director of the Institute of Radiology at the University of Innsbruck from 1931 to 38. His wife, being* Jewish, *they moved to Fordham University (New York). He was naturalized.*

[333] My Faith *is a series of articles published in 1946 in the American Weekly: A Physicist's Faith, by V. F.* **Hess***, published by the Society of Catholic Scientists on March 31, 2021, in Wikipedia;*

[334] Sir Alexander **Fleming** *(UK: Darvel, Scotland 1881–1955 London), was knighted in 1944, for his scientific achievements. Although raised by a Presbyterian father and a Catholic mother he was not very religious. Nevertheless, his ashes are buried at the* Anglican *St Paul Cathedral, his son Robert is Anglican.*

[335] Lord Howard **Florey**, Baron of Adelaide *(Adelaide, S. AUS 1898–1968 Oxford ENG)* Agnostic. *He attended the University of Adelaide where he studied medicine from 1917 to 1921, and continued at Oxford, Cambridge, US. From 1931 he occupied the chair of Pathology at the University of Sheffield, then in 1935 that of Oxford where he worked with E.* **Chain***. In 1941 he was elected a member of the Royal Society (President in 1958), appointed Knight Bachelor in 1944, life peer and Baron of Florey in Adelaide in 1965. 1945 Nobel Prize in Physics,* Agnostic, *a memorial tablet, and a commemorative stone, were unveiled in St Nicholas' Parish, Oxford (1980) and Westminster Abbey (1981) respectively.*

[336] Lord Ernst Boris **Chain** *(Berlin 1906–1979 Castlebar IRL) was an enthusiastic* Jewish *and an ardent* Zionist. *Chemist received his degree in 1930 from F.* **Wilhelm** *University of Berlin. In 1933 The nazis coming to power, and Germany was not safe for a Jew, he moved to London where he obtained a position at University College Hospital. 1935, Lecturer at Oxford, he worked on various research topics until working with H.* **Florey** *and the Biochemist Edward* **Abraham***. He was elected Fellow of the Royal Society in 1948, Knight Bachelor in 1969,* **1945 Nobel** *Prize. In 1965 during the World Jewish Congress Conference of Intellectuals, he expressed clearly his views in a speech entitled "Why I am a Jew."*

of antibiotics, which changed the medical world hugely, it is one of the greatest advances in therapeutic. So, Sir Alexander **Fleming**, Lord Howard **Florey**, and Sir Ernst Boris **Chain** received jointly, the 1945 **Nobel** Prize in Physiology or Medicine *"for the discovery of penicillin and its curative effect in various infectious diseases."* Among these three Nobel Prizes, E. **Chain,** *an enthusiastic Jewish* was the more religious (cf. Chapt. V.5.c). Although he had a Presbyterian background, A. **Fleming** was not very Religious. Nevertheless, he stated [Gc]:

> *"... his life's discoveries were more than chance, it was the invisible hand of God on his life and work."*

Proving, that although having some difficulties with religious institutions, he nevertheless believed in God.

His ashes are buried at the Anglican St Paul's Cathedral, and his son Robert, a medical generalist was an Anglican.

Let's note that, before moving to London, where he will integrate into the Royal Polytechnic Institution, **Fleming** attended the Kilmarnock Academy (a Scottish secondary school) just as did, in 1893, Lord John **Boyd Orr**, 1st Baron of Brechin,[337] a future famous scientific researcher in nutrition, and a *member of the Free Church of Scotland* (dissident Anglicans) was deeply religious until his death and manifested this in his stance in favor of the unfortunate. In 1949, Sir J. Boyd **Orr** was awarded the **Nobel** Peace Prize *"for his lifelong effort to conquer hunger and want, thereby helping to remove a major cause of military conflict and war."* *He donated the whole financial award to institutions devoted to world peace.*

In **1929** Edwin **Hubble** observed, thanks to the telescope of Mt Wilson in California (2,5m in diameter), that the galaxies move away from each other confirming thus, that *"the universe is in expansion,"*[338] as demonstrated by A.

---

[337] Lord John **Boyd Orr,** 1st Baron of Brechin, *(Scotland: Kilmaurs, 1880–1971 Edzell),* member of the Free Church of Scotland. *Scientific researcher in nutrition. He was a successful farmer, teacher, medical doctor, nutritional physiologist, politician, and businessman, ... In 1902, after a Master's degree in Arts, he was posted teacher in a school located in the slums of Glasgow, where he encountered poverty and misery. Realizing that he could not relieve their misery he resigned and returned to the University to study Biology and Medicine. He successfully graduated Bachelor of Medicine in 1912, MD, a postgraduate degree in Medicine in 1914, and D.Sc. (Doctor of Sciences) in 1914. During the First World War, he was posted Medical Officer in an infantry unit. He was awarded the Military Cross and the Distinguished Service Order. Then in 1919, he participated in the expansion of the Rowett Institute where his research was dedicated to animal nutrition and then to human nutrition. Elected Rector of the University of Glasgow, and Member of Parliament. After the Second World War, he contributed as Director General of FAO (Food and Agricultural Organisation) to alleviate food shortages, improve food production, and equitable distribution. Co-founder, the first President of the World Academy of Art and Science, and President of the National Peace Council. He was knighted in 1935.*

[338] The expansion of the universe *(between 5 and 10% per billion years [Hal, p. 77])* does not mean that galaxies are expanding. *Occupying a certain region of the universe, galaxies move, but their size does not evolve, because nuclear forces - more intense than the one producing the expansion - maintain the cohesion between atoms. Moreover, A. **Guth** (ref.391-p.116) demonstrated theoretically the existence of a* "dark matter," *not emitting light, of which nobody*

Friedmann in 1922. He deduced from his measurements the law of "*Hubble-Lemaitre*" according to which the galaxies move away at a speed v proportional to the distance d that separates them; the coefficient of proportionality "*Ho*" was called the "*Hubble constant.*" This is one of the fundamental experimental discoveries of the 20th century.

It is noteworthy that galaxies located sufficiently far from us can appear to move at speeds greater than that of light. However, this phenomenon does not challenge the theory of relativity, as these galaxies are not actually moving through space but are instead following the expansion of space itself. Let's note that the speed limit for moving particles is only valuable within space.

This discovery had a significant impact on A. **Einstein**, leading him to abandon his idea of a stationary universe, which he later admitted was the greatest mistake of his life. The concept of the "*cosmological constant*" was also temporarily set aside.

Despite these setbacks, we can agree with the sentiment expressed by M. **Planck** in [PL1, p1]:

> "... the theory of relativity ... has proved to be the completion and crowning achievement of classical physics...... it has made it possible to take an important step in the direction of the standardization and refinement of its concepts."

The same year the couple of biochemists **Gerty Cori**[339] a *Jewish* converted to *Catholicism* and the *Catholic* **Carl Cori**,[340] published what came to be known as "*Cori's cycle*," explaining how glucose and glycogen are converted in the liver

---

*knows what it is made, keeping the stars at an equal distance from each other. This dark matter induces a gravitational force, preventing the clusters of particles in distant constellations, from moving away from each other. The speed v of the galaxies is all the greater as they are more distant (e.g. a galaxy located 100 million light-years away moves at a speed of 2,300 km/s. If the distance doubles the speed doubles, ...). The name "Hubble-Lemaitre law" refers to the fact that G. **Lemaitre** had predicted this expansion two years earlier.*

[339] **Cory** Gerty, born Radnitz *(Prague, CES 1898–1957 Glendale US-MI). In 1914 she was admitted to the medical school of Prague (unusual for a woman) where she met **Carl Cori**. They graduated and got married in 1920. In 1922, they immigrated to the USA and worked as biochemists at the Roswell Park Cancer Institute of Buffalo (NY). The two were naturalized in 1928. In 1931 they moved to Washington University in St Louis, MI-US. In 1943 she was an associate professor in Biological Chemistry and Pharmacology, then a full professor some months before she won the 1947 **Nobel** Prize in Physiology or Medicine. Elected Fellow of the American Academy of Arts and Sciences in 1953, she received various awards and Prizes. Gerty was an author and coauthor with Carl, of numerous papers. Jews converted in 1920 to* Catholicism.

[340] **Cory** Carl *(Prague, CZK 1896–1984 Cambridge, US-MA). After World War I he returned to the University of Prague where he graduated with a Doctorate in Medicine in 1920. In St Louis Carl obtained a position as a professor in pharmacology and, in 1942 he became a professor in biology. Besides many awards, he shares with his wife a star on the St Louis Walk of Fame and the 1947 **Nobel** Prize in Physiology or Medicine. In 1950 he was elected a Foreign Member of the Royal Society (London).* Catholic.

and bloodstream; answering one of the most fundamental questions about how the human body works. Their subsequent work on carbohydrate metabolism changed our understanding of diabetes. This earned them, the **1947** Nobel Prize of Physiology or Medicine *"for their discovery of the course of catalytic conversion of glycogen."*

Always in 1929, the nuclear physicist Ernest O. **Lawrence,**[341] an *Anglican,* invented the cyclotron, the first cyclical particle accelerator; using magnetic and electrostatic fields to accelerate charged particles. He will improve it in 1931, reaching higher energies. The cyclotron is widely used in the medical field, particularly to produce radioisotopes used in radiotherapy and high-energy charged particles used in proton therapy. He will be awarded the 1939 **Nobel** Prize in Physics *"for the invention and development of the cyclotron and results obtained with it, especially with regard to artificial radioactive elements."* The same year he announced plans for a 100 MeV cyclotron. During the Second World War, he led the Manhattan Project, which built the world's first nuclear weapon. At a conference in early 1945, he declared: *"Atomic bombs will surely shorten the war, let us hope that they will effectively end wars as a possibility in human affairs."*

In **1930** Georges **Lemaitre,**[342] a Belgian *Catholic Canon,* physicist, and astronomer, understood all the interest in the theory of Relativity and proposed the theory of the *"primitive atom"* at the origin of the Big Bang. He predicts that this state, where the density of matter and energy were extreme, should have given rise to the emission of radiation that A. **Penzias** and R. **Wilson** discovered in 1965 (ref.361-p.76). Although, this idea seemed to be a godsend for *Christians* since it

---

[341] **Lawrence** Ernest O *(USA Canton, SD 1901-1958 Palo Alto, CA). Lawrence obtained his doctorate from Yale University, (USCT) in 1925, where, he later became a National Research Fellow and assistant professor. In 1930 he assumed the position of Professor of Physics at the University of California, Berkeley, and subsequently served as the director of the radiation laboratory in 1936. His scholarly contributions include many articles, and he has received numerous esteemed awards. Lawrence was elected in 1934, at the National Academy of Sciences, and was awarded the* **Nobel** *Prize in Physics in 1939.* Anglican, *he was a member of the Episcopal Church.*

[342] **Lemaitre** Georges (BEL: Charleroi 1894-1966 Louvain). He commenced his studies at the Jesuit College of Charleroi in 1904. In 1914, he temporarily halted his education to enlist in the military; where he was honored with the war cross. After returning to the seminary in 1918, he resumed his scientific work. He earned his doctorate in science in 1920 and was ordained as a Catholic priest in 1923. He gained admission to Cambridge as a student researcher in "Cambridge's solar physics laboratory," then worked at MIT, from 1925 to 1927, conducting research on General Relativity and successfully defending his thesis in 1926. Lemaitre *proposed a relativistic model in 1927 and served as a Professor of Astrophysics at the Catholic University of Louvain. In 1931, he formulated the initial cosmological theory of the Universe, positing a dense state, followed by expansion due to an explosion. He was the recipient of the Mendel Medal and Francqui Prize in 1934. He espoused Thomism and advocated for a symbolic interpretation of Genesis, distinguishing between the concept of the beginning and that of creation. In 1960, he was appointed president of the Pontifical Academy of Sciences.*

agreed with the thesis of creation defended by the Bible, **Lemaitre**, a *Thomist canon* that he was, asked not to make any amalgam.

In **1932**, using the Cockroft & Walton generator or multiplier, that they have realized, John D. **Cockcroft**[343] and Ernst T. S. **Walton**[344] performed the first artificial nuclear disintegration in history. Their electrostatic generator creates a high DC voltage from a low voltage input. After a brilliant education in physics at St. John's College, Cambridge, J. **Cockroft**, a *Methodist,* prepared his thesis under the *Anglican* E. **Rutherford**'s supervision, in his Cavendish Laboratory in Cambridge. He met there, in 1927 Ernest Walton *a Methodist who had brilliantly completed a double degree in mathematics and experimental physics.* They will be awarded jointly, the 1951 **Nobel** Prize in Physics *"for their pioneer work on the transmutation of atomic nuclei by artificially accelerated atomic particles."*

In **1934** the Nobel Prize in Physics Louis **de Broglie**, a *Catholic*, developed photon wave mechanics [Bro] demonstrating how the photon theory of light finds its place in the more general framework of wave mechanics.

In **1938** Otto **Hahn**,[345] a *Lutheran,* nicknamed *"Fritz"* and the *"father of*

---

[343] Sir John D. **Cockcroft** *(ENG: Todmorden 1897-1967 Cambridge). He enlisted in November 1915 and was discharged in January 1919. He pursued electrical engineering and eventually earned his doctorate under the guidance of E.* **Rutherford** *in 1928, having achieved the status of a wrangler. During World War II, he played a significant role in the development of the atomic bomb. In 1947 he assumed the position of director of the Atomic Research Establishment at Harwell and was knighted Bachelor. From 1959 to 1967; he served as the inaugural Master of Churchill College, Cambridge. He was honored as a member of the Order of Merit in 1956 and received various prestigious medals from multiple countries. He identified as a* Methodist

[344] **Walton** Ernest *(IRL: Dungarvan, 1903-1995 Belfast). Son of a Methodist pastor, he studied in a Methodist college in Belfast and then at Trinity College in Dublin where he received 7 awards in Mathematics and Experimental Physics. In 1927 he was awarded a scholarship to the Cavendish Laboratory in Cambridge, directed by E.* **Rutherford**, *where he defended his thesis in 1931. During the 1930s, he and J.* **Cockroft** *built an accelerator of particles that bears their name; with which they were the first to break the Li nucleus, producing a He nucleus, allowing them to verify* **Rutherford**'s *atomic theory. In 1942, he was appointed professor of experimental physics. Before receiving the Nobel Prize, he had been awarded the Hughes Medal from the Royal Society of London. After studying Biochemistry at Oxford, he obtained a Ph.D. in Neurochemistry and then, worked for 15 years in various foreign laboratories. As a lecturer in Beirut, he created the National Laboratory of Human Genetics. He joined the Imperial Research Laboratories in London, where he directed a molecular immunity program and a laboratory in Cambridge.*
A committed Methodist. *From 1972 he wrote numerous articles and books on Science and Religion and participated in debates. At his death, the Irish branch of the organization* "Christian in Science" *created* "Lectures on Science and Religion" *to which contributed renowned researchers like the molecular biologist, and* Evangelist.

[345] **Hahn** Otto *(DEU: Frankfurt 1879-1968 Göttingen),* Lutheran. *Worked since 1935 in the laboratory of Lise* **Meitner** *a physicist, on the nuclear fission of uranium. In addition to the 1944* **Nobel** *Prize for Chemistry, O. Hahn received many other distinctions such as the Emile Fischer (1919) and Max Planck (1949) medals; he discovered many radioisotopes, having worked with the greatest of all, such as E.* **Rutherford** *(ref. 252-p.76), the father of Nuclear Physics. Detained by the British at Farm Hall (from the beginning of 1945 to the end of January 1946), with Max* **von Laue**, *Walter* **Gerlach**, *Carl Friedrich* **von Weizsäcker**, *Karl* **Wirtz**, *and* Werner

nuclear chemistry," and his assistant Friedrich **Strassmann**[346] were the first to realize, in December, the fission of uranium nuclei thanks to neutrons. In effect, this discovery, which has been used since 1950 to produce electricity, will unfortunately give rise, first of all, to a frantic race for the development of devastating nuclear bombs.

It should be noted that Lise **Meitner,**[347] a *Protestant* nuclear physicist, and theoretician, director since 1917 of a physics laboratory that had just been created, had involved Otto **Hahn** as soon as 1934 (with whom she had collaborated since 1907) and Fritz **Strassmann** in the study of Uranium, which led to the discovery of fission four years later. Although she converted to *Protestantism* in 1908, she was *Jewish* by birth and had to go into exile in Sweden in 1938 (a country whose nationality she adopted), following the annexation of Austria by the Nazis. She surely deserved to be co-recipient with Otto **Hahn** for the 1944 **Nobel** Prize in Chemistry *"for his discovery of the fission of heavy nuclei."* All the more so because in 1939, together with her nephew Otto **Frisch** (Vienna, AUT 1904-1979 Cambridge, ENG) nuclear physicist, she provided the first theoretical explanation of the nuclear fission of uranium. According to Otto Hahn's biography by Wikipedia, the comity was unfair.

After having noticed the damage created in Japan for which he felt responsible, Otto **Hahn** said:

> *"... I thank God on my knees that we (the Germans) did not make the uranium bomb."*

Numerous have been German scientists who contributed to the process of nuclear fission and the realization of atomic bombs. After the war, Hahn militated against the use of nuclear weapons and served as the first president of the German Atomic Energy Commission.

In **1939,** Robert **Oppenheimer**[348] *a mystic*, and his student George **Volkoff** wrote an article: *"On Massive Neutron Cores"* demonstrating that, when

---

Heisenberg, *it is said that when he learned on 6/08/45 that an atomic bomb had just been dropped on Hiroshima, killing more than 100,000 people, he was deeply affected. So much so that his comrades wondered for several days if he was not going to take his own life.*

[346] **Strassmann** Friedrich *(DEU: Boppard 1902-1980 Mainz), entitled* "The Righteous Among the Nations." *Chemist, appointed Professor in 1946 at the University of Mainz and in 1948, Director of the Max Planck Institute of Chemistry. Enrico Fermi Prize, an asteroid named after him, ...*

[347] **Meitner** *Lise (Vienna, AUT 1878-1968 Cambridge, ENG). A student of* L. **Boltzmann**, *then of* M. **Planck**, *defended her thesis in 1905. In 1912, she became* M. **Planck**'s *assistant and then, in 1917, director of the physics laboratory that was being created. In 1966, she was awarded the Fermi Prize, together with* Otto **Hahn** *and* Fritz **Strassmann**. *Jewish converted to Protestantism.*

[348] **Oppenheimer** Robert *(USA: New York 1904-1967 Princeton, NJ). Noted at an early age for his intelligence, attended the greatest universities in the world and the greatest scientists of the time:* M. **Bohr**, P. **Dirac**, L. **Pauling**, W. **Pauli**, W. **Heisenberg** ... *Theoretical physicist, known for his work in quantum mechanics, nuclear physics (directed the Manhattan Project, hence his*

contracting stars reach a certain critical radius. There is a distance (now called *"horizon event"*) below which nothing can escape its gravitational field, making the star invisible because any light ray or object penetrating inside this surface will be absorbed by the star. We are approaching the term **"black body."**

In **1948** George **Gamow**[349] - whose work significantly influenced the field of astrophysics, cosmology, and nuclear physics explained, among other things, how some elements are formed through nucleosynthesis (ref. 461-p.137) - and his student Ralph **Alpher**[350] proposed, in the line of the *"primitive soup"* model of G. **Lemaitre**, a model of the formation of the universe: a *"dense soup of neutrons and protons"* named *"Ylem,"* that Fred **Hoyle**,[351] coined by derision *"Big Bang,"* during a televised interview in the 1950s. He wanted to mark his skepticism because he was the tenant of the *steady-state theory* of the universe. A theory holding both, the expansion of the universe, and the continuous creation of matter, to keep its density constant. The name *"Big Bang"* has been preserved (cf. Chap.3.2-Pp.145-173). R. **Alpher** and Robert **Herman**[352] were the first to theoretically predict the characteristics of cosmic radiation (Cosmos Microwave Background = CMB), which filled the universe after the Big Bang. We will see in chapter 3, that the theory of relativity has made an enormous leap to science which, at the beginning of the 20th century, was finally able to explain the different steps of the universe formation (since Planck's time), corroborating

---

nickname "Father of the nuclear bomb"), particle physics. Of Jewish origin, his parents were agnostics, but he was deeply mystical and interested in the Hindu *religion, even more so because he mastered Sanskrit, as well as Greek, Latin, French, and German.*

[349] **Gamow** George *(Odessa, UKR 1904-1968 Boulder, US-CO), a theoretical physicist. Emigrating to the USA in 1933, naturalized, he obtained a professorship at Washington University. He became interested in radioactivity in 1928 in Copenhagen with N,* **Bohr***. Joining E.* **Rutherford** *in Cambridge, he gave in 1929 a quantum interpretation of alpha radioactivity, using the "Tunnel Effect." In 1940 he participated in the construction of the atomic bomb in Los Alamos, NM. In 1950 he turned to genetics and collaborated with F.* **Crick** *and J.* **Watson** *who discovered the double helix structure of DNA (1962 Nobel Prize in Physiology or Medicine).*

[350] **Alpher** Ralph *(USA: Washington, D.C. 1921-2007 Austin, TX). Jewish agnostic, cosmologist. During World War II he worked for the US Navy and the Manhattan Project. From 1944 to 1955 he did research at John Hopkins University in Baltimore, MD. In 1948 he defended his thesis: "Big Bang Nucleosynthesis" at the University of Washington D.C. under the direction of G.* **Gamow***. His work was not properly recognized. He nevertheless received many awards including the National Medal of Science in 2005. Author of many scientific papers and science fiction novels.*

[351] Sir Fred **Hoyle** *(ENG: Bingley 1915-2001 Bournemouth). Cosmologist and astronomer, a main detractor of the Big Bang theory because was at the time partisan of a stationary state of the universe. Declaring himself an atheist,* Fred **Hoyle** *promoted the theory of* panspermia *(ref.624-151). Author of many scientific papers and science fiction novels. He was knighted in 1972.*

[352] **Herman** Robert *(USA: New York 1914-1997 Austin, TX) cosmologist. Submitted his Ph.D. in 1940 at Princeton University in molecular spectroscopy. Being forgotten at the time of the discovery of CMB by A.* **Penzias** *and R.* **Wilson***, he nevertheless received in 1993, with R.* **Alpher***, the Henry Draper Medal of the National Academy of Sciences.*

Genesis without having looked for it, and challenging the eternal model of the universe of Greek mythology taken up by Romans, ...

In **1957, B2FH,**[353] Fred **Hoyle** & al. developed the complete theory of nucleosynthesis of elements in stars; a theory that F. **Hoyle** had predicted theoretically in 1946. While the Big Bang theory only explains the formation in the universe of nuclei of atomic mass less than or equal to 7, through this theory its authors demonstrate that the temperature conditions for 3 nuclei of Helium ($^4He_2$) to combine and give a nucleus of carbon: $^{12}C_6$, are met in stars. This is called the *"triple reaction α,"* (α = alpha = $^4He_2$ nucleus). It can then, result in the formation of heavier nuclei such as ^{16}O for example, until iron: ^{26}Fe. The formation of heavier nuclei occurs during the explosion of supernovas. This led Fred **Hoyle,** an *atheist,* to write:

> *"Who could not tell you "...that a superior mathematical intelligence must have designed the properties of the carbon atom?""*

In **1958** Arthur Leonard **Schawlow,**[354] *the "Protestant fairy,"* a *Methodist, and* Charles **Townes,**[355] a *devout Protestant* too, published *"The Theory of Laser;"* Laser is an acronym for Light Amplification by Stimulation Emission of Radiation. Three years earlier they had co-authored a book *"Microwave Spectroscopy."* In **1960** Ch. **Townes** and A. **Schawlow** took out a patent for the *Laser.* Townes will share the **1964 Nobel** Prize in Physics with Nicolay G. **Basov**[356] and Aleksandr M. **Prokhorov,**[357] a *theist "for fundamental work in the field of quantum electronics, which has led to the construction of oscillators and*

---

[353] **B2FH:** *Acronym of Margaret **Burbidge**, Geoffroy **Burbidge**, Fred **Hoyle**, and William* **Fowler.**

[354] **Schawlow** Arthur L. *(USA: Mount Vernon, NY 1921-1999 Palo Alto CA), 1981 **Nobel** Prize in Physics,* Methodist. *Winning a scholarship in Mathematics and Physics (he was interested in radio engineering but there was no scholarship in this discipline) in the Faculty of Arts of Toronto (CAN). In 1945, the war was over, he could resume his studies and prepare a thesis. During his postdoctoral, he worked with* Charles **Townes** *at Columbia University (NY). From 1951 to 1961, as a physicist at Bell Telephone Laboratories, he worked on superconductivity. In 1961 he was appointed a professor of physics at Stanford University (CA) where he remained until his emeritus status. Well known for Laser development and Laser spectroscopy. Member of the National Academy of Sciences (USA), President of various Academies, Granted with a lot of scientific medals.*

[355] **Townes** Charles H. *(USA: Greenville, SC 1915-2015 Oakland, CA).1964 **Nobel** Prize in Physics. Submitted his doctoral dissertation at CAL Tech. Participated during WWII in the development of radar. He joined Columbia University after World War II where he invented the MASER (Microwave Amplification by Stimulated Emission of Radiation), a microwave amplifier, and coinvented the LASER (Light Amplification by Stimulated Emission of Radiation), a visible wave amplifier. Member of the National Academy of Sciences (USA). Professor at MIT (Massachusetts Institute of Technology), Berkeley, CA-US.* Protestant Christian *dixit.*

[356] **Basov** *Nikolaï (RUS: Ousman, 1922-2001 Moscow). Physicist, Academician, University Professor.*

[357] **Prokhorov** *Alexandre M. (Atherton, AUT 1916-2002 Moscow, RUS). Physicist, University Professor, Academician.* Theist.

amplifiers based on the maserlaser principle." In **1981,** A. **Schawlow** and Nicolas **Bloembergen** shared half a part of the **Nobel** Prize in Physics "*for their contribution to the development of laser spectroscopy.*" The "*Arthur Schawlow Prize in Laser Science*" will be established in his honor in 1991. A. Schawlow was raised in his Canadian mother's *Protestant* religion and attended a *Methodist Church.* Cf. epigraph p. 143.

Since its discovery, the laser has been extensively used because the generated light is coherent (all the photons emitted having the same wavelength) and directional (a tiny beam, when other sources produce diffuse beams). It is at the origin of the attribution of Nobel Prizes. Thus, in **1997** Steven **Chu,** Claude **Cohen Tannoudji,** [358] and William Daniel **Phillips** [359] were awarded "*for the development of methods to cool and trap atoms with laser light.*" Although a practicing *Jew,* [360] C. Cohen-Tannoudji leaves these questions at the door of his "*laboratory.*" On the other hand, Bill Phillips is a very committed *apologetic Methodist Christian.* From an early age, he was attracted to the sciences (at 10 he had a workshop in his cellar) but, as he grew older, he realized that his interest in literary subjects had served him well in the scientific field. He stated:

> "*I believe in God. I believe in a personal God who acts in and interacts with the creation. I believe that the observations about the orderliness of the physical universe, and the apparently fine-tuning of the conditions of the universe for the development of life suggest that an intelligent Creator is responsible ... I believe in God because of a personal faith, a faith consistent with what I know about science.*"

---

[358] **Cohen-Tannoudji** Claude *(Constantine, French Algeria 1933)* Jew, *1997* **Nobel** *of Physics. In 1953 he was admitted to the Ecole Normale Supérieure (the best French College with Polytechnique) where he had lectures by the best professors in their disciplines. Once his military service was completed he joined in 1960 A.* **Kastler**'s *team for preparing a Ph.D. submitted in 1962.*
A. Kastler offers him a teaching position (in quantum physics) at the University of Paris. In 1973 he was appointed as Professor at the College de France. Author of many publications he received a lot of awards and honors.
[359] **Phillips** William D. *(Wilkes-Barre, US-PA 1948),* Methodist. **Nobel** *Prize in Physics in 1997. In 1966 he attended Juniata College, Huntingdon PA-US, where he enrolled in an advanced French literature class. Attracted by Princeton he finally chose MIT where he will defend his thesis in 1976. Two years later he was appointed to the National Bureau of Standards (NBS) where he started with laser cooling. Elected to the Pontifical Academy of Sciences in 2004 and to many Academies. Co-founder of the International Society for Science and Religion.*
Methodist. *Raised in the* Christian religion, *by a Methodist father, and a Catholic mother, his faith was maturing and strengthened his respect for other people's values. With his wife, they opted in 1979 for the Fairhaven United Methodist Church in Gaithersburg, MD-US.*
[360] Cohen-Tannoudji, *interviewed on October 21, 1997, by Dominique Leglu for the newspaper* "Libération."

In **1965** Richard **Feynman**,[361] describing himself as a *freethinker*, Sin-Itiro **Tomonaga**[362] and Julian **Schwinger**,[363] *Jewish* shared the **Nobel** Prize in Physics: *"For their fundamental work in quantum electrodynamics, with deep-plowing consequences for the physics of elementary particles."*

The same year, Arno **Penzias**,[364] an *orthodox Jewish*, and Robert W. **Wilson**, researchers at Bell Telephone Laboratories (Murray Hill, NJ), working on new antennas, observed a signal from the atmosphere in the microwave range that they did not expect. After verifications, it turned out that it corresponded to the *"fossil radiation"* predicted by G. Lemaitre, then G. Gamow et al. under the name of *"Cosmic Microwave Background (CMB)"* which earned A. Penzias and R. Wilson to be co-recipients in 1978 of half of the **Nobel** Prize in Physics: *"For their discovery of cosmic microwave background radiation."*

It should be noted that at the same time at Princeton, Robert **Dicke**[365] professor of Physics, and his ex-Ph.D. James **Peebles**[366] tried to demonstrate this

---

[361] **Feynman** Richard *(USA: Queens, NY 1918-1988 Los Ángeles, CA). One of the greatest physicists of the 20th century. Collaborated on the Manhattan Project, taught at Cornell-NY, at Caltech-CA. He authored numerous books on Physics, as well as, the Feynman Diagrams used in string theory. Awards: A. Einstein in 1954, E. Lawrence in 1962, Oersted Medal .... Of Jewish faith he called himself a* free thinker, and he was co-awarded the 1967 **Nobel** Prize in Physics.

[362] **Tomonaga** Sin-Itiro *(Tokyo, JPN 1906-1979) Theoretical physicist, one of the creators of quantum electrodynamics. 1965* **Nobel** *Prize in Physics.*

[363] **Schwinger** Julian *(USA: New York 1918-1994 Los Angeles, CA)* Jewish. *Theoretical physicist, his interest in cold fusion, which was very controversial, earned him, despite his* **Nobel** *Prize, denigration from some of his colleagues.*

[364] **Penzias** Arno A. *(Munich, DEU 1933). Naturalized American. Ph.D. in Physics from Columbia University. Researcher and director of Bell Laboratories (Holmdel, NJ). Winner in 1978 of half of the* **Nobel** *Prize in Physics with Robert Woodrow* **Wilson** *(Houston, TX 1936, Agnostic) for the discovery in 1965 - while they were working on a new type of antenna - of the thermal (3K) cosmological microwave radiation supporting the Big Bang Theory (chapter 3 pp. 147-175). A. Penzias then took an active part in the development of millimeter radio astronomy and the study of interstellar molecules. He is an active conservative Jew, for whom our present knowledge of the universe is in perfect agreement with the Torah, and the Bible, ... During the 1970s A. Penzias was involved in helping Soviet Jewish dissidents (the Refuseniks) who wanted to join Israel. After receiving the Nobel Prize he went to Moscow to meet his colleague:* Viktor **Brailovsky** *(Moscow, RUS 1935) Mathematician, computer scientist,* Orthodox Jew, *member of the Aliyah (meaning "let him come up," implied in Jerusalem), imprisoned for 8 years in Russia for his stance, he was able to emigrate in 1988 to Israel where he held a position as a professor in Mathematics applied to Computer Science at the University of Tel-Aviv. He was also Minister of the Interior, as well as of Science and Technology. See [Hab].*

[365] **Dicke** Robert *(USA: St Louis, MO 1916-1997 Princeton, NJ). Obtained his B.Sc. at Princeton and defended his thesis in 1939 at the University of Rochester, NY. He made important contributions in the fields of Astrophysics, cosmology, gravity atomic physics, and worked extensively on the Equivalence Principle. He was awarded the National Medal of Science in 1970, and the Comstock Prize in Physics in 1973, ...*

[366] **Peebles** James *(Winnipeg, CAN 1935) Cosmologist, theoretical physicist, American winner of half the* **2019 Nobel** *Prize in Physics. Received at Princeton in 1958, he defended his Ph.D. in 1962. Professor in 1972 he is, since 2000, professor emeritus. He contributed a lot to the "Standard Cosmological Model," a pioneer of "cold dark matter" (composed of particles that*

radiation with the help of the *"Dike radiometer"* that R. Dicke had developed at MIT, during the Second World War, and thanks to which he had been able to predict, 20 years earlier, a temperature of the fossil radiation lower than 20K. They could nevertheless theoretically confirm the discovery of A. **Penzias** and R. **Wilson,** as well as the validity of the Big Bang model (see 1992 COBE Experiment (ref. 418-p.124)). J. Peebles proved to be one of the most brilliant theoretical cosmologists of the 20th century, hence his award in 2019 for half of the Nobel Prize in Physics *"For theoretical discoveries in physical cosmology."* The scientific confirmation of the Big Bang has had important theological consequences, causing some scientists to revise their negative ideas about the Old Testament. Emphasizing the fact that our mathematical tools are not usable below Planck time, *Agnostic* astrophysicist Robert **Jastrow**[367] stated [Ja1]: *"At this point, it seems that science will never be able to lift the curtain obscuring the mystery of creation...."*

During this year, Sir Roger **Penrose,**[368] an *Agnostic Jew*, using the Theory of General Relativity, demonstrates mathematically that a star collapsing under the effect of gravity will end up occupying a zero volume where the energy density and the curvature of space-time are infinite, giving what theoretical physicists name a *"singularity"* in a region of space called *"**black hole**."* This hypothesis will become the Penrose Theorem:

---

should move at low speed in front of the celerity c), co-winner in 2005, with J. **Gunn** *and M.* **Rees,** *of the Crafoord Prize in astronomy. This prize is awarded in association with the Nobel Committee and is given to disciplines not eligible for the Nobel Prize. He received many other prestigious prizes.*

[367] **Jastrow** Robert *(USA: N-Y 1925 - 2008 Arlington, VA). Agnostic astrophysicist. Awarded in 1958 a Ph.D. in theoretical physics from Columbia University, NY, and joined NASA, where he will be the president of the commission in charge of establishing the project of exploration of the Moon during the Apollo lunar landing and the chief of the division of theoretical physics. From 1979 to 1992 he was a Professor of Earth Sciences at Dartmouth, NH.*

[368] Sir **Penrose** Robert *(Colchester, ENG 1931).* **2020 Nobel** *Prize in Physics. Professor Emeritus at Oxford. Attended Cambridge where he was awarded his Ph.D. in 1957. Alternating stays between London and the USA he held various positions before accepting a professorship at Oxford where he retired. After having elaborated a theory describing the formation of supernovas and stated - among other things, the "Penrose theorem" - he worked with S.* **Hawking** *on a theory on the origin of the universe. In 2010 he suggested [Pe1] the existence of a universe that preceded the one created during the Big Bang. In this book he answers 3 questions: 1. if something existed before the Big Bang what could it be? 2. what is the source of order in the universe? 3. what is the future? In 2020, he will confirm this hypothesis. He received numerous medals for his work and was knighted in 1994. In 2010 (25/09), during an interview with the BBC, he declared himself an atheist (in the sense that he does not adhere to any religion) and a nonpracticing Jew. On the other hand, in the movie: "A History of Time" he says that he thinks that the universe has a reason to exist, that Man is not the fruit of chance, and that there is something very deep in the reason for the existence of the universe. Fellow of the Royal Society, Wolf Foundation Prize for Physics, ... His list of awards is impressive: Copley's, Einstein's, Dirac's medals.... Author and co-author of many publications, he wrote popular books too.*

"Anybody collapsing under gravitational forces must form a singularity."

He will be awarded, in 2020, half part of the Nobel Prize in Physics "for the discovery that black hole formation is a robust prediction of the general theory of relativity."

At the time, a gifted student: Stephen **Hawking**,[369] discovering Penrose's theorem during a seminar decides, despite the terrible illness that is beginning to take its toll, to dedicate his thesis to the study of time. Reversing the direction of time, he demonstrates that Penrose's theorem is still valid and that **Friedman**'s expanding universe must have begun with a "singularity;" provided that its expansion was fast enough to avoid a collapse of the matter involved. He defended his thesis in 1969 and became one of the most brilliant scientists of our era.

In **1967,** the *Quaker* Jocelyn **Bell**,[370] a graduate student of the renowned *Christian* radio-astronomer Anthony **Hewish**, [371] puts in evidence the first

---

[369] **Hawking** Stephen *(ENG: Oxford 1942-2018 Cambridge). Theoretical physicist, a specialist in black holes, he worked in the fields of mathematics, cosmology, and quantum gravitation. Holder, from 1980 to 2008, of the Lucasian Professor Chair of Cambridge (named after Reverend Henri Lucas, see ref.189-p.58). He attended Oxford in 1959, then Cambridge in 1962; the year during which it was diagnosed that he suffered from amyotrophic lateral sclerosis, a neuronal disease that slowly attacked his muscular nerve functions, rendering him a cripple, able to speak only with the help of a voice synthesizer. Thanks to his enormous strength of character and the support of his family, he defied the prognosis that gave him only 2 years to live. In 1965 he resumed his research and started a family (3 children). He defended his PhD in 1969. Honorary member of the Royal Society of Arts and the Pontifical Academy of Sciences (although agnostic he collaborated until his death with 5 popes in the framework of the Pontifical Academy of Sciences, open to great international scientists, whatever their religion), Fellow of the Royal Society,... His list of awards is impressive: Copley's, Einstein's, and Eddington's medals.... Author and co-author of many publications, he wrote various interesting books.*

[370] **Bell Burnell** Jocelyn *(Lurgan, Ulster, IRL1943). Physicist. President of the Royal Astronomical Society, of the Institute of Physics. Under-graduated in the Quaker Girl's School of York, she graduated with honors her B.Sc. in Physics in 1965 at the University of Glasgow, a Ph.D. in 1969 at Cambridge, and worked with A. **Hewish**, at the building of the interplanetary Scintillation Array at Cambridge, to study quasars. Bell worked at Southampton, London, and Edinburg, and became, in 1986 Project manager for the J. Clerk **Maxwell** Telescope at Hawai. The second woman to become Professor at the Open University UK. She was awarded various medals: Coppley 2021, Royal Medal 2015, ... and honors.*
Active Quaker, *she is more comfortable with the English version of Quaker than with the Northern Ireland version which seems to her* "a bit more fundamentalist, a bit more take the Bible literally, a bit more evangelical, a bit less educated."

[371] **Hewish** Antony *(Fowey, UK 1924-2021). Christian radio astronomer. Attended Cambridge University in 1942 but engaged in war service from 1943-46 at the Telecommunications Research Establishment, Malvern. During this period, he worked with M. **Ryle** on Radar devices. In 1946 he returned to Cambridge, graduated in 1948, joined Ryle's team at the Cavendish Laboratory, he obtained his Ph.D. in 1948. Director of studies in Physics at Churchill College, University Lecturer, and Professor of Radioastronomy from 1971 to his retirement in 1989. He made pioneering measurements of plasma clouds in the ionosphere and received*

pulsars.[372] A. **Hewish** is not convinced, thinking it is a flare star. Some called these observations *"little green men,"* a common English expression for an extraterrestrial civilization. Renewing her measurements several times, she finally convinces **Hewish** who will be awarded together with Martin **Ryle**,[373] the 1974 **Nobel** Prize in Physics *"for their pioneering research in radio astrophysics: Ryle for his observations and inventions, in particular of the aperture synthesis technique, and* **Hewish** *for his decisive role in the discovery of pulsars"*. Although she had been listed second after **Hewish** (her Ph.D. supervisor) on the paper announcing the discovery,[374] she was not co-awarded. Sir Fred **Hoyle** criticized this omission. Some editors said humorously (Nobel = No Bell). J. **Bell** was not affected, saying philosophically that the prize had never been given to a research student. She will discover 200 more pulsars during the following years. In 2018 she will be awarded *"the Special Breakthrough Prize in Fundamental Physics"* which is better rewarded. Interviewed on May 21, 2000 by David DeVorkin[375] about her faith she answered:

"Born and brought up Quaker and still am active in Quakers, probably even very active in Quakers."

We will encounter various quotations of A. **Hewish,** but I retain here:[376]

"As a Christian, I begin to comprehend what life is all about through belief in a creator, some of whose nature was revealed by a Man, who

---

various medals as Eddington's Medal from the Royal Astronomical Society in 1969 and honors. For the benefit of my young readers who might have some difficulties at the university, I have retained the following sentence which should be taken at face value. In an interview for "The War Cry," reprinted by Christian **Evidence** *he said:* "My first year at the University was a bit of a flop … you don't have to be brilliant to get a **Nobel** prize (he had been awarded the 1974 N.P. in Physics), you just have to be doing the rights things at the right time in the right place."

[372] Pulsar. *It is a highly magnetized spinning neutron star assimilated to a magnetic dipole. Bursts of electromagnetic radiations are emitted in the direction of the bipolar axe, we can observe them when the pulsar is directed towards the Earth, i.e., at time intervals ranging between ms to s. The pulsation period is very accurately maintained, meaning that it has a great energy reserve. Another characteristic is that the intensity of its scintillation varies in function of its distance to the sun. Pulsars are very small objects in physical dimensions but big in the sense of mass.*

[373] Sir Martin **Ryle** *(ENG: Brighton 1918–1984 Cambridge). Astronomer, he developed original radio telescopes. He was graduated from Oxford in 1939. During the war, he worked on the development of radars for the R.A.F. in Malvern. In 1945 joins* **Ratcliffe**'s *team at Cavendish Laboratory and worked on the ionosphere. Elected to a Fellowship at Trinity College in 1949 he welcomes A.* **Hewish** *and other colleagues. In 1959 he was the first professor of radio-Astronomy appointed at Cambridge. Awarded lots of medals and honors. From 72 to 82 was a Royal Astronomer. Knighted in 1966.* Agnostic.

[374] **Hewish**, A; **Bell**, S.J.; et al. *"Observation of a Rapidly Pulsating Radio Source."* Nature, Vol. 217, Issue 5130, Pp 709-713 (February 1968).

[375] David **DeVorkin** *"J. Bell Burnell religion,"* for the American Institute of Physics (AIP), Carnegie Institution, Washington DC. 21/05/2000.

[376] *2009 summer session of Does God exist? Cf. www.doesgodexists.org*

was born about 2,000 years ago."

At the end of his life, Sir Martin **Ryle** was more deeply committed to ethical and environmental questions than astronomy. He was the advocate of science and technology for the good of mankind instead of its selfdestruction. In 1983, answering a request from the Pontifical Academy he wrote:

"Our cleverness has grown prodigiously – but not our wisdom."

Seven years after the discovery of J. **Bell Burnell**, the *Quaker* Joseph H. **Taylor** Jr,[377] and his research student Russell A. **Hulse**[378] discover, with the radio telescope at Arecibo, Puerto Rico, the first binary pulsar: two stars in very close proximity (about a few distances Terre-Moon), rotating around each other. They will receive the 1993 Nobel Prize in Physics *"for the discovery of a new type of pulsar, a discovery, that has opened up new possibilities for the study of gravitation."* In 1978, J. H. **Taylor** had demonstrated that the binary pulsar emits gravitational waves according to A. **Einstein prediction.** From 1980 until his retirement J.H. Taylor was a Professor at Princeton University. For him:[379]

"A scientific discovery is also a religious discovery. There is no conflict between science and religion. Our knowledge of God is made larger with every discovery we make about the world" (Brown 2002).

In **1970**, S. **Hawking** [Ha2-p.52], joined forces with the esteemed R. **Penrose,** building upon the latter's theorem dating back to 1965. Together they penned an illuminating article that showcased, under certain conditions, how the presence of singularities need not challenge the validity of the Big Bang model. They demonstrated that black holes are delimited by a surface called an *"event horizon"* beyond which light rays are not absorbed but more or less deflected depending on their proximity. On the other hand, those which cross this limit are absorbed by the black hole and never come out. Hence the famous sentence of R. **Penrose**: *"God abhors a naked singularity"* meaning that the black hole,[380] and

---

[377] **Taylor** Jr. Joseph H. *(Philadelphia, PA-US 1941)* Quaker. *Young, he enjoys building radio telescopes, In 1963 he attended the Harvard Departments of Astronomy, physics, and Mathematics where he was awarded his Ph.D. in radio astronomy in 1968 one year after J.* **Bell** *discovery. He joins the National Radio Astronomy Observatory where he discovers 4 news pulsars. In 1969 he left for Massachusetts University as a Professor of Astronomy and Associate Director of the Five College Radio Astronomy Observatory—1993* **Nobel** *Prize in Physics. In January 2022, there were over 2,000 known pulsars.*

[378] **Hulse** Russell A. *(New York 1950). His fields of research are mainly astrophysics and plasma physics. He is a Professor of Physics at the University of Texas in Dallas. He shared the 1993* **Nobel** *Prize in Physics with Joseph* **Taylor**.

[379] *Cf. 2012Daily.com. Compiled by* **Tihomir** *Dimitrov holder of 2 Masters in Philosophy and Science (nobelists.net, scigod.com/index,php/issue/view/3.*

[380] *Have Black Holes been detected? In 1988,* **Hawking** *said that the probability that the binary system Cygnus X-1 - observed in 1965 by* Louise **Webster** *and then by* Paul **Murdin** *in our Milky Way - was made up of a star orbiting a black hole, was 95%. He also thought that in our*

therefore the singularity, is hidden by the *"Event Horizon."* The collective expertise of these signatories lent credible weight to their findings, adding another layer of fascination to the cosmic narrative.

In **1973**, a pivotal moment in the quest to unveil the secrets of fundamental particles and the mysterious forces that govern them unfolded. David J. **Gross** *Jew*, H. David **Politzer** *Jew,* and Franck **Wilczek**[381] (raised *Catholic* he became a *Pantheist*) made a groundbreaking theoretical revelation about the nature of the *"strong force,"* affectionately dubbed the *"color force,"* that binds quarks together: the elemental constituents of nucleons, which encompass neutrons and protons. Their revelation: the interactive force is proportional to their distance. When they are very close, it is so weak that, they behave as free particles. A discovery that catalyzed the birth of Quantum Chromo Dynamics (QCD), a theory that now plays a central role in understanding the behavior of three forces: the Electromagnetic (introduced in 1865 by J.C. **Maxwell**, ref.234, p.71), the strong and the Weak.

This theoretical breakthrough has contributed significantly to the Standard Model,[382] a fundamental framework in the particle physics realm. In recognition of their pioneering work, they were honored with the **2004 Nobel** Prize in Physics *"for the discovery of asymptotic freedom in the theory of the strong interaction."* As explained by Lars Bergström, Secretary of the Nobel Committee for Physics, during an interview *"This discovery has brought physics one step closer to fulfilling a grand dream, to formulate a unified theory comprising gravity as well, a Theory Of Everything (TOE)."*[383]

---

galaxy there are more than observed stars. The evidence in 2016 of gravitational waves and especially the 1st image in 2019 of Sagittarius A*, TN located in the center of the Milky Way (p. 90) by the EHT project leaves no room for doubt: they exist.

[381] **Wilczek** Frank *(New York, USA 1951). Professor of Physics at MIT since 2000, and a **Nobel** Laureate (2004), he graduated from the University of Chicago, IL with a B.Sc. in Mathematics in 1970; then, studied under David **Gross** at Princeton University, NJ he earned an M.Sc. in Mathematics in 1972 and defended his Ph.D. in physics in 1974. He taught at the Universities of Princeton and California. Besides his studies on quarks and the force that acts on them, he illustrated questions related to cosmology, black holes, and condensed matter physics. Founder and Director of T.D. Lee Institute at the Wilczek Quantum Center in Shanghai. Coauthor of many scientific papers and books, he received many prestigious medals and awards like the Dirac (1994) and Lorentz (2002), and the Prestigious (1.5 million $) Templeton Prize in 2022.* Pantheist. David **Gross** (Washington-DC 1941) and David **Politzer** (New York 1941) are both theoretical physicists and had a Jewish education.

[382] Standard Model*: quantum theory developed in the 1970s, to classify the elementary particles and explain the four fundamental forces.*

[383] The TOE = Theory Of Everything) *should - according to some scientists including S.* **Hawking**- *make it possible to eliminate the initial singularity, which, although at the origin of spectacular advances, is an insurmountable barrier for the Generalized Theory of Relativity. Others doubt it. I do not think that we can scientifically answer G.* **Leibnitz's** *question because, as long as* **Gödel**'s *theorem (ref.613-p.212) will not be invalidated, if ever, we will have to admit that there is a limit to our knowledge of our Universe because we are part of it, and that to cross this limit, it is necessary to be external, which only God can do if other universes exist.*

During the ceremony, Wilczek expressed a sense of wonder and spirituality in the natural world:

"Another thing that shaped my thinking was religious training. I was brought up as a Roman Catholic. I loved the idea that there was a great drama and a grand plan behind existence. Later, under the influence of Bertrand Russell's writings and my increasing awareness of scientific knowledge, I lost faith in conventional religion. A big part of my later quest has been trying to regain some of the sense of purpose, and meaning that was lost. I'm still trying."

In 2022, receiving the prestigious Templeton Prize he will describe himself as a *Pantheist*.

In **1975**, during a conference at the Rutherford-Appleton Laboratory near Oxford, in a fascinating twist, S. **Hawking**, a visionary ahead of his time, made an announcement that stunned the participants:

"Black holes emit radiation."

John G. **Taylor**,[384] chairman of the session, was amazed and said: *"This is nonsense."* But S. **Hawking**, who had until then succeeded in convincing a great part of his scientific colleagues that no radiation, nor particle, could escape from the black hole, had finally accepted the idea put forward in 1973 by the *Jewish* theoretical physicist, Jacob **Bekenstein**,[385] according to which black holes can have entropy and thus a temperature. Drawing upon the reliable realm of quantum field theory, according to which quantum fluctuations[386] are responsible for the

---

[384] **Taylor** John Gerald *(ENG: 1931-2012). Mathematician & physicist obtained his Ph. D. in 1956 at Christ's College, Cambridge, Director of the Centre for Neurological Networks at King's College, London, he was interested in the paranormal. Actor, and writer he published many books for the general public.*

[385] **Bekenstein** Jacob *(Mexico, MEX 1947-2015 Helsinki, FIN). Theoretical physicist, astronomer, practicing* Jew. *Mexican, naturalized American in 1968, obtained a Master of Science degree in 1969 from the Brooklyn Polytechnic Institute, then moved to Princeton, NJ where he defended his thesis in 1972 under the direction of J.* **Wheeler** *(ref. 280-p.85). He completed his postdoctorate at the University of Texas at Austin and published on the entropy of black holes. He emigrated to Israel in 1974, becoming a* "citizen of Israel," *a professor at the Ben-Gurion University of Beersheba (head of the Astrophysics Department), and then at the Hebrew University of Jerusalem in 1990. Elected member of the Academy of Sciences and Humanities of Israel in 1997, Holder of the Einstein Prize in 2015, for his work in the field of gravitational physics, and many other prestigious prizes: Wolf (2012), Weizmann (2011), ..., Rothschild (1988).*

[386] *With the advent of Quantum Mechanics (p. 65), and more particularly, the Heisenberg indeterminacy principle ΔE. $\Delta t = h/2\pi$ it has been demonstrated that the vacuum, is animated by quantum fluctuations. Namely, during a very short time t, one can "borrow" a very great energy E from the vacuum, energy that one gives back to it. Thus, according to Einstein's energy-mass equivalence relation, very massive particles are produced during this short time. It is at the origin of Bekenstein-Hawking entropy's law.*

spontaneous appearance of particle-antiparticle pairs known as virtual because having a very short lifespan, it showed that if such a pair occurs, close to the event horizon (the probability is weak and thus the "*evaporation*" of energy is it also), one of the two elements can remain in the hole while, the other can escape in the infinite, similar to radiation which one called: "*Hawking radiation.*"

Hawking championed the notion, that the Big Bang doesn't necessarily require the crutch of a singularity (required by the General Relativity Theory). His captivating argument invited a fresh perspective, sparking a fervent debate among the scientific community. Here again, S. Hawking makes a total turnaround (which he acknowledges), proving that Reason is not always right about facts and that what seems obvious today can be contradicted tomorrow by recent discoveries. Among other things, he had announced to Etienne Klein, director of research at the CEA, in 1908, during a symposium held in Ajaccio, France that the LHC would not be able to prove the existence of bosons, which was denied in July 2012 (ref.451-p.134).

A prominent figure in this grand cosmic discourse is Ilya **Prigogine,** [387] acclaimed as the father of irreversible thermodynamics, and a laureate of the Nobel Prize in Chemistry in **1977** "*for his contributions to nonequilibrium thermodynamics, particularly the theory of dissipative structures.*" His expertise allowed him to offer a thermodynamic approach to the genesis of the universe. In his revolutionary insights, Prigogine too defied the singularity, presenting an alternative viewpoint that stirred the minds of his contemporaries. [Pr]

In **1978** Vera **Rubin,** [388] *Jewish* astronomer, and astrophysicist published the results of work done with her colleague Kent Ford on about sixty galaxies, confirming the results of their measurements made at the beginning of the decade on the Andromeda galaxy. Namely, that the stars in the outer part and those in the center move at the same speed, which is not the case in the Milky Way for

---

[387] **Prigogine** Iliya *(Moscow, RUS 1917-2003 Brussels, BEL). Of Jewish origin, his family emigrated to Germany in 1921 before settling in Brussels, where he was naturalized in 1949. Polymath he studied Chemistry (defended his thesis in 1941), Psychology, Biology, Biochemistry, Particle Physics, Astrophysics, Cosmology, and Politics. Professor in 1947, he was appointed in 1951 at the Université Libre de Bruxelles (ULB). He also taught at the University of Austin, US-TX. Director of many International Institutes of Physics and Chemistry. He is a specialist in the thermodynamics of irreversible phenomena. He was awarded the title of Viscount in 1990 by the King of Belgium, he has received numerous awards and is a Doctor Honoris Causa of about fifty universities, including the University of Nice.*

[388] **Rubin** Eva *(USA: Philadelphia, PA 1928-2016 Princeton, NJ). From a very young age, she was fascinated by Astronomy. In 1951, she obtained a Master's degree from Cornell University in Ithaca, NY, whose dissertation was controversial, and then, in 1954, a thesis under the direction of G. **Gamow** on the distribution of galaxies (Georgetown University, Washington DC). Successively researcher at Georgetown, then at the Carnegie Institute (Washington), she observed in the 90s that half of the stars in the Galaxy NGC 4550 rotate in one direction and the other half in the other direction. Fought for the right of women to do men's jobs. She was the first woman to be allowed to use the Hale Telescope on Mount Palomar at Caltech. Member of the National Academy of Sciences in 1981, National Medal of Science in 1993, and Gold Medal of the Royal Astronomical Society...*

example, where the most distant move less quickly in order not to leave our galaxy under Newtonian and relativistic theories. Since these theories could not be questioned based on these results[389] alone, it resulted that the presence of Dark Matter (D.M.) reported by Fritz **Zucky**[390] in 1933 was realistic but not proven, contrary to what some people wrote at the time. She calculated at the time that the amount of dark matter should be between 4 and 5 times that of visible matter. In 2019 it was estimated [CC1911] that D.M. intervenes for 25% of the mass-energy of the universe; dark energy, for its part, contributes for 70%, and visible matter (nucleons + electrons), for the remaining 5%. D.M. could be responsible for the acceleration of the expansion rate of the Universe observed in the last 20 years. Many models are currently trying to interpret it but require experiments, involving very high-speed collisions, to be done. They are being prepared at CERN's High-Luminosity (HL), an improved version of the Large Hadron Collider (LHC).

In **1979** Alan **Guth**,[391] a Ph.D. student at M.I.T. breathed new life into Einstein's *"cosmological constant,"* unveiling it as: *"space energy," "dark energy"* or even *"dark matter."* This groundbreaking discovery, sometimes equated with the enigmatic *"**Higgs' field**"* (cf. year 2013, ref.449–p.134), set in motion *Cosmological Inflation* the pivotal starting point of contemporary scientific cosmology, by satisfactorily describing the macroscopic world of stars, galaxies, and the vast expense of the cosmos, without undermining the fundamental contributions of the Big Bang theory[392] (cf. §IV.1.1., p.179). This concept has captivated scientists' imagination, at least for the time being. This constant allows in particular [Ha1 - p.89]:

> *"... to explain why the cosmological radiation has the same temperature in all directions"* ... *"without invoking the act of a God willing to create beings such as us,"*

as S. **Hawking**, who claimed to be an *atheist*, said jokingly. According to A. **Guth**, during the 10^{-45} seconds following the Bang, the universe experienced a fleeting burst of accelerated expansion, multiplying its initial radius by a factor of

---

[389] *Attempts to modify, even partially, the laws of gravity have not been successful.*

[390] **Zucky** Fritz *(Varna, BGR 1898-1974 Pasadena, CA-US) Professor of Astronomy at Caltech who, as early as 1933, realizing that some galaxies were moving faster than expected, predicted the existence of an unknown non-luminous matter (matter that would be called "Dark Matter").*

[391] **Guth** Alan *(New Brunswick, NJ-US 1947). He acquired his master's degree in 1969 and defended his thesis at MIT in 1972. He will work successively at Princeton, Columbia, Cornell, and Stanford. Predicted the existence of magnetic monopolies that have not yet been observed. He is one of the founders of the inflationary theory of the Universe published in 1980 in Physical Reviews "Inflationary Universe" (see chapter 4-p.179). Professor of Physics at MIT, he is the recipient of many prestigious awards among which the 2012 "Prize of Fundamental Physics," the most lucrative.*

[392] *In 2015 the results of measurements made by the Planck satellite launched on 14/05/2009 were published. They allow us to describe the current universe and confirm the theory of cosmic inflation. See on the Internet "Planck satellite" and cosmological background (\Gr] p. 511).*

10^{30}. This extraordinary event fostered a relative uniformity throughout the universe.

In that same year, Sheldon L. **Glashow**,[393] a *Jewish* physicist, Muhammad A. **Salam**,[394] an *Ahmanist Muslim* physicist, and Steven **Weinberg**[395] a cosmologist, were honored with the **Nobel** Prize in Physics *"for their contributions to the unified theory of the weak and electromagnetic interactions between elementary particles, including, among other things, the prediction of the weak neutral current."* Remarkably, more than a century after J.C. **Maxwell** integration of electric and magnetic forces into an *"electromagnetic force,"* these brilliant minds demonstrated that a new force - *"the electroweak force"* - could unify the electromagnetic force (with its near infinite range) and weak nuclear force (with its microscopic range) at high energy (GeV).[396] This remarkable achievement forged a vital *"bridge"* between the realms of the infinitely large and the infinitely small, as physicists embarked on their quest for unification through the theory aptly named *"unified."*

Meanwhile in the material chemistry realm, **1979** proved to be a profoundly fruitful year. John **Goodenough**,[397] an *Evangelist*, and his team postulated that the

---

[393] **Glashow** Sheldon L. *(New York, NY-US 1932) Theoretical physicist. Submits his thesis in 1959 at Harvard (MA), with Nobel Prize winner* Julian **Schwinger** *as thesis advisor (ref.363-p.108). Professor in Houston (TX), Harvard (MA), and Berkeley (CA), holder of numerous scientific distinctions including the 1979* **Nobel** *Prize in Physics. Was a Practicing* Jew.

[394] **Salam** Abdus *(Jhang Sadar, PAK 1926-1996 Oxford, ENG). Obtained a scholarship in 1940 to enter the University of Punjab. In 1952 he was awarded a Ph.D. in Mathematics and Physics, at Cambridge where he taught. Promoted to Professor of Theoretical Physics at Imperial College, London. 1964, Director of the International Centre for Theoretical Physics in Trieste. Co-creator of the Nuclear Research Agency of Pakistan, he became its first director. He was awarded prestigious prizes and elected to the Royal Society in 1959.* Fervent Muslim, *he adhered to Ahmadism* which was banned in Pakistan in 1974. Reason for which he left his country. 1979* **Nobel** *Prize in Physics. the*

* Ahmadism, *a movement created by* Mirza G. **Ahmad** *(Qadiyan, Punjab 1835-1908) who, in 1889, proclaimed that Allah had asked him to restore a pure Islam, which goes against the majority of Muslims for which* **Muhammad** *is the last prophet.*

[395] **Weinberg** Steven *(New York, NY-US 1933) attended Columbia, Cornell, and Princeton (NJ) universities, where he defended his doctoral thesis in 1957. Physicist, cosmologist, and writer. Professor at Austin (TX), Berkeley (CA), Harvard.* **Nobel** *Prize in Physics in 1979 with* S. **Glashow** *and* M. A. **Salam.**

[396] *G, abbreviation of Giga = 10^9; eV = electron volt: energy of an electron subjected to a potential difference of $1V = 1.6.10^{-19}$ joules (J).*

[397] **Goodenough** John *(Jena, DEU 1922-2023 Austin, US-TX). After studying mathematics at Yale University he served as a meteorologist in the US Army during World War II. He integrated into Chicago University in 1946, was introduced to quantum mechanics by Enrico* **Fermi**, *and was awarded a doctorate in physics in 1952. Subsequently, he worked at MIT until 1976 where he developed fast RAM (Random Access Memory) used in laptops and contributed to the development of digital computers. During the late 1970s and early 1980s, he was the head of the Inorganic Chemistry Laboratory at the University of Oxford, London. He has occupied between 1986 and 2018, the chair of Material Science and Engineering, at Texas University in Austin. Evangelist, he was introduced to the Bible in 1945.*

cathode of Lithium (Li) batteries pioneered by Stanley **Whittingham**[398] in 1976, could exhibit even greater potential if constructed using a metal oxide instead of a metal sulfide. Their foresight was vindicated in 1985 when Akira **Yoshino**[399] replaced the Li anode with one crafted from petroleum coke, leading to the commercialization of the 1st Li-ion battery. This groundbreaking innovation would ultimately revolutionize the landscape of laptops, smartphones, and electric cars. In **2019**, Goodenough, Whittingham, and Yoshino were awarded the **Nobel** Prize in Chemistry *"for the development of lithium-ion batteries."* Notably, Goodnough at the age of 97, became the oldest recipient of this esteemed honor, delighting Arthur **Ashkin**[400] who had been awarded in 2018 (at the age of 96) the **Nobel** Prize in Physics for *"the optical tweezers and their application to biological systems."*

---

[398] **Whittingham** *Stanley (Nottingham, ENG 1941). With a passion for Chemistry, he was admitted to Oxford in 1960 where he was awarded his Ph.D. in 1968 before becoming a postdoctoral fellow at Stanford University until 1972. Then he joined Exxon Research & Engineering Company in Linden, US-NJ where he delved into materials science, skillfully crafting innovative materials with unprecedented properties leading to the development of Li-ion batteries. From 1984 to 88 he lent his expertise to Schlumberger; but he embraced a new chapter in his career as a Professor at the Chemistry Department of Binghamton University, US-NY. He dedicated himself to furthering the potential of Li batteries. Before being awarded the Nobel Prize in 2019 his work had already garnered numerous accolades, solidifying his status as a visionary in his field. He was educated as a* Christian. *A teenager, his school days began at St. Paul's Church, and he married Georgina Andai, a* Jewish *Hungarian immigrant at the multi-faith chapel of Stanford University.*

[399] **Yoshino** *Akira (Suita, JPN 1948). During his fourth grade of elementary school, he was fascinated by Chemistry. In 1966, he stepped into Kyoto University, joining the prestigious Department of Petrochemistry within the Faculty of Engineering. There, he delved deep into the field of photochemistry. In 1972, after earning a master's degree, he embarked on a journey as a corporate researcher at Asahi Kasei Corp.. 1981 was a "fateful year" for him; he focused his efforts on polyacetylene, a plastic able to conduct electricity, recently discovered. Aware of* **Whittingham**'s *research he will prove that polyacetylene is the ideal material for creating anodes of Li-ion batteries. In 2005 he was awarded a doctorate in engineering at Osaka University and he had his own laboratory at Asahi Kasei. Since 2017 he has shared his knowledge and passion as a professor at Meijo University in Nagoya.*

[400] **Ashkin** *Arthur (USA: New York, NY 1922-2020 Rumson, NJ).* Jewish *Physicist. As a teenager, his mind is ignited with an unquenchable curiosity for wondrous forces of light. Also in 1940, he enrolled at Columbia College. But in 1945 he was drafted into the army as a skilled technician at Columbia University's Radiation Lab. During this time he built magnetrons for radar. After the war, he pursued his education by completing his bachelor's degree earning in 1947 a BS degree in physics at Columbia University. He then attended Cornell University where he studied nuclear physics and received his Ph.D. in 1952. He was hired by Bell Laboratories where he embarked on groundbreaking research in the field of lasers leaving an indelible mark that resonated throughout the scientific world. His pioneering work brought about a revolutionary transformation in the realm of ocular surgery. Even before being honored with the 2018* **Nobel** *Prize in Physics, his trailblazing efforts had already been recognized with various other coveted medals. His work has been both widely used and widely recognized. He was a member of the National Academies of Engineering and Sciences respectively and a laureate of the National Inventors Hall of Fame…*

These lightweight, rechargeable, and powerful batteries, as expressed by Professor Göran K. Hansson, Secretary General of the Royal Swedish Academy of Sciences, ushered in "*a rechargeable world*" on 9 October 2019.

It is worth mentioning **Ashkin**'s extraordinary achievement with optical tweezers, which enable the manipulation of particles, atoms, molecules, and living cells by coaxing them toward the center of the laser beam. In 1987, Ashkin achieved the remarkable feat of capturing living bacteria without harming them. Today optical tweezers are widely employed to study biological systems, contributing to significant advancements in this field.

In **1980** the "**String theory**,"[401] initiated in 1968, attracted the attention of theoretical physicists. This theory is the fruit of the hard work of two theoretical Physicists, professors John H. **Schwarz**[402] and Michael **Green**,[403] co-initiators of the Green-Schwarz mechanism. In 2013 they were awarded the "*Fundamental Physics Prize*," the most lucrative (USD 3 million). They were convinced since the 1970s, despite the skepticism of the scientific community, that the string theory, the first quantum theory justifying the existence of the **Graviton**, vector of the gravitational waves (resulting from the theory of relativity), was to allow the unification of the theories of the General Relativity and the Quantum Physics. The collaboration of J. **Schwartz** with M. **Green**, et al. allowed them to demonstrate that this theory was mathematically correct which was a big step towards the "*unification theory*" sought by A. **Einstein** et al. A new breath was given to it in 1995 by Edward **Witten**,[404] a practicing *Jewish*, physicist, and

---

[401] String theory. *Subatomic particles are no longer conceptualized as points but as variants of a one-dimensional entity, having different vibration modes from one element to another. This entity called: a* "string," *is a small filament of vibrating energy, having no thickness, only a length. Each mode of vibration, performed by a string produces a mass m growing with the frequency as stated by the relation:* $E = h\nu = mc^2$. *There are 5 variants of this theory (with slight differences) in which the strings can be either closed on themselves or opened (free ends) which is awkward for a theory that wants to unify Gravitation and Quantum! But E.* **Witten** *(below) solved this problem. For example, the graviton would be a vibrating string of length of the order of the Planck length* $(lp = 1.616.10^{-35}$ m), *the smallest length known. The mathematical equations describing string theory require 9 dimensions of space (the 6 additional ones being of the order of magnitude of lp), for describing all the vibrational modes of the so-called* "elementary particles," *which is in line with the geometry of Calabi-Yau space including hexa-dimensional forms that string theory has determined.*

[402] **Schwarz** John, H. *(North Adams, US-MA 1941). Theoretical physicist, at the origin, with M.* **Green** *of the string theory. Student at Harvard-MA, then at Berkeley-CA, he was appointed as a professor at Princeton-NJ then at Caltech. Laureate of many prestigious prizes, among which the most lucrative prize in physics:* "The Fundamental Physics Prize," *shared with M. Green.*

[403] **Green** Michael, B. *(London, ENG 1946) obtained his Ph.D. in particle physics in 1970 at Churchill College, Cambridge. Professor of theoretical physics, appointed in 2005 at the Lucasian Chair of Mathematics at Cambridge (since* **S. Hawking's** *departure). Co-initiator of the Green-Schwarz mechanism. Laureate of numerous prestigious awards, among which the most lucrative physics prize: the* "Fundamental Physics Prize," *shared with J. Schwartz.*

[404] **Witten** Edward *(Baltimore, US-MD 1951), Jewish, stated the importance of religion for him. Professor of mathematical physics at Princeton-NJ., Witten is a specialist in quantum field theory. After a bachelor's degree in history, he worked for George McGovern's presidential*

mathematician of genius, who succeeded in unifying the 5 string theories into one: the *"Superstring Theory"* or *"M-Theory,"*[405] without contradicting the results obtained with the 5 variants of the string theory. If the M theory were verified experimentally, it would not only allow unifying all the particles of matter as well as the messengers in the same string [Gr p. 511] of variable length and frequency (frequency v, energy E, and mass m are related: $E = mc^2 = hv$) those properties determine those of *"objects"* of higher dimensions, called **p-branes** (where p, less than 11, would be the number of spatial dimensions necessary to describe our universe (cf. Chap. 4)); but also, to answer questions that the Standard Model (SM, ref.382-p.113) could not answer. Some cosmologists think it is necessary to introduce new unknown concepts of space and time. Similarly, the prospect that strings larger (of the order of 10^{-18}m) than the primitive ones (of the order of Planck length = 10^{-35}m) could exist and justify the existence of unknown particles of 100 to 1,000 times the mass of the proton, which the LHC could reveal, excites many researchers.

In the pivotal year of **1989,** the field of medical research witnessed a groundbreaking moment. A triumvirate of renowned, independent teams, helmed by distinguished scientists and eminent virologists, left an indelible mark on the medical realm. Thus, Dr. Harvey J. **Alter,**[406] born in a *Jewish* family, Pr. Sir Michael **Houghton,**[407] and Dr. Charles M. **Rice,**[408] accomplished a monumental

---

campaign, and in 1976 he defended his thesis in physics after studying mathematics, where he excelled. One of the major physicists of the end of the 20th century. He received many awards, including the 1990 Fields medal and the most lucrative prize in Physics: the Fundamental Physics Prize."

[405] *The M Theory, using 1 additional spatial dimension, bringing to 10 their number, improves knowledge. It can be considered a real revolution, but it remains to justify it experimentally.*

[406] **Alter** Harvey J. *(New York 1935). A medical researcher, virologist, and physician was born into a Jewish family. In 1960 he obtained a medical degree from the University of Rochester NY-US). Alter co-discovered in 1964 the Hepatitis B antigen. In the Mid-1970s, his research team demonstrated the existence of a hepatitis virus different from the A & B well-known. Their transmission studies on chimpanzees led to the discovery in 1988 of the Hepatitis C Virus (HCV) which was confirmed in 1989. Elected to both the National Academy of Sciences and the Institute of Medicine. He was the former chief of the infection disease section, and associate research director in the department of transfusion medicine, at the NIH in Bethesda (MD-US).* **2020 Nobel** *Prize for Physiology or Medicine.*

[407] Sir Michael **Houghton** *(UK 1949). Scientist, and virologist, he graduated in 1972 from the University of East Anglia (Norwich ENG) with the highest degree in biological sciences, in 1977 he graduated from King's College of London with a doctoral degree in biochemistry. Co-discoverer in 1986 of the Hepatitis D genome and in 1989 of Hepatitis C antibodies. In 2009 he accepted the position of Professor of virology at the University of Alberta (CAN) where his team showed in 2013 that a vaccine for one strain of Hepatitis C is effective against all strains of Hepatitis C. During the interview by G. Rutherford in 2020 (Wikipedia) he says:* "We can save time by transferring technology we developed for the hepatitis C vaccine into the COVID research." *He is knighted in 2021. Besides the 2020 Nobel Prize for Physiology or Medicine, he has a lot of medals and honours.*

[408] **Rice** Charles M. *(Sacramento US-CA 1952) is a virologist well known for his contributions to the development of effective treatments, for HCV infection. In 1974 he graduated with a*

feat: the discovery of the Hepatitis C Virus (HCV). This discovery contributed to a significant leap forward in the future treatment of this viral ailment. Their collective achievement was crowned with the **2020 Nobel** Prize for Physiology or Medicine. The three had already been honored with an array of prestigious awards, underlining the magnitude of their contributions to the field of medical science.

In **1990**, a momentous year unfolded, profoundly impacting Humanity and the delicate balance of life on Earth. The year marked a critical juncture as the IPCC (Intergovernmental Panel on Climate Change), an institution, established and shaped by Sir John **Houghton**[409] in 1988, came to the forefront. J. **Houghton,** a distinguished environmental physicist and devoted *Evangelical Christian,* assumed the helm of this organization. In 1990, the IPCC released its inaugural comprehensive global report, illuminating the profound influence of greenhouse gases on the planet's climate and society. Houghton's remarkable diplomatic finesse ensured the agency's impartiality, adhering to his vision of it being:

"an entirely scientific body not political" [Ho]

as he reiterated during the 25th-anniversary commemoration of the first IPCC report, in 2015.

Throughout his lifetime, Sir J. **Houghton** devoted himself to disseminating the science of climate change, reaching out to a diverse audience, from religious and scientific leaders to foreign dignitaries and fellow scientists. In the words of J. Houghton, the President of the John Ray Initiative:

"In this paper, I first list some of the growing threats to the environment
and introduce the important concept of sustainability. I then explain the
threat arising from human-induced climate change, summarizing its
scientific basis and the most significant impacts. I proceed to outline the
action that is necessary to halt climate change, especially in the energy

---

bachelor's degree in zoology. In 1981 working on RNA viruses, he earned a doctoral degree in biology at Caltech (CA-US) where he remains. His research helped the development of a yellow fever vaccine. He joined the Washington University School in St Louis where in 1989 he identified a DNA clone of the HCV RNA genome. In 1996 he described the HCV genome and then demonstrated its infectious nature. In 2011, appointed at Rockefeller University since 2001, his first drugs blocking HCV were approved by the US Food and Drug Administration. 2020 Nobel Prize for Physiology or Medicine.
[409] Sir John T. **Houghton** *(Wales: Dyserth 1931-2020 Dolgellau). Meteorologist and Environmental physicist. At 16 he went to Oxford to study Physics and was awarded a D. Phil in 1955 by the Jesus College, Oxford, London. There, he became unavoidable in the incipient field, of designing instruments to measure atmospheric temperature and composition, for four of NASA's Nimbus satellites in the 1970s. 2012 Vetlesen Foundation Prize,* referred to as the *"**Nobel** Prize for Earth and Universe Sciences."* Evangelical Christian *and* apologist of Christianity *he co-founded the International Society for Science and Religion. President (2005–2009) of the Victorian Institute. Elder at Aberdovey (Gwynedd, Wales) Presbyterian Church.*

sector. Finally, I emphasize the moral imperative for action and suggest how Christians, in particular, should respond to the challenge."

In 2007, when the IPCC shared half of the Nobel Peace Prize *"for their efforts to strengthen and disseminate knowledge about human-induced climate change and to lay the foundations for the measures that are needed to counteract such changes,"* Sir J. **Houghton**, alongside renowned climatologist and glaciologist Jean **Jouzel**,[410] a Catholic, and member of the French Académie des Sciences, co-presided over the agency. Jean Jouzel was later co-awarded the prestigious Vetlesen Foundation Prize in 2012, often referred to as the *"Nobel Prize for Earth and Universe Sciences."*

However, it is equally crucial to recognize the foresight of a remarkable visionary, Svante **Arrhenius**,[411] more than a century earlier. In 1896, Arrhenius a **Nobel** Laureate in Chemistry in 1903 *"In recognition of the extraordinary service he had rendered to the advancement of chemistry by his electrolytic theory of dissociation,"* published a groundbreaking article in the *"Philosophical Magazine and Journal of Science."* In this foresighted work, he issued a prescient warning to the scientific community, stating that a doubling of carbon dioxide (CO_2) in the atmosphere would result in global warming of approximately 5°C. Regrettably, despite the eclectic and influential nature of Arrhenius' scientific contributions, his visionary insights were not taken seriously then. As one of those scientists who proved to be ahead of his era, Arrhenius faced skepticism and was undervalued for his early understanding of the potential impacts of rising CO_2 levels on global climate. Especially since the previous work of the Mathematician and Physicist Joseph **Fourier**, a *Catholic* (ref. 596-p.198) - who had introduced in the 1820s the concept of the *"greenhouse effect,"* suggesting that the Earth's atmosphere acts

---

[410] **Jouzel** Jean *(Janzé, Bretagne, FR 1947) is a* Catholic *and an international expert in the sciences of the universe. Since 1970, he has been interested in glacial drilling and quickly became a specialist. At the end of the 70s he initiated, the introduction of water isotopes (with which he became familiar while preparing his doctoral thesis in Science, defended in 1974) in a General Circulation Model of the atmosphere. He set up the EPICA project (drilling under the Antarctic) and became its director from 1995 (when he was appointed Director of Research at the CEA) to 2001. At the same time, from 1997 to 2001, he held positions of responsibility in environmental and glaciology laboratories. From 2001 to 2008, he was director of the Pierre-Simon Laplace laboratory (grouping 7 climatology laboratories). Having joined the IPCC in 1994, he became vice-president in 2002, the year in which he was co-recipient with* Claude **Laurius** *of the CNRS Gold Medal for their contribution to the international Vostok project to study climate series; among various honors, he received the 2012 Vetlesen Foundation Prize,* referred to as the ***"Nobel** Prize for Earth and Universe Sciences."*

[411] **Arrhenius** Svante *(SWE: Vik 1859-1927 Stockholm). A precocious child raised* Lutheran, *he was a pioneer in various branches of Physics, the author of the Arrhenius law in 1889, correlating the speed of a reaction and the temperature. He entered the University of Uppsala, where he studied mathematics, chemistry, and physics, he was awarded a thesis in 1884. After various professorships, he became in 1895 professor in physics in Stockholm, then rector. Author of many publications, books, and lectures he received many awards and distinctions: the Society's Davy Medal, the Faraday Medal of the Chemical Society, 1903* **Nobel** *Prize in Chemistry.*

like a greenhouse, allowing sunlight to enter but trapping some of the outgoing heat, thus warming the planet - could have caught the attention of scientists earlier.

Notably, in **1990**, Joseph E. **Murray**,[412] a *Catholic* surgeon, jointly received the **Nobel** Prize in Physiology or Medicine with Edward D. **Thomas** *"for their discoveries concerning organ and cell transplantation in the treatment of human disease."* These pioneering researchers unveiled the means to manage organ transplant rejection, a phenomenon that had stymied earlier attempts, as recognized by the 1912 **Nobel** Prize Alexis **Carrel**, *Catholic* too (ref. 292-87) who referred to it as a *"biological force"* hindering the successful transplantation of organs between individuals. Today, we are well aware, of the necessity of taking specific measures to counter the immune system's rejection of foreign bodies.

In 2005, he was awarded the **Laetare Medal**[413] by the University of Notre Dame (IN-US) in recognition of outstanding service to the *Catholic Church* and society.

The **1990** year, was also a pivotal one, for the knowledge of the matter. The Nobel Prize in Physics was awarded jointly to Jerome I. **Friedman**,[414] Henry W. **Kendall,** [415] and Richard E. **Taylor** [416] *"for their pioneering investigations*

---

[412] **Murray** Joseph E. *(US-MA: Milford 1919-2012 Boston)* Catholic. *A pioneer in the transplantation of kidneys. He was a star athlete at Milford High School when he had to make a choice that turned out to be the right one for hundreds of thousands of patients in the world who could not otherwise be cured. Graduating in Medicine at Harvard Medical School he began his internship in Bakersfield, VT-US. During World War II he served in a plastic surgery unit. Nobel Prize in Physiology or Medicine. Many awards.*

[413] Laetare Medal *is announced to the recipient on the Laetare Sunday (the fourth Sunday of the Lent period. A period celebrating the 40 days, beginning on Ash Wednesday, during which Jesus endured satan's temptation in the desert). It is deserved to an American Catholic or a group* "whose genius has ennobled the arts and sciences, illustrated the ideals of the Church, and enriched the heritage of humanity"

[414] **Friedmann** Jerome I. *(Chicago, US-IL 1930), 1990* **Nobel** *in Physics. Admitted in 1950 to the Physics Dept. of the University of Chicago where E.* **Fermi** *was his supervisor. In 1956, he defended his Ph.D. then in 1957, worked with Robert* **Hofstadter***, *H.* **Kendall,** *and Richard* **Taylor**. *In 1960 he was hired as a member of the Physics Dept. of MIT, then Director of the Lab. for Nuclear Physics in 1980, and head of the Physics Dept from 1983 to 1988. Raised Jewish he claims to be an Atheist.* * *R.* Hofstadter *(New York 1915-1990 Stanford, CA) was raised Episcopalian, but he identified more with his Jewish roots and shared the 1961* **Nobel** *Prize In Physics with Rudolf* **Mössbauer** *(DEU: Munich 1929-2011 Grünwald)* Atheist.

[415] **Kendall** Henry *(Boston, MA 1926). He entered Amherst College, MA in 1946, then in 1950 at MIT where he studied Physics, he had M.* **Deutsch** *as a supervisor for his Ph.D. defended in 1955. Joined* **Hofstadter** *R's team, worked with* **Friedmann**, *and known* **Taylor**. *Environmental activist, he was in 1969 a co-founder of* "Union Concerned Scientists (UCS)" *an independent group pressing for control of technologies that may be harmful or dangerous. Was laureate of the 1990* **Nobel Prize** *in Physics.*

[416] **Taylor** Richard E. *(Medicine Hat, CAN 1929-Stanford, US-CA 2018). He attended the University of Alberta, CAN where he received a master's degree in 1952. He moved to Stanford University, US-CA where he defended a Ph.D. in 1962, then joined the SLAC where, he became a full professor in 1970, and conducted with Friedmann and Kendall the MIT-Experiment. 1990* **Nobel** *Prize in Physics. Director of the Lab. for Nuclear Physics in 1980, and head of the Physics Dept from 1983 to 1988. Raised Jewish he claims to be an Atheist.* * R. Hofstadter *(New York*

concerning the deep inelastic scattering of electrons on protons and bound neutrons, which have been of essential importance for the development of the quark model in particle physics."

Let's return in 1968. In this pivotal year, the 1990 Nobel laureates made a breakthrough, in our understanding of the inner structure of matter. Their series of investigations, called the *"SLAC-MIT Experiment,"* because it was realized with the two-mile-long SLAC Accelerator (Stanford Linear Accelerator) put into evidence the first indications that the nucleons have an inner structure with point-like scattering centers, according to the *"Standard Model"* (ref.382-p.113). These were later interpreted as being quarks. The quarks that had been searched in vain since the 1950s represent the building blocks of nucleons, and leptons (spin [angular momentum] = +/- ½, including electrons, muons, tauons, neutrinos ... and their antiparticles). They constitute 99% of the Earth's mass including our bodies. Quarks are bound by *"gluons"* massless particles. The concept of quarks to explain the observed behavior of strongly interacting particles had been proposed in 1964 independently by the physicists Georg **Zweig** and Murray **Gell-Mann** [417] who received the 1969 **Nobel** Prize *"for his contributions and discoveries concerning the classification of elementary particles and their interactions."*

In **1992** the scientists in charge of the COBE (COsmic Background Explorer)[418] satellite officially announced that the mission, whose aim was to study the *"fossil radiation"* or *"Cosmic Microwave Background"* (CMB), had confirmed experimentally that the structures (galaxies, clusters of galaxies) that we now observe were in their germ state 380,000 years after the Big Bang (cf. p.179) validating the *"Standard Cosmological Model"*[Pee] of which J. **Peebles**, the 2019 Nobel Prize in Physics, is one of the pioneers. J. Peebles has also worked a lot on, the temperature conditions necessary for galaxies to form. The **Nobel** Prize in Physics will be awarded in 2006 to the two designers of the instruments and

---

*1915-1990 Stanford, CA) was raised Episcopalian, but he identified more with his Jewish roots and shared the 1961 **Nobel** Prize In Physics with Rudolf **Mössbauer** (DEU: Munich 1929-2011 Grünwald) Atheist.*

[417] **Gell-Mann** Murray *(USA: New York 1929-2019 Santa Fe, NM). He earned his Bachelor's Degree in Physics from Yale University and subsequently, held various academic positions at different Universities in the US. From 1955 to 1993, he served as a Professor at the California Institute of Technology, with occasional breaks for Sabbatical periods at the College de France, and numerous engagements at CERN. Throughout his career, Gell-Mann collaborated with renowned physicists such as R. Feynman, and R. E. Block, yielding significant contributions to elementary particles.*

[418] COBE *satellite was placed in 1989, on an orbit at 900 km altitude, after having embarked among others the DMR (Differential Microwave Radiometer), to detect the anisotropy in temperature of space and the FIRAS (Far Infrared Absolute Spectrometer) intended to make a comparative measurement of the Cosmic Microwave Background (*CMB, detected by **Penzias** and **Wilson** in 1965, cf. p. 108) with the radiation of the black body.*

initiators of the project: Georges F. **Smoot**[419] and his colleague John C. **Mather**[420] for *"their discovery of the shape of the spectrum of the black body and the anisotropy of the cosmic microwave background."* At the announcement of the news S. **Hawking** declared:

"... it is the discovery of the century, perhaps even of all times."

Also in **1992**, during the Meeting of the American Physical Society in Washington, G. **Smoot** declared:

"If you are religious, it is like seeing God,"

Stephen **Maran**, editor of the Astronomy and Astrophysics Encyclopedia, said:

"It's like Genesis."

and I will finish with the front page of Newsweek:

"The handwriting of God."

On **October 6, 1995**, the Haute-Provence Observatory in France (04) received honors for its groundbreaking achievements. Using the ELODIE spectrograph, built by a team of engineers and scientists led by the astrophysicist Michel **Mayor**,[421] this latter and his student Didier **Queloz**,[422] made a momentous announcement: the discovery of the first exoplanet (extra-solar planet), orbiting a star still alive located 51 light-years from the Sun.[423] Their remarkable

---

[419] **Smoot** George F. *(Yukon, US-FL 1945) astrophysicist and cosmologist. Joins MIT where he graduates in Mathematics and Physics and then obtains a Ph.D. fellowship. Works with:* David **Frisch** *(1918-1991, who participated in the realization of the 1st atomic bomb),* Louis S. **Osborne** *(Rome, ITA 1923-2012 Lexington, US-MA),* Christian, *nicknamed "gentle giant of physics," because he participated in the training of many researchers who became famous, then the Nobel prizes in Physics in 1990:* Jerome I. **Friedman** *and* Henry **Kendall** *and in 1976* Samuel C.C. **Ting** *(Ann Arbor, US-MI 1936).*

[420] **Mather** John C. *(Roanoke, US-VA 1946) astrophysicist. Designer of the FIRAS. Early on, he was interested in science. So, he joined Berkeley as a student, he was taught by* Henry **Frisch**, *David's father (see Smoot). In 1970 he started a thesis on CMR - under the direction of the 1966 Nobel Prize in Physics* Charles H. **Townes** *(ref. 355,-p.106) - which he defended in 1974. In 1976 NASA agreed to integrate his spectrometer (FIRAS) and that of G. Smoot in one of its projects. He was laureate of the 2006 **Nobel** Prize in Physics.*

[421] **Mayor** Michel *(Lausanne, CHE 1942) defended his thesis in 1971 in Geneva where he studied astrophysics. Professor in 1984 at the University of Geneva, he studied double stars. Director of the Geneva Observatory from 1998 to 2004. He invented various spectrometers. Holder of numerous awards including the A. Einstein Medal, in 2004, ... Wolf Prize in 2017, and the 2019 **Nobel** Prize for Physics.*

[422] **Queloz** Didier *(CHE 1966) astronomer. After graduating from the University of Geneva prepared at the Geneva Observatory, under the* **Mayor's** *direction, a thesis that he defended in 1995. Professor at the Geneva Observatory and at the University of Cambridge, he published the same year in "Nature" the discovery made with* Michel **Mayor** *of the exoplanet 51 Pegasi a. Wolf Prize 2017. 2019 **Nobel** Prize in Physics.*

[423] *The exoplanets discovered previously were discovered around the Pulsar PSR B1257+12 (neutron star [1/50th] of the Earth] having a rotation period on itself of about 6.22ms). This is*

contribution earned them half of the 2019 Nobel Prize in Physics with the citation: *"For the discovery of an exoplanet: 51 Pegasi b, orbiting a solar-type star on 6/10/1995."* As of November 2023, the Extrasolar Planets Encyclopedia (EPE) reports 5,482 discovered exoplanets across 4,045 systems, with 875 identified as part of multiple planetary systems.

In **1996**, *Christian* Richard E. **Smalley**[424] was bestowed with the **Nobel** Prize in Chemistry, a joint honor shared with Sir Harold W. **Kroto** and Robert F. **Curl** *"for their discovery of fullerenes,"* a new form of carbon distinct from diamond and graphite. These incredibly stable molecules, despite being made up of 60 or more carbon atoms, arranged in a hollow sphere or tube, sparked up a revolution in chemistry. Their discovery has led groundbreaking research and innovations across multiple scientific fields, [425] paving the way for a more sustainable, technologically advanced, and medically promising world. The name *"Buckminsterfullerene"* was thoughtfully proposed by Sir Harold **Kroto** from the visionary architect Richard Buckminster **Fuller**,[426] a *Unitarian* minister, known for his environmental activism and advocacy for recycling to minimize waste.

Fuller's geodesic domes bore a striking resemblance to the new atom structure. Throughout his brilliant career, **Smalley** demonstrated an exceptional concern for pressing environmental issues. He was a driving force in global warming, affordable and clean energy, and the quest for sustainable alternatives to fossil fuels. In October 2005 at an alumni banquet at Tuskegee University (US-AL) he said:

---

the result, of 1 year of data collected by ELODIE a spectrograph of the Haute-Provence Observatory located in Saint-Michel-de l'Observatoire, FR.

[424] **Smalley** Richard E. *(US: Akron, OH 1943-2005 Houston, TX)* Christian, *He spent two years at Hope College before enrolling at the University of Michigan where he graduated in 1965. He will have various positions until 1969 when he moved with his wife to Princeton to begin his Ph.D. In 1973, he began a postdoctoral period at the University of Chicago. In 1976 he was appointed as Assistant Professor in the Department of Chemistry at Rice University in Houston where he built various generations of supersonic beam apparatus used to study large molecules and clusters and discover the Buckminsterfullerene or C60, presenting a great interest in the realization of eyewear and optical sensors because they attenuate the transmission of light. Smalley was the Director of the Center for Nanoscale Science and Technology from 1991. Besides the 1996* **Nobel** *Prize in Physics, he received various distinctions.*

[425] Fullerenes *are new forms of Carbon obtained in an inert gas atmosphere submitted to an intense pulse of Laser light. They possess exceptional physical and chemical properties (high tensile strength, remarkable thermal stability, conducting electricity, ...) making them valuable for numerous technological advancements in nanotechnologies, medical applications where they are suitable carriers for drug delivery, and space studies: fullerenes have been discovered in space...*

[426] **Fuller** Richard B. *(USA: Milton, MA 1895–1983 Los Angeles, CA) architect, inventor (28US patents), designer, author, philosopher. From Milton Academy, a renowned school, twice expelled from Harvard as a non-conformist misfit he held various jobs before finding his way in 1927. He will develop many architectural designs. Elected at the American Academy of Sciences and the National Academy of Design in 1968, The Presidential Medal of Freedom from R. Nixon in 1983,... Many honorary doctorates....*Unitarian Minister.

"My short two years at Hope starting as a freshman in 1961 were immensely important to me. I went to chapel, studied religion, attended church more than I had ever done before, and was with people who took these issues seriously. I valued that greatly back then. Recently I have gone back to church regularly with a new focus to understand as best as I can what makes Christianity so vital and powerful in the lives of billions of people today, even though almost 2,000 years passed since the death and resurrection of Christ."

2.11. 21st century Human Genome: The Language of God? Will gravitational waves replace Galileo's telescope?

Note. Readers unfamiliar with biological terms can consult Appendix 1, p. 140.

"... today we learn the language in which God created life..."

It is with this sentence that on **06/26/2000**, Bill **Clinton**,[427] the 42nd President of the USA and a *Baptist Christian*, presented to the international community, groundbreaking results from the Human Genome Sequencing Project. Over the past decade, two independent teams of physicians and biologists diligently have worked on sequencing the Human Genome (Appendix 2, p. 141) to identify all its protein-coding genes. This scientific feat is comparable to the presentation of the American West map by Meriwether **Lewis**[428] to President Thomas **Jefferson**[429] who declared:

---

[427] **Clinton** William S. *(Hope, US-AR 1946). 42nd President of the USA from 1993 to 2001. Raised in a non-religious family he joined the Baptist church in his town at a very young age. Very early on he became interested in politics, joining the "Boys State" from 1963 to 1964 to prepare future civil servants. A brilliant student at Hot Springs High School, he revealed his talent for the saxophone. He graduated from Georgetown in 1968, entered Yale Law School in 1970, and graduated in 1973. Elected Governor of Arkansas in 1978. Just elected President of the United States, he introduced the "Family and Medical Care Act" in 1993, which aimed to provide universal coverage. It was rejected by the Senate, which did so again in 1994 with the "Family Medical Act." He was behind the meeting between Israeli Prime Minister Yitzhak Rabin and Palestinian leader Yasser Arafat at Camp David on September 13, 1993. Created in 1997 the "Clinton Foundation," with humanitarian, philanthropic, and ecological goals.*

[428] **Lewis** Meriwether *(USA: Charlottesville, VA 1774-1809 Tennessee, TN). Berkeley-CA University. Enrolled in the militia in 1794, then from 1795 to 1801 in the U.S. Army where he obtained the rank of captain. He was then private secretary to Thomas Jefferson and prepared the Lewis and **Clark** (1770-1838) expedition which, leaving Hartford, IL in 1804, led them to the Pacific coast in 1805. The goal was to find waterways allowing the crossing of North America from East to West. They brought back geographical maps, observed and described nearly 200 plants and 122 animal species, identified about fifty Amerindian tribes with whom they established commercial links, and favored the colonization of the country by European immigrants. It seems that they owe a lot to Sacagawea, the Amerindian interpreter who saved the expedition on two occasions.*

[429] **Jefferson** Thomas *(USA: Shadwell 1743-1826 Monticello, VA). Highly educated. 3rd President of the USA (1801-1809) was vice-president from 1797 to 1801 and secretary of state*

"Without a shadow of a doubt this is the most important, the most wonderful map ever produced by mankind."

The results of the Human Genome Sequencing project became available in September 2003; they compiled genetic data from diverse individuals worldwide, represented as a 3-billion letter text coded using the 4 letters A, T, G, and C (for the nucleotides: Adenine, Thymine, Guanine, Cytosine). This remarkable achievement sheds light on the intricate complexities of our genetic makeup and its impact on various traits and genetic diseases. However, it is now clear that only a small portion (1,5%) of the genome codes for proteins hence the necessity to launch in the crowd the ENCODE (ENCyclopedia Of DNA Elements) project, which aims to shed light on the complex regulatory mechanisms that govern the expression and function of genes. This knowledge can provide valuable insights into human biology, disease mechanisms, and potential therapeutic targets. Two competing teams have started. The first one was *"Celera Genomics,"* a private, for-profit company financed by PerkinElmer[430] and led by the scientist-turned businessman: Craig **Venter**,[431] and the second one *"the International Human Genome Project"* an international initiative created under the auspices of the American *"National Institutes of Health"* (NIH). The latter's initial director was James **Watson**,[432] who, along with Francis **Crick** and Maurice **Wilkins** won the

---

of G. **Washington** *from 1790 to 1793. Magistrate participated in 1776 in the Declaration of Independence of the US and was ambassador to France. He founded the Republican-Democratic Party,. Negotiated with Napoleon the sale of Louisiana. Very influenced by* Christianity and Deism.

[430] PerkinElmer *consulting company was founded in 1937 by* Richard Scott Perkin *and* Charles Elmer *and headquartered in Wellesley, MA. Manufactures equipment for medical diagnostics, chemical analysis, agronomy, and genetics, ... 10,000 employees worldwide. In charge of polishing the main mirror of the Hubble Space Telescope, it was discovered that it did not conform to the specifications, resulting in blurred images.*

[431] **Venter** *Craig (Salt Lake City, US-UT 1946). Biotechnologist and businessman. Obtained in 1975 a Ph.D. in the mode of action of Adrenalin. He then began a scientific career at the University of Buffalo, NY; then, in 1984, he joined the NIH (National Institutes of Health) and started automating genome sequencing. He filed patents, which triggered the ire of biologists, including Nobel Prize winner* James **Watson**. *Venter resigned in 1992 and set up a private foundation* "The Institute For Genome Research (ITGR)" *which he left in 1998 to found* "Celera Genomics" *with PerkinElmer. In 2002, he left to found the* "Graig Venter Institute."

[432] **Watson** James *(Chicago, US-IL 1928). Biochemist and geneticist, atheist. After defending his thesis in Biology in 1950, he went to Copenhagen to learn the methods used in biochemistry and became interested in the structure of nucleic acids. At the end of 1951, he joined the crystallography laboratory of Cambridge. Appointed Professor of Biology at Harvard in 1961. Holder of numerous distinctions of which he was deprived for racist remarks and eugenics, he was co-winner of the 1962* **Nobel** *Prize in Physiology or Medicine. Note the important part played by* Rosalind **Franklin*** *whose DNA images were hijacked without her knowledge.* Watson *even went so far as to minimize his role in this discovery, an argumentation which was refuted by* F. **Crick**, L. **Pauling**, ...

***Franklin** Rosalind *(ENG: Notting Hill 1920-1958 London),* Jewish. *Chemist, molecular biologist, crystallographer, and geneticist, she obtained her Ph.D. in Physics and Chemistry at*

Nobel Prize in Physiology or Medicine in 1962 *"for their discoveries on the molecular structure of nucleic acids and its significance for transmission of information in living matter."* Following a dispute with the NIH director, J. Watson resigned in 1992. Francis **Collins**[433] a renowned geneticist and *Christian apologist,* who would be awarded the prestigious **Templeton**[434] Prize in 2020, succeeded him after careful consideration. He wrote [Col1] that as a *believing scientist,*[435] he had to accept because the realization of this project *"was a chance to know the language of God and to contribute to the healing of his contemporaries."*

Cambridge in 1945. Learned the techniques of X-ray diffractometry in Paris from 1947 to 1950 and joined King's College in London in 1951. The pictures she made of the structure of DNA were very useful for the confirmation of the double helix structure of DNA. In 2003 the Royal Society created the Rosalind-Franklin Prize, and in 2008 the Louisa-Gross-Horwitz Prize was awarded posthumously by Columbia University (in recognition of an outstanding contribution to fundamental research in biology or biochemistry).

[433] **Collins** Francis S. *(Staunton, US-VA 1950), scientist, physician, geneticist. Initially educated by his mother, he entered public school at age 10 and, at 14 - discovering with wonder the power of science - decided to become a chemist. At 16 he entered the University of Virginia and became an* Agnostic *"not wanting to know." He then prepared for a Ph.D. at Yale in mathematics and quantum physics. After 2 years, not seeing what quantum mechanics could open up for him, he took a course in biochemistry, and it was a revelation: "DNA has the elegance of mathematics" for him. At 22 years old, married, with a daughter, about to finish his thesis, but wishing to do something for humanity, he turns to medicine and is admitted to the University of North Carolina. This is his path, he can combine his love of math with his love of medicine. A top-level geneticist and director of a university laboratory, he has been President of the National Institute of Health since 2008. Member of the National Academy of Sciences, the Academy of Medicine, and the Pontifical Academy of Sciences. Recipient of numerous awards: Biotechnology Heritage Award in 2001, the William Allan Award in 2005 by the American Society of Human Genetics, the National Medal Award of Sciences in 2008, ... the Pontifical Key Scientific Awards, the Warren Alpert Foundation Prize in 2018, the Templeton Prize in 2020.....*

[434] Sir John **Templeton** *(Winchester, TN-US 1912-2008 Nassau, BHS).* Presbyterian *businessman and philanthropist devoted most of his fortune to religious and scientific associations. He majored in economics in 1934 at Yale University and was awarded a scholarship to Balliol College in Oxford, a school that prepares future world leaders (Bill Clinton, Wasim Sajjad (Pt of Pakistan), prime ministers, ambassadors...). Holder of the* "Chartered Financial Analyst" *international professional recognition, he created in 1973 a foundation awarding the Templeton Prize* "for the progress of research and discoveries concerning spiritual realities," *endowed with 1 million pounds sterling, It is the largest annual prize given to an individual, making no distinction between religions.*

[435] **Collins** F. S. Christian *apologist. It was at Yale, upon discovering that Einstein did not believe in Yahweh, the God of his Jewish people (whom he nevertheless defended), that he became an atheist. At the age of 26, while caring for terminally ill patients, he discovered that many of his patients found in faith the strength to fight the suffering they endured, without having done anything to deserve it. Deciding to find out if God existed, he contacted a Methodist priest who lent him* "Mere Christianity" *by* **C.S. Lewis** *(ref.734-p.269). His reading was a revelation, he realized that science did not have the tools to lead to belief in God, only faith could do that; he became a* Christian Evangelist. *Belief in the God of Abraham, the Creator, seemed to him more rational than nonbelief. His stance in favor of Christianity earned him enmity in the scientific world, which is not new. A* Christian apologist, *he has written and edited many books, of which* "The Language of God: A Scientist Presents Evidence for Belief" *[Col] is the best known.*

Among the many esteemed scientists who contributed to this remarkable endeavor, it is worth mentioning Georgia M. **Dunston**, [436] a *Baptist* Afro-American scientist, with an inspiring career. Despite facing racial segregation, she secured scholarships and grants to become the first member of her family to graduate from college. Driven by her passion for studying the human genome that she eloquently describes as *"the most elegant living information and communication system known to science"* Dunston focused on researching health disparities affecting African descent. In an interview with J. Stout, [437] she shared her belief that:

"... the human genome encodes biblical truths on the laws of life revealed in and through research on human identity and population diversity. Moreover, I perceive and pose the genome human as a type of sacred text *...."*

As a scientist, she claims that: *"the concept of race has no place in biology. .... We all belong to Homo sapiens united by our shared humanity;"* and *"As a believer, I am convinced that the genome has been sequenced in our time to provide solid evidence from "big data science" for this generation, requires sense-based knowledge, to believe biblical truths of our identity and genomic inheritance in relationship to ourselves, others, and our God."*

It is essential to enlighten the significant contributions of Frederick **Sanger**, [438] a twice **Nobel** Prize laureate in Chemistry (1958 and 1980). His work on accurately and rapidly sequencing long stretches of DNA greatly influenced the realization of the Human Genome Sequencing Project. Additionally, Walter

---

In 2020, on the sidelines of the ceremony organized in honor of the awarding of the Templeton Prize, he was interviewed by N. Ganesh for The Times of India on the reasons for his conversion from atheist to Christian. *He declared:* "I was already pursuing a scientific career when, as a medical student, I needed to understand why people believe in God. I approached faith with suspicion that reason, would have to be sacrificed to accept the spiritual worldview. I found out that was not at all the case. Atheism, the assertion of a universal negative turned out, to be the most irrational of the choices. To my surprise, reason was very much on the side of faith. I ended up being converted to Christianity."

[436] **Dunston** Georgia M. *(Norfolk, US-VA 1944) Baptist. In 1968 she earned her BS in Biology at Norfolk University, a Master's degree from Tuskegee University (AL-US), and defended his Ph.D. in immunology in 1972 at Michigan University (MI-US). Associate Professor at Howard University, then full Professor of Microbiology, and director of the Human Immunogenetics Laboratory she worked for the National Institute of Health, created the National Human Genome Center, and formed a collaboration with F.* **Collins**, *Washington D.C.*

[437] **Stout** Jeremiah and **Dunston** Georgia M. *"Scientist Spotlight: Georgia M. Dunston." Biologos Feb. 22, 2019.*

[438] **Sanger** Frederick *(ENG: Rendcomb 1918–2013 Gloucestershire). Ph.D. in 1943.* **Nobel** *Prize in Chemistry in 1958* "for his work on the structure of proteins, especially that of insulin," *and in 1980 shared half part with* Walter **Gilbert** "for their contributions concerning the determination of base sequences in nucleic acids." *Sanger brought up a Quaker but became* Agnostic.

Gilbert's[439] efforts, a biochemist, and physicist who joined J. **Watson**'s team in 1960, were pivotal in advancing the groundbreaking scientific endeavor. Gilbert shared half of the 1980 **Nobel** Prize in Chemistry with F. Sanger:

"for their contributions concerning the determination of base sequences in nucleic acids."

In **2007** the Nobel Prize for physics was awarded jointly to Albert **Fert**[440] and Peter **Grünberg**,[441] *Catholic, "for the discovery of Giant Magnetoresistance* (GMR)." The GMR discovery changed the way our computers look and function. It triggered the development of powerful magnetic hard disk reading heads, which have been used and improved since 1997. The performance of these heads has made it possible to multiply by a hundred the volume of information stored on the same surface. Although **Gründberg** declares himself as a *conservative catholic,* he is an observer of other religions and, does not intend to classify them because, according to him, no one knows which is the best.

"Per se, they are all equivalent. What counts is how religion is practiced, for example with tolerance. And yet I believe that there is more than what we see, hear, etc., or can detect with instruments. But it is a feeling born out of many details of my personal experience and therefore impossible to share or communicate."

In **2009** Svante S. **Pääbo**,[442] a *Christian,* sequences the Neanderthal genome

---

[439] **Gilbert** Walter (Boston, US-MA 1932). *After a degree in Physics in 1953 at Harvard he graduated with a Ph.D. in Mathematics in 1957 with Abdus **Salam** as thesis advisor. Lecturer at Harvard in 1958, then Assistant Professor in Physics in 1958, and appointed Professor in Biophysics in 1968. In 1980 he shared with F. **Sanger** ½ the **Nobel** Prize in Chemistry.*

[440] **Fert** Albert *(Carcassonne, FR 1938) 2007 **Nobel** Prize in Physics. At the age of seventeen, he obtained "the French baccalauréat." Two years later he entered the "Ecole Normale Supérieure" (ENS) in Paris. He defended his thesis in 1970. Followed a post-doctoral in Leeds. Assistant Professor at the University of Paris-Sud in 1970. Professor in 76. His research activities were diversified, developing lots of international cooperations very fruitful. Made lots of publications and received a great number of prestigious awards.*

[441] **Grünberg** Peter *(Pilsen, CZ 1939-2018 Jülich, DE). Begins to study physics at Frankfurt University in 1958. Obtains his Ph.D. at the Technical University of Darmstadt in 1968 for his application of optical spectroscopy to the* "determination of crystal field split energy levels of rare earth ions in garnets." *Works a postdoc in Ottawa. In 1972, he took a position as a research scientist in Jülich where he investigated some semiconductors. In 1988, he discovered independently of A. **Fert**, but simultaneously, the Giant Magnetoresistive (GMR) effect. Before being awarded the 2007 **Nobel** Prize in Physics, he had received prestigious awards, including the Wolf Prize, and the nomination in 2006 as European Inventor of the Year .... Catholic.*

[442] **Pääbo** Svante *(Stockholm, Sweden 1955), "the father of paleo genomics" revealed the genetic differences existing between hominins and Homo sapiens. He defended his Ph.D. in 1986 at Uppsala University, made his post-doc from 1986 to 87 at Zurich then until 1990 at Berkeley where he joined Allan Wilson, a pioneer in evolutionary biology, and started to develop methods to study Neanderthal DNA. Professor of biology at the University of Munich in 1990 he became the founding director of the M. Planck Institute for Evolutionary Anthropology in Leipzig. In*

from samples found in the Vindija cave in Croatia. It is now well established that modern humans (Homo Sapiens) have 99.5% identical genomes to Neanderthal. He would be awarded the 2022 **Nobel** Prize in Physiology or Medicine *"for his discoveries concerning the genomes of extinct hominids and human evolutio*n."

In **2011** the Nobel Prize in Physics for *"the discovery of the accelerated expansion of the universe through the observation of very distant Supernovas"* in 1998, was awarded to Saul **Perlmutter,**[443] Brian **Schmidt**[444] & Adam **Riess.**[445] Until 1998 physicists majority thought that the speed of expansion of the universe was decreasing under the effect of gravitational forces.[446] An explanation of this expansion is found in the presence of Dark Matter (cf. p.81-p.90) constituting 25% of the total mass of the Universe that maintains, thanks to the force of gravitation, these clusters of galaxies together.

In the same year two famous scientists, covered many important distinctions, Emmanuelle **Charpentier,**[447] a *Catholic,* and Jennifer **Doudna,**[448] a *Christian,*

---

*1997 Pääbo's team realized the sequencing of Neanderthal DNA and in 2009 the first version of the Neanderthal genome was obtained in his Institute. His list of honors and awards is great. He does not mix religion and science. Agnostic according to me, although TG Time (www.tvguidetime.com) calls him Christian. He was laureate of the 2022 **Nobel** Prize in Physiology or Medicine.*

[443] **Perlmutter** Saul *(Champaign-Urbana, US-IL 1959). Berkeley-CA University. He studied at Harvard-MA and defended his thesis at Berkeley-CA in 1986. Winner of half of the **Nobel** Prize in Physics in 2011 (the other half awarded to B. **Schmidt** and A. **Riess**), recipient of numerous other prizes.*

[444] **Schmidt** Brian *(Missoula, US-MT 1967). Astronomer, cosmologist, and Professor at the Australian National University in Canberra. He studied at Harvard-MA. Co-recipient with **Riess** Adam of half of the **Nobel** Prize in Physics in 2011 "for the discovery of the acceleration of the expansion of the Universe," and recipient of many other awards.*

[445] **Riess** Adam *(USA: Washington-DC 1969), astrophysicist and cosmologist. Trained at Harvard-MA, Berkeley-CA, MIT, Professor at Johns-Hopkins-MD, Berkeley-CA. Co-recipient with **Schmidt** Brian of half of the **Nobel** Prize in Physics in 2011, recipient of many other prizes.*

[446] *During an interview with Panos Charitos of CERN (cf. CERNCOURIER 2019, Vol 59, N° 5, pp. 47-48,* **Riess** said that when their discovery was announced, many physicists thought, they had made a mistake in interpreting the experimental results. This was not the case, as other experiments corroborated theirs. Moreover, this acceleration was susceptible to provide answers to the observation that some stars seem older than the universe. It becomes clear that is not sufficient to appeal to dark energy, but also to improve the precision of the value of the *"Hubble constant"* cf. p 72.

[447] **Charpentier** Emmanuelle *(FRA: Juvisy-sur-Orge 1968). Director of the Max Planck Unit for the Science of Pathogens, Berlin. Microbiologist, geneticist, and biochemist she studied at the P. and M. Curie University (Paris, Fr), and graduated in 1995 from the Institut Pasteur (Pari). She did postdoctoral work at Rockefeller University (NY), University Medical Center (NY), and University of Vienna (Aus) she moved to Sweden where she became associate professor and lab head, and then in Germany at the Max Planck Institute for Infection. Member of various Academies, including the French Académie des Sciences (2017), and the Pontifical Academy of Sciences (2021), she was awarded the 2020 **Nobel** Prize in Chemistry. She is a* progressive Catholic.

[448] **Doudna** Jennifer *(Washington US-DC 1964). She is the Li Ka Shing Chancellor's Chair Professor at the University of California, Berkeley, US-CA, where she directs the Innovative*

met during a research conference, and began a collaboration that materialized in 2020 by obtaining the **Nobel** Prize in Chemistry: *"for the development of a method for genome editing."* The new tool: The CRISPR/Cas9 genetic scissors is revolutionary. Better known as the *"Swiss Army knife of genome editing"* because the Cas9 protein has the property of cutting DNA at specific sequences with high precision, it gives a practical way to rewrite the code of life! Many important discoveries -in cancer therapies, inherited diseases, crops...- can be made with this tool. We are entering a new era that can bring the greatest benefit to humankind but the worst too, like the acceleration of mutations in a population. It is the object of ethical questions.

On **October 09, 2013**, the **Nobel** Prize in Physics *"For the theoretical discovery of a mechanism contributing to our understanding of the origin of mass of subatomic particles, which has been recently confirmed by the discovery of the predicted fundamental particle (the Higgs boson) by the ATLAS and CMS experiments conducted at the CERN Large Hadron Collider"* is attributed jointly to the brilliant theoretical physicists Peter **Higgs**,[449] and François **Englert**[450] a *Catholic.* Indeed, on 4/7/2012 the scientific community had learned that a 126 GeV particle, with the characteristics of the Higgs boson[451] or **BEH** (an acronym

---

Genomics Institute. Her interest in sciences was nurtured at Hilo High School (Hawaii). Bachelor's degree in Biochemistry at Pomona Collège, Claremont, US-CA (1985), Ph.D. in Biological Chemistry and Molecular Pharmacology in 1989 at Harvard Medical School, Boston MA-US in 1989. She went to the University of Colorado Boulder to study the underlying mechanisms of RNA enzymes. In 1994 she moved to Yale University, New Haven, US-CT where she was promoted to Professor in 2000. She moved to Harvard University in Cambridge US-MA, then to Berkeley. She was elected in 2002 at the National Academy of Sciences, and in 2003 at the National Academy of Medicine, ...Member of various Academies including the Pontifical Academy of Sciences (2021), she was laureate of the 2020 Nobel Prize in Chemistry.

[449] **Higgs** *Peter (Newcastle, ENG 1929) is a theoretical physicist specializing in particles, string theory, and cosmology. Recipient of numerous awards: Dirac in 1997, Wolf in 2004, Sakurai in 2010, Nobel in Physics in 2013,... Professor of Physics at the University of Edinburgh. The scientists gave his name: Higgs' field to a modern variant of the ether, with one important difference: it does not affect the theory of special relativity, nor the speed of light and does not intervene in its propagation. At the origin of the Universe, the "Higgs' field" should have properties identical to the other fields; but, as the temperature decreased, instead of canceling itself, it took a non-zero value, forming the "Higgs ocean." It would give mass to the fundamental particles and would be at the origin of the inertia that any mass presents to an acceleration. It would act directly on the electrons and quarks of atoms. The presence of the Higgs field in all the Universe leads us to re-examine our conception of nothingness; what becomes of the notion of vacuum? Philosophically it is interesting because it could go in the direction of a divine spirituality filling all the so-called "empty" space.*

[450] **Englert** *François (Etterbeek, BEL 1932). His parents, Polish Jews, had emigrated to Belgium. He converted to Catholicism during the German invasion to escape persecution. Theoretical physicist. Professor at the University of Cornell-NY, Université Libre de Bruxelles, and Tel Aviv. Awards: Francqui in 1982, Wolf in 2004, ... the Belgian Kingdom confers on him in 2014 the title of Baron. He had shared in 2013 the Nobel Prize in Physics with Higgs.*

[451] *Higgs' Boson. The bosons are particles carrying energy and forces across the universe. The Higgs boson was expected according to the Standard Model (ref.382-p.113); it is an elementary*

for **Brout, Englert,** and **Higgs**), had just been demonstrated experimentally at the LHC which is the most powerful particle accelerator in the world. The BEH, called colloquially the "messenger" of the Higgs wave, as are the photon, vector of the E.M.Ws, and the graviton are that of the Gravitational Waves- was predicted theoretically in 1964. The Atlas and CMS experiments bring together in the world more than 6,000 people. The discovery of the Higgs boson, which has since been confirmed,[452] is a major step in the history of physics. It confirms the *"Standard Model,"* classifying all known and predicted elementary particles, it strengthens the existence of the Higgs' field, which theorists in particle physics and cosmology have been talking about for years, despite the absence of experimental evidence. The study of the BEH boson should lead to a better understanding of the infinitely small [Ge]. Leon **Lederman,**[453] co-winner of the 1988 **Nobel** Prize in Physics *"For the neutrino beam method and the demonstration of the doublet structure of leptons by the discovery of the muon neutrino,"* called the Higgs boson: *"The God particle"* [Le]. One should not draw any conclusions about Lederman's faith but, only a reference to the fact that the Higgs' field would be at the origin of all the massive fundamental particles. It is not the case for all non-elementary particles; although it interacts with photons and gluons, it does not give energy to them. According to Matthew **McCullough** [454] Higgs' boson specificities *"make it a formidable tool in the hunt of Dark Matter,"* particularly for the search of its constitution. Do not forget that D.M. represents about 25% of all the matter contained in the universe.

The same year, the ESA (European Space Agency) provided the scientific community with a map of the fossil radiation in the sky, available thanks to the measurements made by the European satellite Planck, put into orbit in 2009. This map shows slight anisotropies and small temperature variations (3rd decimal). It completes and refines the results provided by the American space observatory WMAP (acronym of: Wilkinson Microwave Anisotropy Probe), put into orbit in 2001.

---

particle, unstable, uncharged, and without spin. We have spoken (ref.236–p.72) of the 4 fundamental forces their vectors are 4 other bosons that are "force carriers" called: "gauge bosons": 1. γ photon, the vector of E.M.Ws, 2. G gluon, the vector of the strong nuclear force, 3. Z neutral and W+/-, charged weak bosons, vectors of the weak nuclear force, 4. G graviton, vector of gravity, not observed only predicted by the S.M.. For those interested, there exist other bosons: atoms, nuclei,...

[452] *Not a month goes by without the CERNCOURIER journal mentioning an experiment - usually an interaction - confirming its existence, satisfying the Standard Model. In April 2018, for example, CERNCOURIER [At] published p.13, under the title "ATLAS illuminates the Higgs boson at 13TeV," that the study of the Higgs Boson annihilation into 2 photons has led to a better understanding of it.*

[453] **Lederman** Leon, Max *(USA: New York 1922-2018 Rexburg, ID). Physicist, mathematician, University Professor at Chicago, Columbia), director of Fermilab, discovered the bottom quark, ... Holder of numerous prizes: Wolf in 1982,* **Nobel** *in 1988 with Melvin* **Schwartz** *and Jack* **Steinberg***, Enrico Fermi in 1992,*

[454] **McCullough** Matthew (1985). *Senior in the CERN Department of Theoretical Physics. The quote is from CERNCOURIER Vol. 57 N° 6, July/August 2017, Pp. 34-39.*

On **11/06/2016**, CERN announced to the scientific community the dawn of a new era: the first gravitational waves (G.W.) named "*GW150914*" (Gravitational Waves, year, month, day) were revealed on September 15th, 2015, by **LIGO.**[455] These waves result from the merger -at 1 billion light-years from our Earth- of 2 black holes of masses respectively equal to 36 and 29 times the solar mass giving 1 black hole of mass equal to 62 times that of the sun. The 3 lost solar masses (36 + 29 = 65) are converted into gravitational waves (G.W.s) according to the equivalence relation predicted by Einstein as early as 1916. Just like the E.M.W.s, the G.W.s move through space.

On **9/29/2017**, CERN announced that VIRGO installed near Pisa-It, detected gravitational waves on August 14th, for the first time. It had just started working again. This discovery is important because the same gravitational waves were detected respectively 6ms and 14ms before by the LIGO detectors in Hanford (Washington) and Livingston (Louisiana) which were at their 4th detection (the 3rd one having taken place on January 4, 2017). These waves were due this time to the encounter of 2 black holes of 30 and 25 solar masses which, after swirling around each other, merged at 1.8 billion light-years from us, releasing an energy of about 3 solar masses. In addition, by triangularization, the researchers were able to specify the area of the universe where this encounter took place: in the southern hemisphere towards the Eridanus constellation. Barry **Barish,**[456] leader of the high energy physics group at Caltech, and chief scientist of LIGO from 1994 to 2005 is quoted as saying:

> "*Like Galileo looking through his telescope, this is a breakthrough in astronomy.*"

The announcement that year by the **Nobel** committee that half of the 2017 physics prize was awarded to Rainer **Weiss,**[457] a research professor at MIT, the

---

[455] LIGO: *Laser Interferometer Gravitational-Wave Observatory, includes various laboratories in the world, gathering 900 scientists exploiting the data of two giant Michelson interferometers of 4km long, built at the Hanford Nuclear Complex (Washington State), and at Livingston (Louisiana State) distant of 3.600km. Since 2007 it has been collaborating with the VIRGO detector (Italian name for Virgo) a Michelson interferometer made of two 3km long tubes built in Cascina near Pisa. After an interruption, for improvements, it had just been put back into service a few months earlier. France and Italy are at the origin of the VIRGO project, in which the Netherlands, Poland, and Hungary are associated. It brings together 320 researchers.*

[456] **Barish** Barry C. (Omaha, US-NE 1936). Co-winner (1/4) of the 2017 **Nobel** Prize in Physics. His Polish parents of Jewish origin immigrated to the USA. After studying at Berkeley, he joined Caltech in 1963 as an experimental physicist in the field of particle *physics where he was appointed Linde Professor of Physics. From 2005 to 2013 he worked on the ILC (International Linear Collider) project which will be complementary to the LHC (Large Hadron Collider) at CERN.*

[457] **Weiss** Rainer *(Berlin, DEU 1932), Co-Winner (1/2) of the 2017 **Nobel** Prize in Physics, astrophysicist, a physicist specializing in gravitation. His family, of Jewish origin, emigrated to the USA (St Louis) in 1938. He obtained American nationality. After brilliant studies at Columbia, MIT, and Princeton, he taught at MIT from 1964. A specialist in cosmic ray*

other half being attributed to B. **Barish** and his colleague from Caltech, Kip **Thorne**,[458] for "*their decisive contribution to the realization of the LIGO/VIRGO interferometers and the observation of the first gravitational waves*" predicted in 1916 by A. **Einstein**, that is to say, one century before, confirms that it is a major advance for all those who want to know more about the origins of the Universe. Moreover, it is a further confirmation, if it were necessary, of the importance of the General Relativity Theory and the continuous contributions that improve its power.

It is noteworthy that in January 2019, the journal CERNCOURIER [CC1911] revealed that to date LIGO/VIRGO allowed the detection of 11 gravitational waves (G.W.s), 10 of which result from black hole mergers and neutron star collapses (ref.490, p.150). The last one observed GW170729 would have released, about 5 billion years ago, the equivalent of 5 solar masses in the form of gravitational radiation. According to Professor Dr. Alessandra **Buonanno**,[459] a specialist in gravitational waves, some of these waves may have been formed at the very beginning of the creation of the universe, following cosmic inflation for example. Their detection will give us information on the beginning of the Universe, information that we would not have otherwise.

With the study of G.W.s, scientists have a new technique to probe the Universe. It is a revolution comparable to the discovery of electromagnetic waves. Indeed, the G.W.s., resulting from the deformation of space-time, whose messengers are gravitons moving in gigantic numbers, are responsible for the gravitational force which has the particularity of being among the 4 known forces, the weakest and thus interacting the weakest with matter. It can pass through opaque media to the E.M.W. for example. Thus, offering a new vision of the universe to its explorers. Moreover, any element being able to generate a G.W.s - which is not the case for EMWs, because only charged particles can generate photons - we can hope to have a means of detecting unknown particles, or even universes invisible until now but whose existence **Epicurus** (ref.61–p.20) had envisaged. Tim Gershon, professor at the University of Warwick-UK, says [Ge] that the coupling of the results obtained with the telescopes detecting the G.Ws (studying the infinitely large) and those obtained with the Higgs bosons -that he describes as a "*new microscope*" because it allows the study of the infinitely small- are promising.

Nevertheless, as our knowledge progresses, new questions arise, which may call into question what we thought we had acquired; but, showing that celestial

---

spectrometry, he is at the origin of interferometers dedicated to the study of gravitational waves, and co-founder of LIGO.

[458] **Thorne** Kip *(Logan, US-UT 1940). Co-recipient (1/4) of the 2017 **Nobel** Prize in Physics. After his studies at Caltech and Princeton, he joined Caltech as a teacher in 1967. Specialist in the Theory of General Relativity and astrophysicist. Theoretical physicist he is an authority in the field of gravitational waves. Co-founder of LIGO.*

[459] **Buonanno** Alessandra "The dawn of a new era" *CERNCOURIER 2019 Vol. 57 N° 1, pp. 16-20.*

mechanics is very elaborate although based on a small number of fundamental constants and that, if it all works in our Universe, it is not an effect of chance. The universe can only exist because the elementary particles, divided into 3 families, have the properties required for this. In the same way, if the Sun produces heat and light necessary to our life on the Earth it is because $4,6.10^9$ kg of matter[460] is converted each second into luminous energy E (according to the relation: $E = mc^2$) following nuclear fusion reactions.[461] The answer to the question asked at the beginning of this paragraph: will gravitational waves replace Galileo's telescope? is yes according to Barry **Barish** [Bar] because, since the discovery by G. **Galileo** of the moons of Jupiter and although scientists have learned a lot about the Universe thanks to astronomers who have tracked down the electromagnetic waves emitted by a tiny part of the stars that populate the Universe, scientists expect a lot from the "*Einstein telescope.*" Although this project is currently at the conceptual stage, the main steps of its realization are set by the GWIC (Gravitational Wave International Committee): LIGO should increase from 4 to 40km, LISA (Laser Interferometer Space Antenna): 3 satellites constituting a triangle of 2.5km on each side, a project of the ESA (European Space Agency) should begin 2030, ... allowing to study the Big Bang itself by going back to the beginning of history.

The same year the Event Horizon Telescope (EHT[462]) obtained images of a black hole in the center of Messier 87 (MS87), a giant elliptical galaxy spotted for the first time in 1781 by the *Christian* Astronomer Charles **Messier.**[463] It is located about 55 million light-years from Earth.

---

[460] The Sun, *located $149.5.10^6$ km from the Earth, has a mass $M = 1.991.10^{30}$ kg (332.10^6 times that of the Earth). Its diameter is $1.392.10^6$ km (116 times that of the Earth).*

[461] Nuclear fusion reactions *can occur when the temperature is higher than $10^7 °C$ because, from then on, the kinetic energy of the nuclei is sufficient for two nuclei to overcome the coulomb barrier that pushes them apart and for fusion to occur, giving rise to a new heavier nucleus. Thus, in the center of the sun, the temperature of the plasma (p, n, nuclei, radiation, ...) is about $10^8 °C$, and atoms are ionized (electrons have been torn)). Two main reactions occur (for the nucleic notation cf. Soddy 1910 p. 87):*

a)   *4 protons ($^1H_1$) fuse producing a Helium nucleus ($^4He_2$), 2 positrons (e+, positive electron), 2 neutrinos (ν), and energy in the form of radiation in particular:* $4\ ^1H_1 => {}^4He_2 + 2\ e^+ + 2\nu$ + *great energy.*

b)   *2 helium atoms fuse to give an unstable beryllium atom ($^8Be_4*$) ($T_{1/2} = 10^{-16}$s) which fuses with another helium atom to give carbon 12 ($^{12}C_6$):* $2\ ^4He_2 => {}^8Be_4* + \gamma$ + Energy followed by: $^8Be_4* + {}^4He_2 => {}^{12}C_6 + \gamma$ + Energy, and $^{12}C_6 + {}^4He_2 => {}^{16}O_8 + \gamma$ + Energy.

[462] EHT (Event Horizon Telescope): *International network linking 8 radio telescopes, of very high resolution, located all around the Earth, on about twenty sites, to detect Black Holes. It forms a single "Earth-sized" virtual telescope and brings together 300 researchers from 60 scientific institutes. In addition to other facilities, the EHT network includes the Atacama Large Millimeter/submillimeter Array (ALMA) and the Atacama Pathfinder EXperiment (APEX) in the Atacama Desert in Chile, co-owned and cooperated by ESO.*

[463] **Messier** Charles *(FRA: Badonviller (31) 1739–1817 Paris)* Christian *Astronomer. He was about 14 years old when he observed his first comet. In 1771 he published the* "Nebulae and

On **April 9, 2019,** the professor of Radio Astronomy Heino **Fackle,**[464] a lay *Pastor* in a *Protestant* Church in the Netherlands, and chairman of the scientific council of the EHT's project, announced that he had obtained the first *"image"* of a black hole (Sagittarius A*) located in the center of our galaxy about 25,640 light-years away. In reality, only the dust and radiation at the outer limit of the event horizon are visible and delimit a dark disk containing the black hole.

The three Laureates of the **2020 Nobel** Prize in Physics made groundbreaking work studying compact and supermassive objects. But, as said David Haviland, chair of the Nobel Committee for Physics: *"These exotic objects still pose many questions that beg for answers and motivate future research."* We explained p.77 why Sir **Penrose** (ref.368-p.109), was awarded one-half of it. The other half was awarded jointly, to two leading figures in the field of Astronomy: Andrea **Ghez,**[465] a professor at UCLA (CA-US) since 2002, and Reinhard **Genzel,**[466] the Director of Max Planck Institute for Extraterrestrial Physics, *"for the discovery of supermassive compact object at the center of our galaxy."* Using the Keck telescopes in Hawaii and the ESO (European Southern Observatory)[467] in Chile, the two teams could peer separately *"The Monster at the Heart of our Galaxy,"* as they call it, through the heavy dust. Its scientific name is Sagittarius A* (cf. chap. III.2), and its mass is about 4.2 million times that of the Sun.

The same year Professors Svante **Pääbo** (ref.442, p.192) and Hugo **Zerberg** correlated the virulence of COVID-19 and the greater vulnerability of certain patients, who needed hospitalization and ventilation three times more than others, with their greater **Neanderthal** DNA heritage, which is known to vary between 1

Star Cluster Catalog" *containing 110 objects. He was elected fellow of the Royal Society of England, and a member of the Royal Swedish Academy of Sciences, and the French Academy of Sciences. He was awarded the French Cross of the Legion of Honor…*

[464] **Fackle** Heino *(Cologne, DEU 1966). Professor of Radio Astronomy and Particle Physics since 2003. Physicist, he obtained his Ph.D. in 1994 at the University of Cologne, worked at the Max Planck Institute for Radio Astronomy in Bonn, and later at the universities of Maryland and Arizona. Winner of numerous awards: Ludwig Biermann Award in 2000, the Academy Award of the University of Berlin-Brandenburg in 2006, rewarding a researcher who has opened a promising new path in research (on Black Holes in this case), the Spinoza Prize in 2011, the highest scientific award in the Netherlands, … Member of the Academia Europaea (2013), the Royal Netherlands Academy of Arts and Sciences (2014) …*
Christian, *he exercises the mission of a* Protestant lay pastor *in his spare time. In 2020 he was named Honorary Doctor of the Protestant Theology Faculty of Brussels, in recognition of his successful combination of his high-level work in cosmology and his Faith in Christianity.*
[465] **Ghez** Andrea *(US-NY 1965). Astrophysicist, Ph.D. at California Institute of Technology in Pasadena, US-CA in 1992. She received numerous honors and awards and was Elected to the National Academy of Sciences in 2004. 2020* **Nobel** *Prize in Physics.*
[466] **Genzel** Reinhard *(Bad Homburg vor der Höhe in Hesse 1952). Max Planck Institute for Extraterrestrial Physics, Garching, DE and University of California, Berkeley, US-CA. 2020* **Nobel** *Prize in Physics.*
[467] ESO *was created in 1962 by 5 European governments: Belgium, Germany, France, the Netherlands, and Sweden. They are now 16. It employs about 730 staff members. Its apparatus gives information in the wavelength range from 20 μm to 200 nm.*

and 4% in modern humans (cf. Chapter 3). This effect is added to age and certain pathologies.

On the 12th of May **2022,** Geoffrey C. **Bower**[468] unveiled the first image of the accretion disk around Sagittarius A* confirming it. It is 1000 times smaller than M87. Its image is an average of various images captured by the EHT extracted from its 2017 observations. Although we can't see the event horizon itself, because it can't emit light, the glowing gas orbiting the black hole reveals a telltale signature: a dark central region (called a shadow) surrounded by a bright ring-like structure.

Will this century be the one of the unification of the four fundamental forces of the universe into one, according to the Theory Of Everything (TOE, ref.383-p.114)? this would open new perspectives for the physicists but also open up broader philosophical discussions. S. **Hawking** [Ha], who declared himself an *agnostic*, an eminent ambassador of science, and one of the most brilliant contemporary astrophysicists, and cosmologists, said:

> "... *if we manage to discover a unifying theory, it should be understandable by everyone...... all* (the *"classical"* philosophers, in particular, who since the 20th century have given way to scientists capable of assimilating the mathematics and physics necessary for the development of modern cosmology) *will be able to take part in the discussion of why we exist and why our universe exists. And if we ever find the answer, it will be the triumph of human reason which will allow us to know the mind of God."*

Let's note that S. Hawking remained always cautious about making definitive statements about the existence or nature of God within the context of scientific exploration.

I'll give the word to Alister **McGrath** (ref.739, p.272), a world-renowned expert on the relationship between science and religion, who expresses the same point of view:

> "*The Theory Of Everything says there is one big thing that explains everything; one place where the buck stops. But, that's exactly what Christians have been saying about God: the buck stops with Him. The Theory Of Everything: the method of making sense of the world, the theories that we know, could simply be an explanation of the mind of God"* [CEA].

Indeed, if we believe that God created the world, there is an intrinsic rationality to the world that reflects God's wisdom and justice.

---

[468] **Bower** Geoffrey C *"Focus on First Sgr A* Result from the Event Horizon Telescope,"* The Astrophysical Journal 12th May 2022. *ESO (cf. www. eso.org") reveals the first direct visual evidence of the presence of Sagittarius A*. Cosmologist.*

Appendix 1.
Illustration of a misuse of the 3 first proposals of the
Kalam. (ref.90, p. 27)

Proposition I. *The simple substance is the "parcel."* It can be compared to the atom, even if it does not correspond to the state of our knowledge. *It has a zero mass (but when two parcels meet, they acquire a quantity and form a "body" having a quantity), is unlimited in number because God creates them when he wants, is the seat of the accident*[469] *which does not last 2 instants (cf. proposition III): which is re-initiated or not, regularly, by God according to his will.* Argument against causality and in favor of God's will.

Proposition II. *Existence of the void.* It is necessary to justify the parcel and the movement which would be impossible without a vacuum. These first two propositions disagree with Aristotle's ideas.

Proposition III. *Time is composed of instants* (cf. Aristotle: instants = small times divisible up to a certain value (h, mn, s, ...).

I will simply reproduce one example of demonstrations made by Salomon **Munk**[470] the commentator of Moses **Maimonides** [Ma1]. Let two runners A and B run the same distance in different times. Suppose that A comes first, and B comes second. We say that A is faster than B. The Mu'tazilites said that there is no one faster than the other. Indeed, applying proposition III by saying that the very small running instants are interspersed with resting instants which are shorter for A than for B, it follows that A and B could run as much time if we omit the resting instants and, that there is not one faster than the other!

Their sophist reasonings were not appreciated by the *Sufis* who had a mystical vision of the Koran, a vision that is found among the *Sunnis* and the *Shiites*.

---

[469] Examples of accidents *are taste, color, movement, and time, ... There are opposite accidents: death and life, knowledge and ignorance, power and powerlessness... For G.* **Leibniz** ([Lz2]-point 8) *the accident is a being whose notion does not include everything that can be attributed to the subject to whom one attributes this notion.*

[470] **Munk** *Salomon (Głogów, POL 1803-1867 Paris), Jew. Specialist in medieval Jewish-Arabic literature. Translator, curator at the imperial library in Paris (oriental division).*

Appendix 2.

DNA (Deoxyribonucleic Acid) holding the fundamental genetic material serves as the blueprint for life. The DNA results of the association of two very long polymeric chains, that wound each other, constituting a regular double helix. Each chain is a regular assembly of nucleotides (10 per turn of the helix) whose sugars are linked by a phosphate group. There are four main nucleotides, each is characterized by a different purine base (Adenine = A, and Guanine = G) or pyrimidine base (Thymine = T, and Cytosine = C). The coupling between the two helices can only be done if A-T and G-C pairings are satisfied.

Gene: the sequence of DNA, is the fundamental unit of heredity. It plays a crucial role in determining specific traits in living organisms (humans, animals, and plants) and can be classified into two main types:

- coding genes containing the instructions for producing a specific protein. Proteins are the main functional molecules in cells and play essential roles in cell structure, function, and regulation. Through a process named: "*translation*," DNA is transcribed into messenger RNA (mRNA), which is translated into a chain of amino acids to form the specific protein.

- non-coding genes don't directly code for a protein. Their roles are increasingly studied and understood; particularly the regulatory genes, which intervene in cell development, and various biological processes.

Chromosomes contain the DNA, a human has 22 pairs of chromosomes called "*autosomes*" and one pair of chromosomes, known as "*gonosomes*" which differentiates males (XY) from females (2X).

Human genome or **genetic material**. In individuals, we refer to the genotype, which encompasses the complete set of genes they possess. The observable characteristics, encompassing morphology, physiology, and behavior are collectively known as the phenotype. Remarkably, the human genome contains approximately 25,000 genes, contributing to the diversity of life as we know it.

International Genome Human Project: A Scientific Triumph of Collaboration and Discovery. The idea of this project, which was based on the full human genetic information, previously estimated at 150,000 genes, but revealed only 25,000, was launched in the late 1980s. At the time, it was not unanimously supported by the scientific community. It was James D. **Watson** who, at the end of the 1980s, succeeded in convincing the American Senate to embark on this adventure and finance the project. Rightly appointed to head the project, he recruited the best teams (20) in the world, representing 6 nations (Germany, England, China, France, Japan, and the USA), i.e. over 20,000 researchers. In 1996, the teams were reorganized, and they all agreed that all the researchers in the world would have the opportunity to work on the project. This open access had allowed the private group *"Celera,"* to raise funds from pharmaceutical companies including PerkinElmer, and to declare that they could complete the work well before **Collins'** team. In return, they would sell the results of the work. The U.S. Senate, comparing *"Venter's yacht to Collins's motorcycle,"* was about to suspend the appropriations arguing that the public group was too slow and bureaucratic. Nevertheless, he granted Collins et al. an extension... An ideological battle ensued. After 18 months of hard work 24/7, Collins et al. had collected 90% of the information. Meanwhile, Venter et al. used Collin's team's results without communicating theirs. In 2000, a secret meeting was held between Collins and Venter, who agreed that President Clinton should be informed of their results in a joint meeting. Over the next 3 years, Collins et al. completed the work. The final result was announced in April 2004, fifty years after Watson and Crick publicized the DNA double helix structure.

Chapter 3.

Unraveling the Mysteries of the Universe

Not enough experiments have been made yet to be able to determine exactly the laws according to which this universal spirit acts.

Isaac Newton

"It seems to me that when confronted with the marvels of life and the universe, one must ask why and not just how .... I find a need for God in the universe and in my own life."

Arthur Schawlow[471]

3.1. Warning.

"I can understand that it is possible to look at the Earth and be an atheist, but I cannot understand that one can look up at night at the sky and say that there is no God" [Al].

Abraham Lincoln[472]

We have seen in Chapter 2, that our scientific knowledge has evolved step by step, discovering constants, laws, fundamental forces, and fields, ... thanks to the unceasing work of researchers: experimenters and theorists who improve, day after day, their tools that become more and more sophisticated because the task is

---

[471] **Schawlow A.** *(ref.351, p.106), 1981 **Nobel** Prize in Physics made this statement during an alumni banquet, according to Henry **Margenau***.*
***Margenau** Henry (Bielefeld, DE 1901–1997 Hamden US-CT), Ph.D. in 1929 at Yale University wrote extensively on the connections between religions and science. In his thesis, he argued that Quantum Mechanics plays an important role in our brain, and our being in general; what is, according to him, responsible for an indeterminism comparable to chance. But, having been created free of our decisions the choice is imputable to us. It is difficult to classify religiously; I would opt for deist. Author of various books, the quote attributed above to Schawlow is taken from* "Cosmos, Bios Theos: Scientists Reflect on Science, God and the Origins of the Universe, Life, and Homo sapiens." *Open Court Publishing Company, 1992.*
[472] **Lincoln** Abraham *(USA: Hardin County, KY 1809-1865 Washington, DC (assassinated)) 16th President of the USA, helped abolish slavery.*

immense. These efforts were not in vain, they allowed the description of limited domains of events; and, on the condition of neglecting certain effects or making approximations, scientists were able to arrive at the Big Bang Theory that we are going to synthesize to allow the reader of books - trying to interpret Genesis, according to scientific discoveries[473] - to have some basis.

But no theory is immutable, it can be questioned, thus Newtonian mechanics despite its invaluable contribution welcomed the theory of relativity and quantum mechanics, which have broadened the scope of our knowledge, but which are still insufficient. To go further it is necessary not only to make the two mechanics (relativistic and quantum) compatible, but that one of the two, or even both, must be profoundly reworked [CC1911] given the increase observed over the last 20 years in the universe's expansion acceleration and the various questions raised. We could remark that not only theories evolve but also their role. The definition proposed in the 19th century by the *Catholic* Academician, Pierre **Duhem**,[474] one of the founders of Modern Physics and Chemistry:

> "*A physical theory is not an explanation. It is a system of mathematical propositions, deduced from a small number of principles, which aim to represent as simply, as completely, and as accurately as possible, a set of experimental laws,*"[Du]

is still valid but in the 20th century the positions are reversed: the experiment verifies the validity of the models proposed by theoretical physicists, and cosmologists, which require important means.

Adherence to the advanced theories on the origins of the universe requires from the non-specialist in cosmology, astronomy, and theoretical physics, a greater faith in science and scientists, or at least comparable to that which the average believer has in the Gospels, the testimonies of the apostles and the holy martyrs who gave their lives, sources of his faith in God. Because the picture of the universe that scientists give us, calls for very elaborate mathematical tools that few physicists master and, for notions of n-dimensional spaces that it is unthinkable to represent. This representation of the world that scientists give us is what M. **Planck** (ref.255-p.78) called: the "*world of physics or the image of the world proposed by physics*"[475] created by the human mind, whose function is to

---

[473] *Which I will not attempt to do and would advise against because Genesis is not a scientific book, it is often allegorical.*

[474] **Duhem** Pierre *(FRA: Paris 1861-1916 Cabrespine). Theoretical physicist, a specialist in thermodynamics (Gibbs-Duhem equation), chemist, one of the pioneers of irreversible processes, and co-founder of Modern Physics-Chemistry. Philosopher of science, He entered the ENS in 1882 and taught physics at the Faculty of Sciences of Lille, then in Bordeaux. A member of the Académie des Sciences from 1900.* Catholic, *follower of Blaise* **Pascal**.

[475] Max Planck defines ([Pl1] Pp. 5 & 6) the 3 worlds:
  - *the* world of the senses "that we discover better every day thanks to the laws that scientists establish;"

lead us to a piece of knowledge, as complete and simple as possible, of the objective real world; and does not leave any place to the world of the senses that each of us can interpret in its way and of which B. **Pascal** (ref.170-p.48) said:

> *"The testimony of the senses cannot be misleading; it is on it that scientific experimentation is based. But one can make, in the judgment, bad use of the sensory data"* [Pa-Note 9].

Let us recall that the Principle of Indeterminacy (ref.316, p.93) and Gödel's theorem (ref.184, p.56) support the thesis according to which the world of the senses is different from the real world - because on the one hand, any observation of the object, has an action on it and thus disturbs its state, which is verified experimentally; and that on the other hand, to be able to study it objectively, we would have to place ourselves outside of it. however, E. **Schrödinger** (309-p.91), considering that the mind and the world are made up of the same elements stated:

> *"The world is giving to me only once, not one existing and one perceived"* ([Sc2], p. 127).

He leaves the door open for others' prospects:

> *"I consider it possible that our models can be modified in such a fashion that they do not exhibit at any moment properties that cannot in principle be observed simultaneously"* ([Sc2], p. 125).

3.2. Our universe history through the Big Bang theory.

In the intricate tapestry of our cosmic history, we embark on a journey that reveals the sublime qualities of our universe: its majestic order, breathtaking beauty, ethical imperative, and the profound sense of worship it inspires. As John **Polkinghorne** (ref.571-p.182) eloquently expresses:

> *"...the priceless characteristics of the world, its order, its beauty, its ethical imperative, its worship experience, reflect the personal traits of the Creator who is rational, joyful, good and holy."* ([Po], p. 56).

- *the* real world: "enjoying an autonomous existence, independent of man, and which we can never apprehend directly, but only through our perceptions, thanks to the signs it communicates to us;"
- *and the* "world of physics or the image of the world proposed by physics" *created by the human mind.* "Its function is to lead to a knowledge as complete as possible of the real world" *on the one hand,* "to lead to a description as simple as possible of the world of the senses" *on the other hand.*

In the spirit of Sir Isaac **Newton** who humbly declared in a letter to R. **Hooke**, in **1965**:

"If I have seen further, it is by standing upon the shoulders of giants."

Our narrative begins not in that year, but in 1930 when the visionary astronomer and *Catholic canon* Georges **Lemaître** (ref.342, p.102) proposed the theory of the *"primitive atom."* This concept, which was the precursor of the Big Bang theory, evolved thanks to the insights of Georges **Gamow** (ref.349, p.105) and Ralph **Alpher** (ref.350, p.105), culminating in the publication of the first model in **1948**. The BBC unveiled it to the public' in 1949, scientifically narrating the evolution of the Universe from Planck's Time ($\tau_p = 10^{-43}$s), approximately 13.80 billion ($13.80.10^9$) years ago on our time scale.[476] The Big Bang theory invites us to ponder the cosmic decision that triggered this monumental explosion. Who or what initiated it? The question *"What was God doing then?"* echoes through time, yet, much like St Augustine's playful response to the timelessness of God:

"God was preparing hell for those who ask this question,"

we find ourselves unable to answer definitively. As we traverse the enigmatic span of 10^{-43}s before Planck's Time (τ_p), a moment void of scientific insight, we encounter a canvas waiting for the brushstroke of favorable conditions. The *"how"* and *"why"* of these conditions elude scientific understanding, leaving us with tantalizing questions that persist to this day.

In Chapter IV, we delve into the different stages proposed by the Big Bang theory, exploring the lanes scientists navigate to refine this model and address the inquiries that naturally arise. A spotlight shines on

Alan (ref.391, p.116) **Guth**'s 1979 inflationary theory, offering a lens that aligns with the earliest moments of cosmic evolution. Additionally, we venture into Roger **Penrose**'s Conformal Cyclic Cosmology (C.C.C.), an intellectual odyssey honored by the **2020** Nobel Prize in Physics.

According to the Big Bang theory, at the moment of the cosmic explosion,

---

[476] *According to the results published in 2013, following the processing of data collected by the Planck satellite, our Universe would be aged $(13,799 + /- 0,021)10^9$ years. The initial moment should not be assimilated to the instant 0. To convince ourselves of this we can refer to the Celsius scale of temperatures t that we adopt for reasons of convenience but whose current 0 has not always corresponded to the temperature of solidification of water. Scientists use the absolute temperature T expressed in Kelvin (K), such as $T(K) = t(°C) + 273.15$. We see that in this scale our 0 °C is 273.15 K. Everything is relative. 0K is the lowest temperature that can exist. Everything would be stationary there, the energy would be zero. In 2015 researchers at MIT succeeded in cooling a molecule to 500nK ($500.10^{-9}K$), or 500 billionths of a degree above absolute 0!*

The universe existed as an infinitesimally small, [477] infinitely dense [478] *"singularity"*[479] cloaked in an unimaginable high temperature;[480] an awe-inspiring prelude to the symphony that unfolded afterward, beyond anything we can imagine.

Expansion of the universe after the Planck time

- 10^{-43}s ATB (After The Bang = after the explosion), the universe is an incandescent magma extremely dense whose temperature T is about 10^{32}K. The space swells, its size increasing the temperature decreases. All elementary particles: quarks,[481] leptons, bosons, photons, ... have a zero mass.

- 10^{-35}s ATB, T $\approx 10^{28}$K. Up to this temperature according to the Grand Unification Theory (G.U.T.) the 4 forces are one, and the symmetry is perfect. Below, according to Glasgow and al. the three nongravitational forces (cf. 1865 **Maxwell**'s work refs. 234-p71 & ref. 236-p.71) are one.

- 10^{-11}s ATB, T $\approx 10^{15}$K. There is a cosmological phase transition: the Higgs field condenses to reach a non-zero value (until there, its value oscillates around 0 like the other fields) and fills the whole universe. It results that, all the masses of particles such as leptons, and quarks, ... take a non-zero value, hence a breaking of the symmetry, and the electromagnetic and weak nuclear forces mainly.

---

[477] *The sphere containing all the initial energy had a diameter equal to the Planck wavelength (= 10^{-35}m), that is to say, 10^{21} (= 1,000 billion of billions) times smaller than the diameter of the nucleus of the hydrogen atom (equal to 10^{-14}m). We can deduce that its energy density was infinite. Below this distance, not only does the notion of space as we conceive it no longer make sense, but quantum mechanics is currently in conflict with the theory of general relativity. M.* **Planck** *was the first to point out that, when the gravitational field is too large (almost infinite), our theories do not allow us to describe the behavior of atoms and light. This was the case before Planck's time. We, therefore, speak of a* "**Planck Wall**" *that cannot be crossed until theorists succeed in unifying the theories of gravity and quantum into a* "quantum theory of gravity."

[478] *According to the Buddhist astrophysicist* Trinh Xuan **Thuan** *([Tn], p. 42-43) the initial density of the universe must be set with a precision of 10^{-60}. In other words, if we change the initial density by one number after sixty zeros, the universe would be sterile: neither you nor I would be here to discuss it.*

[479] Singularity: *concept proposed by Sir R.* **Penrose** *(ref.368-p.109), adopted by S.* **Hawking** *during his thesis defense and confirmed by an article co-signed with R.* **Penrose** *in 1970 (p.112 & 113) before S. Hawking himself discarded it in 1975 (p.114) to justify his theory of black hole radiation.*

[480] *The higher the temperature, the greater the thermal agitation energy of the particles (Brownian motion), so their propensity to repel each other is all the greater. During the shocks, they produce pairs of particles and antiparticles that annihilate when they meet, their energy is found in the form of radiation.*

[481] Quarks *(constituents of protons and neutrons) and leptons (spin [angular momentum] = +/- ½, including electrons, muons, tauons, neutrinos ... and their antiparticles) are the basic materials of our universe. Lepton comes from the Greek meaning* "light mass."

- 10^{-7}s ATB, the universe is so dense and hot that we have a plasma[482] of elementary constituents. Very quickly the nucleons (protons and neutrons) will be formed.

- 1s ATB, T $\approx 10^{10}$K. The universe then contains photons that, are stopped by light particles such as electrons (e⁻), positrons (e⁺), neutrinos (v), antineutrinos (v*), protons (p = $^1p_1$), neutrons (n = $^1n_0$) with which they coexist. *Cf. Soddy 1910- p.87 to understand the representation of a nuclei ^AX_Z.*

As the universe continues to expand[483] the temperature T continues to decrease, and so does the production of antiparticles; electrons, and positrons which annihilate for a large part, producing photons. Some electrons will remain. S. **Hawking** points out [Ha2] that, if at that time the expansion rate had been even 1.10^{-16} times smaller than it was,[484] the universe would have collapsed on itself and we can guess what would happen next.

- 100s ATB, T $\approx 10^9$K (this is the temperature at the heart of stars). The energy of thermal agitation having decreased, the strong nuclear force is sufficient to retain the protons and neutrons that will combine to produce deuterium nuclei[485] D (= $^2H_1$, hydrogen said *"heavy"* because its nucleus contains not only 1 proton but also 1 neutron). The deuterium nuclei D will combine with protons and neutrons to form helium nuclei ($^4He_2$) consisting of 2 neutrons (n = $^1n_0$) and 2 protons ($^1p_1$), as well as heavier nuclei such as lithium ($^6Li_3$) involving 3 protons and 3 neutrons and beryllium ($^7Be_4$) consisting of 4 protons and 3 neutrons. This is what is commonly called *"primordial nucleosynthesis."* On the other hand, cosmic radiation, which is important will destroy, by *"spallation,"*[486] the heavier nuclei like Carbon ^{12}C, and Oxygen ^{16}O which are broken giving lighter nuclei.

- A **few minutes ATB** ($\approx$ 3'): the universe was filled with a hot gas - composed of light atoms: 75% of hydrogen gas (made of 2 hydrogen nuclei $^1H_1$), 23% of helium ($^4He_2$), deuterium (D = $^2H_1$) and Lithium ($^6Li_3$) - practically uniformly distributed, in a state of very low entropy,

---

[482] Plasma: *4th state of matter (the first 3 being: solid, liquid, and gaseous) in which the atoms disintegrate to make room for the particles that compose them.*

[483] *We will see in Chapter 4 how* **Guth**'s *inflationary model can explain this inflation which occurs despite the gravitational force that should prevent it.*

[484] 1.10^{-16} = 1/10 million billion! *When scientists are astonished by the precision of the conditions that have been necessary to get to where we are, we don't have to hesitate, we can reject the theory of randomness in all certainty ([Po], p. 56).*

[485] *At these temperatures the electromagnetic force responsible for the association nucleus-electron is insufficient to counterbalance the energy of the particles.*

[486] Spallation: *the nuclei have such a high energy that they explode, giving various particles.*

at the origin of the order observed today. This makes B. **Greene** [GR] say that:

"*...the present order is a cosmological relic.*"

- A **few hours ATB**, the production of helium and heavier elements has ceased.

- **300.000 years ATB.** T = few 10^3K, the electromagnetic forces are sufficiently important to overcome the energy of agitation of electrons and nuclei and allow the formation of electrically neutral atoms. The expansion of the Universe will continue for more than 700,000 years, in an almost uniform manner, and the temperature will continue to decrease isotropically.

- **380.000 years ATB.** The radiation can propagate freely in the universe (which lights up), it will dilute in space, whose temperature continues to decrease. The universe ceasing to be opaque our most powerful telescopes will be able to receive the visible radiation emitted from this date. The structures (galaxies, clusters of galaxies) we now observe were yet in their germ.

- **1.10^9 (1 billion) years ATB**, the temperature is 10K. In this homogeneous space existed some inhomogeneities that quantum mechanics justifies from the inhomogeneities prevailing in the initial subatomic world (before the Bang). [487] These inhomogeneities, extending during the expansion form in various points of space nebulae of gas turning on themselves which, under the effect of the attractive force of gravitation, will condense, and form galaxies made up of stars[488]

---

[487] *Scientists demonstrate that, when the star is created, the entropy decreases, because a certain order of matter is obtained. This explanation follows from the work on inflationary theory, see [Ge] for more details.*

[488] *A star is formed from a gaseous cloud (mainly hydrogen). When its mass is large enough, it contracts under the effect of gravitational forces. Then, the shocks between particles produce a temperature rise, inducing nuclear fusions (ref.461-p.137) during which 4 hydrogen atoms give rise to 1 helium atom. This operation releases, not only, luminous radiation* justifying the luminosity of the star; but also heat, producing an increase in pressure of the gas inside the star; enough to prevent it from contracting more. As long as the pressure of the gas is, counterbalanced by gravitational forces, it is stable. But, as the hydrogen reserves are depleted, the temperature decreases and so, does the gas pressure. The compression due to the gravitational forces then produces a contraction, responsible for new heating of the heart of the star inducing the fusion of helium nuclei into heavier elements: C, O_2, Fe, ... This results in an expansion of the star which can give rise to a "supergiant" star: mass greater than 20 Solar Masses. On the other hand, as it becomes denser, its temperature increases. With time (counted in billions of years) the heart of the star, known as the first generation star, contracts again, exhausting its fuel, and finally collapses, the outer layers breaking up into elements that are released into the interstellar medium, giving rise to a supernova (ref.490).*

around which are formed planets[489] and their derivatives... According to the second law of thermodynamics, the energy released by the stars contributes to the universe's entropy increase. This is how the Milky Way began to form, from diffuse gas provided by **supernovas.**[490] Without these massive stars made of carbon among others, Earth's life would not exist. The residual diffuse gas has a mass of about 10^9 MS (1 billion MS, where MS = Solar Mass), growing by merging with other galaxies. At the heart of the Milky Way is Sagittarius A*, a supermassive black hole (BH) that has grown in mass by accretion of gas and dust evolving in its immediate environment and then by absorption into the *"Event Horizon."* During the accretion the matter swirls around the BH, compresses under the effect of gravity, and heats up, reaching thousands of billions of degrees. Hence the emission of radiation.

Astrophysicist Caleb **Scharf**[491] claimed [S&V]:

"The Milky Way harbors a black hole that is just active enough to moderate the formation of 2nd generation stars, but not active enough to smother it completely. A happy medium without which life could probably not have developed."

The team of astrophysicists led by Chuck J. **Hailey** (Columbia University-NY) identified [Hai] & [Ko] a dozen black holes (the size of a star), smaller than Sagittarius A*, orbiting around it within a radius of three light years. Given that these are the

---

* The emission of radiation *in the universe increases its entropy more than it decreases the one inside the star, satisfying the 2nd Principle of Thermodynamics (any irreversible transformation sees its entropy increase).*

[489] Planet *comes from the Greek* "wandering star." *It must satisfy, since 2006, 3 criteria:* **1.** *to orbit a star,* **2.** *to be practically spherical, and* **3.** *to have eliminated all debris other than moons, around its orbit. The dust around a star that has not aggregated to it gathers under the effect of the gravitational force in planetesimals, sets of some 10km in diameter, larger sets: oligarchs resulting from the attraction of other planetesimals. The planetesimals located near the Sun, have an energy (function of the temperature) more important, and during shocks between them, some can form planets. Debris resulting from these shocks aggregates and forms the* "moons" *of these planets.*

[490] Supernova. *As a star exhausts its fuel, it reaches a critical moment when it cools and collapses on itself, leading to a supernova. For the Sun, this process will occur in approximately 4.5 billion years. Under the intense gravitational forces, the core of the star contracts so dramatically that a black hole forms, while the outer layers are expelled into space as debris. This event emits intense radiation. Smaller stars, however, conclude their life cycles differently, becoming either a white dwarf (a cold, stable star) or a neutron star, described by Landau using the principle of indeterminacy.*

[491] **Scharf** Caleb (England alive). *Defended his Ph.D. at the University of Cambridge, England, and Post-doctorate at NASA. Astronomer, British cosmologist, who is particularly interested in exoplanets and astrobiology, and director of the* "Columbia Astrobiology Center" *in NY. Author of many popular books.*

brightest[492] they have observed, they predict the existence of 300 to 1,000 others. A. **Generozov**'s team, from the same university, has proposed from this work a model predicting a range between 10,000 and 40,000! The total mass of our galaxy would currently be of the order of 10^{12} MS if we included the dark mass (DM).

- At **7 billion years (ATB)** the speed of expansion of the universe stops decreasing and starts to increase. The universe is substantially homogeneous, the stars and planets "*dissolve*" in the whole. Under the effect of the gravitational force, the galaxies group themselves in swarms of galaxies.

- **8.2 to 8.7.10⁹ (8.7 billion) years ATB** that is to say 5.0 billion years ago, our galaxy the "*Milky Way*," came to maturity.

- **9.2 10⁹ years:**[493] Following the collapse of a nebula of gas and dust from the debris[494] of supernovas a star - among the 200 billion that the Milky Way contains - the protosun (flat disc in rotation) is formed under the effect of the gravitational force. When it begins to shine it is surrounded by a "*protoplanetary*" disk of small bodies rotating around it. Shocks, and mergers, will follow one another for a few tens of millions of years and, under the influence of the gravitational force, always it, the debris that revolves around it will collide with him and merge giving the Sun that we know, whose diameter is then equal to 1,391,400 km, its mass is approximately 2.10^{30} kg or about 99.9% of the mass of the solar system. It is located at about 30,000 light-years from the center of our galaxy and rotates around it with a period comprised between 225 and 250 million years.

The remaining: 0.1%, of the mass of the Solar System is distributed among its planets, their 175 moons, and debris, forming around the Sun, the Solar

---

[492] Brightest *is the term used to say that they are the easiest to detect. Indeed, black holes do not shine, unlike a star. On the other hand, they are very often surrounded by stars from which they tear out (we use the verb "to accredit") matter thanks to their strong gravitational constant, which increases their entropy. This increase is compensated by heat emission and therefore by radiation that these researchers have succeeded in separating from others.*

[493] Since 2005 *(cf. Science et Vie N° 1055-08/2005.) a hypothesis adopted by the scientific community has been elaborated at the*
*Nice Côte d'Azur Observatory (06 Fr) by 4 researchers: Alessandro **Morbidelli**, an Italian working at the Nice Côte d'Azur Observatory received the Brazilian Rodney **Gomes**, the Greek Kleomenis **Tsiganis** and the American Harold **Levison** to work on the origin of the Sun.*

[494] This debris *contains, among other things, heavy atoms (2%) such as carbon (C), without which life on Earth could not have existed. Few known galaxies contain stars massive enough to produce heavy atoms.*

System.[495] Starting from the Sun we have respectively: Mercury, Venus, the Earth (located at 149.6 million km from the Sun = 1 AU),[496] Mars, Jupiter, Saturn, Uranus, and Neptune (at 30 AU), It is interesting to note that the planets are perfectly aligned, that their rotations around the Sun are done for 6 of them in an anti-clockwise direction called "*direct direction.*" Only Uranus and Venus turn on themselves in the clockwise direction. This particularity of Uranus and Venus is attributed to a collision with a body of very large mass during their formation. Specialists think that a 9th planet exists in our solar system.[497] It would be located at about 1.000 AU from the Sun and its mass would be 10 times that of Earth. Among these planets, we distinguish:

- **the giant planets**: Jupiter (70% of the 0.1%) at 5 AU from the Sun, Saturn (20% of the 0.1%) at 9.5 AU. The two are gaseous, made up mainly of hydrogen and helium; Uranus (4% of the 0.1%) at 19 AU and Neptune (5% of the 0.1%) at 30 AU, are called "*ice planets*" and made up essentially of water, methane, ammonia...

- **The rocky or telluric planets**, the closest to the Sun are respectively: Mercury, Venus, Earth,[498] and Mars which will appear 50 million years later[499] and will share the remaining 1% or 1/1000 of the mass of our solar system. They result from collisions and mergers of massive protoplanetary bodies. The fusion of two of these bodies gives birth to one of these planets which, under the combined effects of heat, gravitational force, and rotation will take an almost spherical shape. The debris of matter resulting from the collision will form after a few hundred years the satellites of our planets: the moons. We counted 175 of them.

---

[495] *The study of* "young" stars *such as HL Tauri, which formed only 100,000 years ago, should provide important information on the formation of the solar system. It is 450 light-years from Earth.*

[496] Astronomical Unit (AU): *distance between the Sun and the Earth, about 150 million km.*

[497] *Remember that Pluto, which was long considered as the 9th planet of the solar system, was demoted in 2006 to the rank of a* dwarf planet *because in recent decades many planets of similar sizes or even larger (eg. Eris) have been discovered in the solar system. Which led in 2006 to a new definition of planets.*

[498] *The* Earth *has the highest density (5.5 = 4 times that of the Sun) of the planets in the solar system. Although 80% of the surface of the Earth is covered with water (Oceans, Seas, ...), it is considered as a rocky planet because the average thickness of water is equal to only 1/1000th of its diameter.*

[499] *The most recent* dating *of the Earth: 4.54 10^9 years was made in 1953 by* Clair **Patterson** *(USA: Mitchellville, IA 1922-1995 Sea Ranch, CA), a geochemist specializing in the dating of meteorites, using the method called "Uranium (U) - Lead (Pb)." Based on the fact that among the 26 known isotopes of U, the two most important are: $^{238}U_{92}$ (<99.28%, $T_{1/2}$ = 4.5.10^9 years), and $^{235}U_{92}$ (<0.72%, $T_{1/2}$ = 0.7.10^9 years). The decay of $^{238}U_{92}$ results in stable $^{206}Pb_{82}$, while that of $^{238}U_{92}$ results in stable $^{207}Pb_{82}$. The measurements of the ratios $^{238}U_{92}$ /$^{206}Pb_{82}$ and $^{235}U_{92}$ /$^{207}Pb_{82}$ allow to know with precision the date of birth of the studied samples. Following the Apollo 15 mission (1971) researchers from Stony Brook University, NY dated pieces of lunar rock: 4.15 +/- 0.25 billion years, approximately the age of the Earth.*

It's noteworthy that our Moon stabilizes the Earth's axis of rotation itself, preventing chaotic variations in temperature and pressure. It is, in fact, one of the two gravitational forces, the other being that of the Sun, which, combined with the Coriolis force (inertial force linked to the rotation of the Earth on itself), are at the origin of the tides. These latter contribute (according to NASA geodynamicist Bruce Gordon Bills) to the circulation of heat from tropical waters to the poles.

At the very beginning of its formation, the Earth was very hot and without an atmosphere, intense seismic and volcanic activity reigned. The gases contained in the rock evaporated and, as the Earth cooled, some of the gases condensed in the atmosphere forming a large cloudy layer that filtered U-V rays and allowed the appearance of life that could not be human as long as the cloudy layer did not contain oxygen. The appearance of human life is a fact that appears for the moment as unique in our Solar System, but ......

Remark. The distinction between gaseous and telluric is due to their distance from the Sun. The more one moves away from it, the more the temperature decreases. Below the "*line of the ices: T = -120°C*" that is to say approximately 700.10^6km of the Sun = 4,7 AU), the light elements: Hydrogen (H_2), Helium (He), ... evaporate and remain in the gaseous state around their planet, within which persist mainly the metals, the rocks from where their name of telluric planets. Beyond the "*ice line*": H_2, He, carbon (C), nitrogen (N_2), and oxygen (O_2) come together, forming water H_2O, methane (CH_4), and ammonia (NH_3), which solidify. The mass of these planets is sufficiently important to attract gases that will solidify and orbit these "*gaseous planets.*"

If the first **exoplanets** (planets outside the Solar System) were discovered about 30 years ago, hundreds of them are currently being discovered every year. Xinyu **Dai** and Eduardo **Guerra** (Oklahoma University) have detected the existence of exoplanets in a galaxy located 3.8 billion light-years away, located between the black hole and us (see CERNCOURIER Vol. 58 N° 3 April 2018 p.21). The James Webb Space Telescope (JWST) jointly realized by NASA, ESA, and CSA (Canadian Space Agency) that an "*Ariane*" launcher put into orbit, at 1.5 million km from Earth, in January 2022 (it was launched on December 25, 2021) could lift the veil on the existence - still hypothetical, although mathematically confirmed - of exoplanets called "*Ocean Planets*" [FZ]. Intermediate in size between telluric and gaseous planets, they could exist around Trappist 1, an ultracold red dwarf star located at about 40 light-years from Earth. Their particularity: they have a large mantle of liquid water enveloping a core of icy dust. JWST is the largest optical telescope in space with high-sensitivity and high-resolution instruments. The public could see the first image on 11v July 2022.

-    **10.10^9 (10 billion) years**[500] ATB or **$3.8.10^9$ (10 billion)** years ago, the

---

[500] **Helmi** et al. *Nature 2018 Vol. 563 p. 85.*

Milky Way merges with **Gaia Enceladus**[501] a smaller galaxy (1/4 of its mass).

Let's now take the countdown. There are:

- **3.8. 10^9 years**. Jupiter and Saturn modify their orbits, inducing a destabilization of the solar system: Neptune and Uranus move away from the Sun, and a rain of asteroids falls on all the other planets. It will be necessary to wait another 600 million years for the solar system to settle down and start to stabilize[502] so that the orbits of the planets initially circular become elliptic. The Solar System was then half as large as it is today and was surrounded by a thick belt of "*planetesimals*" (a few hundred km in diameter).

All the conditions were met for elements with atomic mass greater than 7 to form inside the Sun.

Merlin **COLE**, Dept of Particle Physics, Univ. of Geneva reports [CE1707] that Sarah **Sadavoy** and Steven **Stahler,** her director of thesis, and professor at (the University of California, Berkeley, CA)[503] have shown that the sun, belonging to a second or third-generation of stars, had at its birth a twin which disappeared for an unknown reason, at a date that is not known.

- **3,7.10^9 years: The Step of the Life or passage from the molecule to the cell**. Scientists believe that the first bacteria, as well as the DNA responsible for their reproduction, appeared. But where did these organisms come from? The hypotheses of Panspermia[504] which have been revived, by prominent scientists but, without clear evidence are evocated in Chapter V.3.

---

[501] Gaia, *a satellite put into orbit in 2013 by the European Space Agency, aims to map part of our galaxy and measure the characteristics of more than a billion stars (galaxies, stars, asteroids, ...). It allowed the discovery that more than 50% of the stars of the Milky Way were formed about 3 billion years ago.*

[502] Apollo 11 *(takeoff on July 16, 1969 - landing on July 20) learned to us, thanks to the instruments deposited by N. **Armstrong**, 1st man to have walked on the Moon, and B. **Aldrin**, that the Moon is still moving away from the Earth at the speed of 4cm/year.*

[503] Merlin **COLE**, *Dept of Particle Physics, Univ. of Geneva reports that Sarah **Sadavoy** (Max Planck Institute for Astronomy, Heidelberg) preparing for a Ph.D. in astronomy, and Steven **Stahler** (Professor of Astronomy at the University of California at Berkeley), having studied the formation of stars in the Perseus cluster, thanks to the Very Large Array (VLA) radio telescopes of New Mexico and the James Clerk Maxwell Telescope (JCMT) of Hawaii, have deduced that all the stars in the Perseus cluster have a very high speed.*

[504] Panspermia *is the addition of "pan" meaning "universal" to "Sperma" the seed. It is a biological paradigm according to which life on Earth has been seeded from space. With the emergence of new diseases such as the Coronavirus, this paradigm is experiencing a new boom. What's more, like Hoyle some scientists believe they can invalidate the Darwinian theory of evolution.*

For Dan **Vasilescu** [Vu2] "*... to place at the origin the aminoacyl adenylate, which is, at the same time a nucleic acid and a protein in miniature, is an attractive hypothesis.*" In his "*Molecular and Thermodynamic Approach to the Origin of Life*" he says that some questions "*must be answered; namely:*

- Do primitive biological objects exist today that could be at the inert-living border?

- Is contemporary science able to describe the functioning of the biological object?"

While waiting for experimental proofs, we can reflect on the proposal of Jesuit paleontologist **Teilhard de Chardin**[505] for whom, within the framework of the permanent evolution of the universe, the appearance of Life results from the maturation of a unique germination point in the history of the universe, comparable to the appearance of atoms 10 billion years earlier (cf. [Te2] p. 89):

"Without a long maturation, no profound change can occur in Nature. On the other hand, given such a period, it is inevitable that something new will occur."

Prokaryotes, single-cell organisms without a nucleus, such as bacteria, are produced. These microscopic and innumerable cells are, as he says [Te2]: natural

---

[505] **Teilhard de Chardin** Pierre (Orcines (63) Fr 1881-1955 New-York, USA). Since 1920 is one of the greatest paleontologists and geologists. Priest, Jesuit, his scientific field of action covers geology (China), mammalian paleontology, and human prehistory - his team demonstrated that the fossil man of "Choukoutien" or "Sinanthropus" (-500,000 years), which he helped to discover, near Beijing, is a Homo Faber (works iron). Philosophy and Theology (at the origin of the Law of Complexity/Consciousness, including *the cosmic Christ (end of time))* *give all its meaning to evolutionary biology. As a scientist, he was convinced that the theory of evolution opened a new vision of the universe fully compatible with the Christian dogma, of which he was the guarantor as a priest and Jesuit, which before Vatican II was not well seen by the Church. He graduated in philosophy in 1887, became a priest in 1911, joined the paleontology laboratory of the Museum of Natural History in Paris in 1912, was a stretcher-bearer during the Great War (he refused to become a chaplain in order to remain close to the men), and published* "La vie cosmique" *in 1916. When the war was over, he obtained three licenses (geology, botany and zoology). In 1922, he became a lecturer and defended his thesis:* "The mammals of the French Lower Eocene and their statigraphy." *In 1923 he discovered China where the Roman curia kept him away, and stayed there from 1926 to 1946. In 1931 he discovered the evidence that the sinanthrope was a homo faber. He was awarded the Mendel Medal in 1937 at the Philadelphia Congress for his research in human paleontology. He was sent to New York in 1951 by his hierarchy, participated in excavations in Africa and put forward the hypothesis of the African origin of man. His conceptions of original sin and human thought, among others, made him an heterodox until* Joseph **Ratzinger** *rehabilitated him. Elected* "Pope **Benedict XVI**" he will *confirm* [Ben]*, and praise his conception of the* **noosphere***: a thin layer surrounding the Earth and grouping together the consciousnesses of humanity, a place where Man and the world are one with God [Te&Te2]. He wrote several essays. Officer of the Legion of Honor, Holder of the Military Medal ...Nominated in 1950 at the French Académie des sciences.*

grains of Life as the atom is the natural grain of inorganic Matter.

Remark. On August 5, 2012, the *"Mars Science Laboratory"* landed on Mars. It has been launched on the 26th of November 2011. On its board *"Curiosity,"* the largest, and most sophisticated rover ever sent to Mars. It was designed to answer a crucial question: did Mars ever have the necessary environmental conditions to be habitable? The 6/08/22 the astrophysicists say *"Yes, it was 3.6.10⁹ years ago."* Curiosity scientific tools found minerals and evidence. Since then, the situation has deteriorated. We will return to its possible consequences in § V.3. p.217.

- **3.10^9 years**: the appearance of the blue algae which will produce oxygen by photosynthesis. Essential element.

- **$2,1.10^9$ years**: the appearance of the eukaryotes: unicellular and multicellular organisms having a nucleus.

- **700.10^6 years** ($10^6 = 1$ million): formation of soft-bodied beings, without shell or skeleton (worms, jellyfish, ...). The forms of life of low entropy are present.

- **445.10^6 years** (the Ordovician): The **first mass extinction** was brutal and short. There were no living beings on Earth because oxygen was rare. 70% of marine life was destroyed. Followed by the appearance on Earth of the first plants (moss, lichens), scorpions, and mites, followed 20.10^6 years later by fish, invertebrates (no backbone), and insects. The cause is a supernova, releasing very intense radiation.

- **365.10^6 years** (the Devonian): The **second longer mass extinction**, about 3 million years. The vegetation on the Earth was then luxuriant, the humidity and the heat that reigned there favored the presence of insects. 75% of the marine species disappeared. The cause is Global warming, leading to augmentation in volcanic eruptions, including a major volcanic explosion under Siberia.

- **250.10^6 years** (Permian): The **third mass extinction**, the most important: 95% of animal species disappeared, and the continents reunited. Cause: The impact of meteorites leading to an important volcanic explosion thus the emission of CO_2 and Sulphur, responsible for acid rain destroying all the fauna.

- **230.10^6 years**: appearance of crocodiles, dinosaurs, and birds,

- **225.10^6 years**: appearance of mammals, vertebrate animals with lungs, feeding their babies with milk produced by the mammary glands of females. They are mostly viviparous (gestation of newborns is done inside the mother).

- **200.10^6 years** (Triassic-Jurassic), **fourth mass extinction:** the fauna is destroyed, and only some reptiles, mammals, and dinosaurs remain. The cause is controversial.

-    **65.10⁶ years** (end of the Cretaceous, beginning of the Tertiary), **fifth mass extinction**:[506] 70% of animal species disappeared, including the dinosaurs. The <u>cause</u> is an intense volcanic activity that would have lasted 1.5 million years, and the impact of a huge meteorite (15 km in diameter) on Yucatan.

-    **60/55.10⁶ years** before our era [Hn1], there were **Primates**[507] on the whole of Earth. The organization chart (fig.3 – p.158) shows the phylogeny from the Primates to the Homos Sapiens families from which modern man emerged, as demonstrated by recent work carried out by paleo geneticists. Some minority currents may contest this representation, but it is currently the most recognized one. The dates depend upon the scientific discoveries, some of which may be modified. Chronologically we know that:

-    **55.10⁶ years** ago they divided into **Strepsirrhinians**[508] with a raised nose and **Haplorrhines** characterized by a flat nose, which gave rise:

-    **45.10⁶ years ago**, the **Tarsiiformes**[509] in North America, Europe, and Asia, which formed a single continental block at the time (today they are found only on a few islands in Southeast Asia); and another branch constituted by the **Simiiformes**[510] or **Anthropoids,** our distant ancestors in Africa, who:

-    **34.10⁶ years** before our era, gave way in the "*old world*" to the

---

[506] *At the end of the chapter, we made a modelization intended to predict when the 6th mass extinction will occur.*

[507] Primates: *A subset of the placental mammals (the fetus is wrapped in a placenta) to which we belong. They have two prehensile hands with five fingers and five toes on each foot, three kinds of teeth, two pectoral teats, and two well-developed cerebral hemispheres (see Robert Vol. 7 p. 764). The oldest fossils today, dating back 70.10⁶ years, were discovered in the USA in Montan* [Hn1]. *The Tree of Life (Figure 9) shows us how primates differentiated from other species. It should be noted that 40.10⁶ years ago, resulting from ice ages, and various meteorite impacts, primates were only found in sub-Saharan regions.*

[508] Strepsirrhinians: *A subset of Primates including lemurs, lorises, galagos, aye-aye.*

[509] Tarsiiformes: *tree-dwelling, jumping, nocturnal primates, such as tarsiers with large, bulging eyes, large feet, long tails, and ears pointing in all directions. Do not leave their tree. Measure 15 cm without a tail and weigh about 120g. Emit and receive ultrasound. Protected species.*

[510] Anthropoids or Simiiformes: *large monkeys resembling humans, such as Orangutans, Chimpanzees, Gibbons, Gorillas, etc., without tails, with large brains, relying on the backs of the phalanges of the hands and the soles of the feet to walk. Depending on the climatic conditions some of them will survive for thousands or even millions of years.*

Catarrhinians,[511] anthropoids without tails, with noses pointing downwards, and then, **9 million years later,** in the *"new world,"* [512] to the **Platyrrhines** (wide noses), their long-tailed whorl.

According to Professor Jean-Jacques **Hublin,** [513] (cf. [Hn1]), *the first anthropoids to appear in the New World may have been carried on plant rafts, crushed trees, etc., because the gap between Africa and South America was smaller than nowadays.*

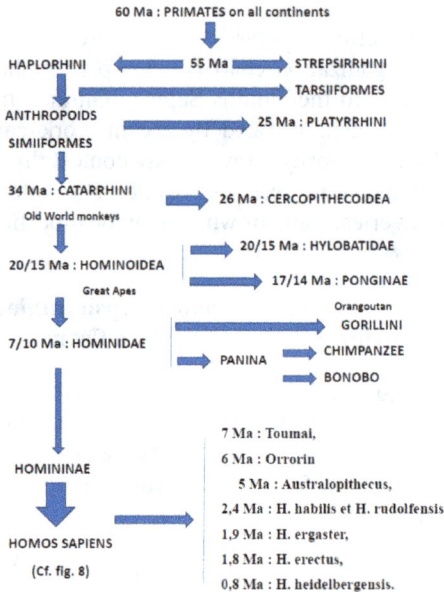

60 Ma : PRIMATES on all continents

HAPLORHINI   55 Ma   STREPSIRRHINI

TARSIIFORMES

ANTHROPOIDS   25 Ma : PLATYRRHINI

SIMIIFORMES

34 Ma : CATARRHINI   26 Ma : CERCOPITHECOIDEA
Old World monkeys

20/15 Ma : HYLOBATIDAE

20/15 Ma : HOMINOIDEA   17/14 Ma : PONGINAE
Great Apes

Orangoutan
GORILLINI

7/10 Ma : HOMINIDAE   CHIMPANZEE
PANINA

BONOBO

7 Ma : Toumaï,

6 Ma : Orrorin

HOMININAE   5 Ma : Australopithecus,

2,4 Ma : H. habilis et H. rudolfensis

1,9 Ma : H. ergaster,

HOMOS SAPIENS   1,8 Ma : H. erectus,

(Cf. fig. 8)   0,8 Ma : H. heidelbergensis.

Figure 3. Evolution of Primates.

Always, according to Hublin [Hn2], it is at the end of the Oligocene; approximately:

---

[511] Catarrhinians, *anthropoids that appeared in Africa at the beginning of the Oligocene (-33.9 to -23.03. 10^6 y); only reached their maximum size and diversification at the end of the Pliocene (-5.3 to -2.6 Ma) in Africa and India, in tropical and subtropical areas.*

[512] According to J.-J. **Hublin** (cf. [Hn1]), *the first anthropoids to appear in the New World may have been carried on plant rafts, crushed trees, etc., because the gap between Africa and South America was smaller than nowadays.*

[513] **Hublin** Jean-Jacques (Mostaganem, Algeria Fr 1953) *Paleo anthropologist. Director of the* "Evolution of Man" *department of the Max Planck Institute for Evolutionary Anthropology, Leipzig, DEU. Professor at the Collège de France since 2014, he always refused during interviews to mix science and religion. For me he is Agnostic. Ph.D. in 1978 at P. & M. Curie, Paris, Fr; Doctorat ès-Sciences in 1991 in paleoanthropology at Bordeaux, Fr. Joined the CNRS in 1981 and remained there until 1992 when he left to teach in the USA at Berkeley, Harvard & Stanford. 1991 professor in Bordeaux, at the same time teaching in Leiden and Leipzig. Author of numerous articles and books.*

- **23.10^6** years before our era, that happened the division of the **Catarrhines** into: **Cercopithecoidea**, long-tailed monkeys: baboons, macaques, ... and: **Hominidae,** [514] a name given by **Teilhard de Chardin,** meaning *"looks like a man."* They are also called *"Great Apes."* There are 3 species:

a) **Hylobatid or Gibbons, 15 to 20.10^6 years before our era.** Small monkeys with long arms that facilitate brachiation (swinging from one branch to another at high speed (80 km/h)). They were arboreal, monogamous..., and found mainly in Asia, South-East India, and Indonesia. They can be divided into 4 species according to their number of chromosomes (38, 44, 50, or 52).

b) **Pongidae**[515] **17 to 14.10^6 years ago,** including the **Orangutans.** This subgroup is controversial, some place it in the Hominidae subgroup.

c) **Hominids 10 to 7.10^6 years before our era,** which are divided into 3 groups:

- **Gorillines** (Gorillas),

- **Panines** (Chimpanzees), close to man. **1.10^6 years ago,** they gave rise to two sub-groups: the socalled *"common"* chimpanzees and the *"pygmy"* chimpanzees or Bonobos,

- *and the* **Hominins,** *a genus* **Homo** comprising several fossil groups that disappeared **40.10^3 years** before our era, absorbed by **Homo Sapiens** or **Modern Man,** who came from the Far East but had African ancestry. According to J.-J. **Hublin** [Hn4] Homo Sapiens migrated, following environmental changes.[516]

The oldest known fossil of the **hominin genus** is "Toumaï" who lived **7.10^6** years ago. His skull was discovered in 2001 by Michel[517] **Brunet's** team in Djourab (Chadian Sahara). Later, other bones belonging to it were found.

---

[514] Hominidae (-16.10^{6}a): *Family of primates descended from the Great Apes such as the Australopithecines, which gave rise to the genus Homo, of which only Homo Sapiens remains, to which we will return later.*

[515] Pongo or Pongidae: *an anthropoid with a large brain, to which the bonobos, chimpanzees, and gorillas were associated until recent discoveries.*

[516] Over the last 500,000 years, *dramatic climatic changes have occurred. Africa experienced a period of "green Sahara," river systems, and wide lakes, up to the size of Germany. The same is true for Saudi Arabia. The first traces of Homos outside Africa appear in the Levant between 177 and 199 thousand years ago.*

[517] **Brunet** Michel *(Magné, Fr 1940) Chair of Paleontology at the Collège de France, Paris, Paleoanthropologist. Ph.D. in paleontology in 1966 in Paris. State doctorate in 1975, Professor in 1989. Participated in and directed numerous excavations.*

Orrorin, a million years younger, was discovered by Brigitte **Senut,**[518] Martin **Pickford,**[519] et al. in 2000 in the Tugen Hills region of Kenya. It has both human and simian features.

But, it is **Australopithecus**[520] that is the precursor of the genus **Homo**. It is a bipedal hunter, making tools, standing upright, and knowing how to climb trees. According to J.J. **Hublin**, it may have appeared **6.10⁶** 6.10^6 years ago, but the oldest fossils discovered to date do not go beyond **5 million years.** J.J. **Hublin** distinguishes two African australopithecines (figs. 4 & 5):

a) A **Robust Australopithecus**, about 1.60 m tall, weighing between 50 and 70 kg, with a cranial cavity of 500 cm³. It appeared in southern and then eastern Africa. Examples: **Olduvai** (Tanzania), **Zinjanthrope** discovered in **1959**... **P. Boisei** (fig. 5) and **P. Robustus** were part of the **Paranthropus** family, which is still under debate.

Fig. 4. Australopithèque afarensis, Lucy.[521]	Fig. 5. Australopithèque boisei.	Fig. 6. Homo habilis
3,2 My. Hadar, Ethiopie	1,7 My Olduvai Tanzanie	1,5 My Gorges Olduavi, Tanzanie

---

[518] **Sénut** Brigitte *(Paris, Fr 1954). Paleo primatologist and paleoanthropologist, Professor at the National Museum of Natural History in Paris. She participated in important discoveries including that of* "Orrorin," *and is laureate of numerous awards.*

[519] **Pickford** Martin *(Trowbridge, ENG 1943). Palaeo primatologist, involved in important discoveries including that of* "Orrorin." *Held positions at the Museum National d'Histoire Naturelle in Paris, Professor in Mainz...*

[520] Australopithecus (Australopithecus africanus): *Hominian, able to carve stone and make fire. The oldest fossil dates from 4.2.10⁶ (millions of)* **years** *and was discovered in East Africa. It became extinct about 2.10⁶ years ago (see figs. 2 & 3). Its largest (1.25 - 1.5m) and most robust descendant, Paranthropus, was discovered in the Transvaal in 1948. According to some it lived between 2.7 and 1.10⁶ years and, includes three very muscular species:* P. **Boisel** *(brain between 500 and 550 cm³),* P. **Aethiopicus,** *and* P. **Robustus.** We have classified (figs. 5 - 10) the fossils according to the years they were discovered.

[521] Lucy *(some scientists believe she was a man). Discovered in 1974, it is the first complete* Australopithecus *fossil. She was 25 years old, had a small brain, comparable to that of a chimpanzee, was 1.1m tall, and weighed 30kg, associating walking with climbing ability.*

b) Gracile Australopithecus, from which **Homo Sapiens** descended, in East Africa (1m10 - 1m30, 30 - 40 kg, skull 400 to 550 cm³). Vegetarian at first, then hunter of small animals and occasional scavenger. **Lucy,** discovered in 1974 in Ethiopia, is an example. They became extinct 1 million years ago.

In **1995** a team led by Ron **Clark** [522] discovered in Sibelberg Cave (Sterkfontein) near Johannesburg, South Africa, fossilized members of *"Little Foot"* belonging to the *"Australopithecus Prometheus"* lineage identified in 2015, distinct from **Africanus**.

Today, all specialists agree that **Homo habilis**[523] (fig. 6), which appeared in Eastern and Southern Africa **2.4.10⁶ years** ago, with a skull larger than 600 cm³ (due to its food, including meat rich in calories), is the first link in our chain, despite the discovery of **Homo rudolfensis** in 1972 near Lake Turkana between Kenya and Ethiopia. Although they lived at the same time the latter differs from Homo habilis in having a larger cranial volume (650 - 750 cm³). **Homo ergaster**[524] succeeded Homo Habilis **1.9.10⁶ years ago**. He may have been the ancestor of **Homo Erectus**[525] (fig. 7) which first appeared in Asia **1.8.10⁶ years** ago. Not all paleo anthropologists agree on this separation of the two genera, although it was adopted by the majority in 1975. **Homo erectus**, in the broadest sense of the term, includes two large families: the **"Pithecanthropus,"** [526] discovered in Java, which has a face with an *"Afro-Asian"* shape, and the **"Sinanthropus"**[527] discovered in China, whose existence was initially estimated to be between **780 and 300,000 years** before our era. All these species are fossilized.

---

[522] **Clark** Ron *(South Africa 1944) participated in numerous excavations.*

[523] Homo habilis (skilled man), fig. 6. *Had a cranial cavity of between 600 and 700 cm3, measured between 1.3 and 1.5m and weighed between 35 and 45kg. Lived in East Africa between (-2.4 and -1.6 Ma), cohabited for some time with* Australopithecus, Paranthropes *and other* homos *cf. figure 8.*

[524] Homo ergaster: *skull 750 to 1,050 cm³, height 1.55 to 1.70m, weight 50 to 65 kg.*

[525] Homo Erectus, fig. 7, *walked straight. He appeared 1.8.10⁶ years ago in Asia and disappeared about 300.10³ years ago. Measurements: skull 800 to 1,200 cm³, height 1.50 to 1.65 m, weight 45 to 55 kg. It is currently known that at least one variant has been found in Pretoria (South Africa). It was bipedal, yet moved easily through the trees.*

[526] Pithecanthropus: *Former name for* "Homo Erectus." *Has features like the present-day* Anthropoid Apes *and* Homo Sapiens. *Skull (800 to 1,000 cm³), smaller than modern man (1,400 cm³), stood upright and knew how to light a fire. That of* Java *or* Solo *man (name of the river near which fossils have been found) is dated between -250 and -150 thousand years, that of* Terra Amata, Nice (06, Fr): *-400 thousand years, that of* Vallonnet, Cap Martin (06, Fr): *- 900 thousand years.*

[527] Sinanthropus or Peking man *(780 to 300 kans).* P. **Teilhard de Chardin** *contributed to its discovery and naming:* Homo Faber *man, as a species capable of technique, of making tools. According to* **A. Leroi-Gourhan** *("Le Geste et la Parole" t. 2, p. 266 Albin Michel 1964), until 1963 it was believed that the* Sinanthrope *was the oldest Homo in China. But that year the* Lanthian man *(1.63 to 1.15 million years old) was discovered and in 1989 the* Yunxian man *(936 million years old).*

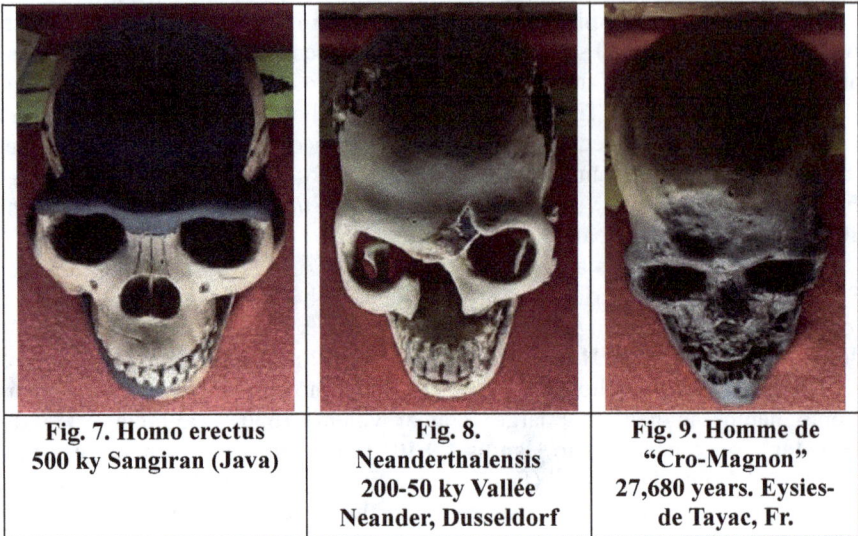

Fig. 7. Homo erectus 500 ky Sangiran (Java)	Fig. 8. Neanderthalensis 200-50 ky Vallée Neander, Dusseldorf	Fig. 9. Homme de "Cro-Magnon" 27,680 years. Eysies-de Tayac, Fr.

In 1907 "**Homo Heidelberg**" was discovered near the German town of the same name. J.J. **Hublin** [Hn3] and others claim that he had an African origin although living in Europe from **800 to 250,000 years** before our era, making, according to some - including one of the pioneers of paleogenomics (evolutionary genetics), the Christian S. **Pääbo** (ref 442, p.131), 2022 Nobel Prize in Physiology or Medicine - the ancestor of **Neanderthal**[528] (fig. 9), **Nesher Ramla**,[529] and **Denisova**.[530] He had a skull between 1,000 and 1,300 cm³, was 1.65 m tall, and hunted large game such as rhinoceros. Some researchers distinguish it from the

---

[528] Neanderthal man: *(-400 to -40 thousand years B.C.). The first fossil was discovered in 1856 in the Neander Valley near Dusseldorf in the Rhineland, Germany, a species closest to modern man. Maximum height: 1.65m. Larger brain (about 1.5dm3). He was a vegetarian. A complete skull fossil, dated between 50 and 20 ka BC, was discovered at* "la Ferrassie," *near La Bugue in the Dordogne, France. Since, other fossils have been discovered in the Middle East (Iraq, Syria, Lebanon, Israel), Central Asia: Uzbekistan, Siberia, Caucasus ... European and Asian Modern men have 1 to 4% of Neanderthal DNA.*

[529] Nesher Ramla *(400,000 to 130,000 years ago), discovered in Israel in the 2010s, and confirmed by the work of* Israel **Hershkovitz** *et al. in 2021. Fossils found in the Misliya cave, Mount Carmel, Israel attest to the arrival of Homo sapiens in the Levant 200,000 years before our era.*

[530] Denisova Man *was identified in 2009* by Svante **Pääbo** (ref.442, p.132), from a phalanx found in the Denisova cave in Altai, Siberia. It is thought to have lived from 200,000 to 40,000 years before our era. Mitochondrial DNA analyses carried out by S. **Pääbo**, revealed that it was genetically distinct from Neanderthals and modern humans (see [SP] p. 30); and also, that gene transfer had occurred from Denisova to Homo sapiens out of Africa around 70,000 years ago. In Southeast Asia individuals carry up to 6% of Denisova DNA, the specific gene EPAS1 is common among Tibetans.

genus "**Homo Rhodesiensis**" [531] which lived at the same time in Northern Rhodesia.

- - **400,000 years before our era**, the "**Neanderthal Man**" appeared in the Middle East and in Europe (fig. 8). The first fossil discovered in Europe was in the Neander Valley, near Düsseldorf, in the Rhineland, Germany. It became extinct about 40,000 years before our era. In terms of DNA, the similarity with modern man - initially called '**Cro-Magnon**[532] **man**' (fig. 10), then "**Homo Sapiens**"[533] (Modern man) - is striking. Neanderthals have 23 pairs of chromosomes (like us), whereas great apes have 24.

Since the work of S. **Pääbo**, a *Christian*, it is well established that modern humans (Homo Sapiens) have 99.5% identical genomes to Neanderthal. Yet, individuals can have 1 to 4% Neanderthal DNA.

Now, we know [SP] that 6 species of hominins coexisted (cf. fig. 11), and mated, for a time, with the lineage that appeared 300 thousand years ago to end up with Homo sapiens (modern man), which is known to have been more prolific than them, absorbing definitively the other 6, by the theory of evolution. Some Y.E.C. see it as a confirmation of the genesis, which is not the case of S. Pääbo for whom religion and science are two different realms.

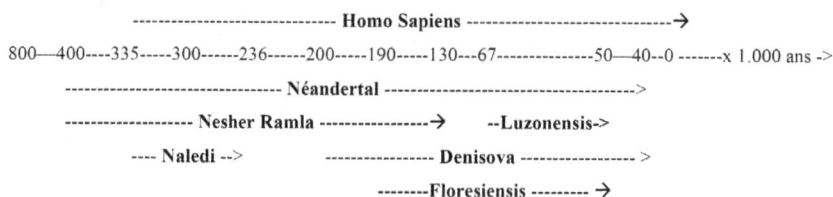

```
-------------------------------- Homo Sapiens -----------------------------→
800—400----335-----300------236------200-----190-----130---67--------------50—40--0 -------x 1.000 ans ->
-------------------------------- Néandertal ------------------------------->
-------------------- Nesher Ramla ----------------→      --Luzonensis->
---- Naledi -->            ---------------- Denisova ----------------- >
                           -------Floresiensis --------- →
```

Figure 10. Coexistence of the 7 Homos Sapiens genera. (From [SP]).

The six "cousins" genera of **Homo sapiens***:* **Neanderthal, Nesher Ramla,**

---

[531] Rhodesian Man *or Kabwe Man, the name of the Iron and Zinc mine (now in Zambia) where a skull (1,230 cm³) 110,000 years old, and various bones, were discovered in 1921.*

[532] Cro-Magnon Man *(fig. 9) is a Homo sapiens discovered in 1868 in the Cro-Magnon shelter in Les Eysies (Périgord, Fr) by* Louis **Lartet**. *Traces of his presence in Europe 35,000 years ago have been discovered. Size between 1.72 and 2m. Maximum lifespan: 35 years. He had a brain as large as ours. It is attributed to an important artistic production: Chauvet, Lascaux, Altamira, etc. Mastered hunting with the manufacture of throwing weapons. Contemporary of the Neanderthals and witnessed their disappearance*

[533] Homo Sapiens *is no longer called* "Homo sapiens sapiens" *since paleontologists have agreed to differentiate him from the Neanderthals and the 5 other species of Homo with whom he cohabited and perhaps even mated for nearly 260,000 years before taking over.*

Naledi,[534] **Denisova, Floriensis,**[535] and **Luzonensi**[536] are not, according to S. **Pääbo,**[SP] ancestors of Homo Sapiens; nevertheless, a transfer of genes took place from these now extinct hominins to Homo sapiens after the migration out of Africa, about 70,000 years ago. This affects the way our immune system reacts to infections.

Professor Alysson[537] **Muori's** neurobiology team also presented evidence in February **2021** that we share 99.5% of genes and 1- 4% of DNA with Neanderthals and demonstrated that by introducing the Neanderthal variant of the NOVA1 gene into the brain tissue of developing Homo Sapiens, their neuronal connections are altered. For A. Muori, Neanderthals and Denisovans have one version of the NOVA1 gene while Homo Sapiens has another that allowed them to survive and develop their brains.

- **50,000 years before our era,** after several incursions into Europe, Homo Sapiens, the modern man, spread all over the planet. The team led by Professor J.J. **Hublin** reported, in 2020 [Hn5], the discovery of 43,000-year-old fossils in the Bacho Kiro cave in Bulgaria. Not only, they adapted them to the environment, by exploiting it more efficiently, but they also knew how to create more extensive networks, going beyond the family and even the neighborhood.

After this state of affairs, which has evolved considerably thanks to the work of Svante **Pääbo**'s team, let's take a step back to see how ideas have evolved about Homo sapiens, modern man.

According to the *Jesuit* paleontologist **Teilhard de Chardin**: with Homo Sapiens "*it is a definitively liberated Thought which makes explosion... appears Art, naturalistic art makes us enter of full foot in the conscience of the disappeared beings*" (cf. [Te2] p. 218). Likewise, if we want to understand how from the great apes endowed with Intelligence, we arrived at the Modern Man, who alone is

---

[534] H. Naledi *(335 to 236,000 years old), was discovered in 2015 in the Rising Star Caves near Johannesburg in South Africa.*

[535] H. Floriensis *(190 to 50,000 years old), was discovered in 2003 on the Indonesian island of Flores. He is small: 1 to 1.10 m, with a skull of around 420 cm³ but an evolved brain.* Pygmies currently live there and have Neanderthal *and* Denisovian *but not Florian ancestry. Their small size may be the result of insufficient food resources.*

[536] H. Luzonensis or Luzon Man *(60 to 40,000 years ago), was discovered in 2007 in the Philippine Callao Cave on the island of Luzon. The skeleton of a rhinoceros, a species extinct on the island for 100,000 years, was discovered there; confirming a human presence on the island dating back 700,000 years, as indicated by the fossils found at the neighboring site of Kalinka.*

[537] **Muori** Alysson *(Sao Paulo, BRA ~1974). Professor of Neurobiology at the University of California, San Diego School of Medicine since 2008. Obtained a BSc in Biology in 1995 from the State University of Campinas in Sao Paulo, Brazil; then defended a Ph.D. in Genetics in 2001 at the University of Sao Paulo followed by a postdoc in 2002 at the Salk Institute (La Jolla, San Diego). He is a laureate of numerous awards.*

endowed with Consciousness of himself,[538] it is necessary to look for the reason in the evolutionary perfection of the brain because *"In the living, the brain is a measure of the consciousness."* The *Russian Orthodox Christian* geneticist Theodosius **Dobzhansky**[539] will corroborate this assertion two decades later [Ec2]:

"Self-awareness is a fundamental character of the human species. In evolution, this character represents a novelty; the biological species from which humanity is descended possess only rudiments or even a total absence of self-awareness. Awareness, however, has brought with it sinister companions, fear, anxiety, and the realization of one's death. The consciousness of one's death is a handicap for man."

The idea that this consciousness could disappear with the body displeases E. **Schrödinger**, a *Christian*, for whom each body has its consciousness, accepting however the Hindu concept of Brahman according to which: each consciousness is a manifestation of a universal consciousness or Soul spread in the Universe, assimilated to God [Sc2]. According to him ([Sc2] p. 89) the idea of the plurality is proper to the Western religions. For the philosopher Karl **Popper:**[540]

"The appearance of any consciousness, capable of self-reflection is in truth one of the greatest miracles."

This is what distinguishes the human person from superior primates, such as the chimpanzee, endowed with consciousness but not with self-consciousness.

Renowned *Christian* biologist and geneticist Charles **Birch**[541] introduces a captivating perspective on human knowledge. He posits that our understanding

---

[538] Self-awareness *designates all kinds of the psyche, leading to Reflection, a characteristic of the Modern Man. For* P. **Teilhard de Chardin**, *speaking about Consciousness ([Te2] p. 148), the Spirit [which he assimilates to Energy] is essentially the power of synthesis and organization. Consciousness characterizes a high-level experience involving the memory of past events and conscious anticipation of future events.*

[539] **Dobzhansky** Theodosius *(Nemyriv, UKR 1900-1975 San Jacinto, US-CA). Biologist, geneticist, Russian Orthodox Christian.*

[540] **Popper** Karl *(Vienna, AUT 1902-1994 London, ENG). One of the most influential philosophers of the 20th century. Raised in the Protestant religion by Jewish parents, he was attracted to Marxism during his university studies and turned away from it to adopt the ideas of social liberalism. For him, metaphysical theories are irrefutable by experience.*

[541] **Birch** Charles *(AUS: Melbourne1918-2009 Sydney). Biologist, geneticist, and* Anglican theologian. *He attended Scotch College in Melbourne, was awarded a Bachelor of Agricultural Sciences from the University of Melbourne in 1939, and went to the Wait Institute in Adelaïde where he was awarded 1948 a Doctorate Degree in Sciences on "Insect Ecology." Meantime, he moved to Senior Lecturer at the University of Sydney in 1948 he became a Professor of Zoology in 1960 and Professor of Biology from 1963 to 1983 the year he was elected Emeritus professor until his death. Raised as an Anglican he moved away from the too-rigid evangelical outlook of Anglican Melbourne Church when he encountered the* "Student Christian Movement" *in* Adelaïde, "a liberal organization that had an alternative way of looking at Christianity, science, and the world" *which offered a way to get rid of the feeling of sinful, and wickedness and*

comprises two facets: the externally observable realm, scrutinized by science, and the internal landscape of consciousness, a profound realm that has evolved since time immemorial. This internal domain encompasses a sophisticated interplay of memories from the past and a conscious anticipation of future events.

Immersed in the inspiring writings of Alfred **Whitehead**[542] a prominent mathematician and one of the greatest metaphysicists of the 20th century, the father of the *"process thought."*[543] Birch rejects the dualistic viewpoints of figures like Sir John **Eccles** (ref.164, p.76), the 1963 **Nobel** in Physiology or Medicine, and Theodosian **Dobzanski**. Instead, he advocates for a more all-encompassing role for consciousness in the universe, transcending the boundaries of living organisms. In contrast to the materialistic leanings of his peers, Birch aligns himself with panexperientialism[Bc], a doctrine positing that mentality and

---

discover Christian forgiveness. He integrated the Uniting Church of Australia (a union of Methodist, Presbyterian, and Congregational churches); in 1977 he became vice-chairman of the Church and Science Committee. Author of scientific publications and books on Science, Religion, and Ecology. 1990 **Templeton** *Prize.*

[542] **Whitehead** Alfred N. *(Ramsgate, ENG 1861-1947 Cambridge, US-MA), was a distinguished mathematician, philosopher, and metaphysicist. Renowned as the co-author with his student Bertrand Russell, of* "Principia Mathematica" *a seminal work on the foundations of mathematics, he also ventured into proposing a gravity theory as a logically viable alternative to Einstein's general theory of relativity,... He began his academic journey at Trinity College, Cambridge in 1880 where he focused on mathematics, earning his B.A. in 1884. Subsequently, he was elected a fellow and enrolled in teaching mathematics and physics until 1910. Then he taught mathematics and mechanics at University College London. Later, he held the position of professor of mathematics at Imperial College, London until 1924. His intellectual journey led him to Harvard University, in Cambridge, US-MA, where he was elected professor of philosophy.*

Whitehead's recognition extended to prestigious institutions, with his election to the American Academy of Arts and Sciences in 1925 and the American Philosophical Society in 1926. In the late 1910s and early 1920s, he shifted his focus to the philosophy of sciences, culminating in his influential work, "Process and Reality" *(1929). This book stands as a cornerstone of the development of* "process philosophy," *particularly emphasizing a philosophy of organisms. His intellectual contributions spanned the realms of mathematics, physics, and philosophy, leaving an indelible mark on each discipline.*

Despite being raised and married in the Anglican Methodist tradition, his personal beliefs evolved. I identify him as a Theist or a Deist, the distinction being elusive.

[543] Process thought. *It proposes that reality is fundamentally characterized by processes or events rather than fixed, static substances (like fundamental particles (cf. Chapter II). He emphasizes the dynamic nature of existence, suggesting that everything is in a state of becoming and is interconnected through a network of processes.* **Whitehead** *introduced the concept of* "actual occasions" *or* "actual entities," *which are the basic units of reality. These occasions are not enduring substances but momentary* "events" *that constitute the flow of experience. Process thought rejects the Cartesian dualism of matter and mind and instead sees the world as a continuing process of becoming, with everything interconnected and interrelated. It is opposed to the* **Darwinian** *evolution theory.*

This process philosophy has found applications in various fields. Besides philosophy and metaphysics searching to understand God, His nature, His relation to the World, it has been applied to educational philosophy, environment, and quantum physics.

physicality are two aspects of a singular phenomenon. According to Birch, matter and mind have evolved harmoniously since the dawn of time, fostering a soulful connection akin to **Eccles**'s concept.

Birch's conceptualization of the world diverges from conventional physical substances, favoring the term "*entities*," or "*things*." Rejecting the overly materialistic notion of the "*body*," he delves into the realms of quantum physics and duality (cf. chapter II), coining the term "*events*" to describe these fundamental building blocks. Drawing inspiration from **Schrödinger**'s musing on life (ref.309, p.91) involving quantum principles,[Sc2] Birch associates consciousness not with specific material forms but with "*events*," a perspective reminiscent of **Leibnitz**'s monads (Cf. p53). These elemental "*things*," possess a subjective aspect termed "*experience*" or "*feeling*." Birch discerns between individual entities or events, such as elemental particles, electrons, and nucleons on one hand and compounds such as atoms, and cells on the other hand, which exhibit feeling, and "*aggregates*" like as rocks, chairs, which lack this capacity. His colleague Dobzanski said jokingly that "*his atoms have a brain*."

After having demonstrated that only the branch of the chordates (cf. Tree of Life, Fig. 11) & could lead to Man, **Teilhard de Chardin** demonstrates that from "*layer to layer*"[544] the brain grew. He notes that in mammals "*the brain is more voluminous and more folded than in any other group of vertebrates*" (cf. the interpretation of A. **Muori** on the previous page). At the end of the Pliocene[545] (-2.58 My), the mammalian nappe is steady, which presages a change of state that he calls a "*step*"[546] because evolution has not stopped:

> "*...it is only with blows of chances that the Life proceeds; but with blows of chances recognized and seized, - that is to say psychically selected,*"

There is, according to him, somewhere in the Biosphere, a point of singularity that is being prepared.[547] If among the animals of the Pliocene many seem to arrive at their term, the **Primates** on the contrary, having forms similar to the Man, constitute "*a set of fans or whorls interlocked*" more advised than the others; they were able "*to rise by successive spurts until the borders of the Intelligence.*" Among these, he retains the **Catarrhinians** (cf. p. 159). Thought appears with the consciousness of oneself, we tend towards a hominization of the species.

---

[544] *Large periods (of the order of 80 Ma) during which life evolves and* "Time erases each weak line in the drawings of Life" *(cf. [Te2] p. 130).*

[545] Pliocene: *5.332 ± 0.005 to 2.588 ± 0.005 million, succeeded the Miocene. 5.3 million years ago the Strait of Gibraltar opens again.*

[546] *To picture* the step *he takes the example of a change of state in physics at a given point. When water is heated at a pressure of 1 atmosphere, it gradually changes from a liquid to a gas at 100 °C. During all the time that water changes state the temperature remains equal to 100°C: we have a step.*

[547] *Using a different language from that of* Ch. **Darwin**, *while expressing almost the same thing,* P. **Teilhard de Chardin** *thinks that* "Without a long period of maturation, no profound change can take place in Nature. On the other hand, given such a period, it is fatal that something new should occur": *the jumping of a step.*

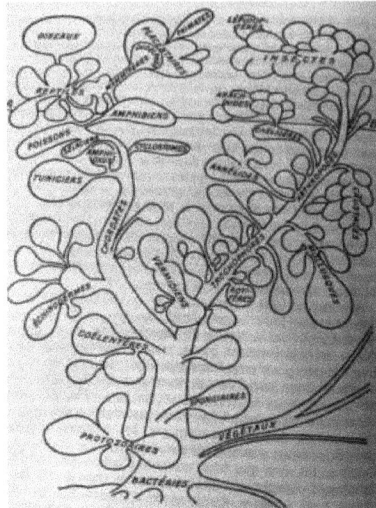

Fig. 11. Tree of life ([Te2] p. 146)

When did this happen? Difficult to say according to him, because:

"When, in all domains, a new thing begins to dawn around us, we do not distinguish it - for the good reason that we would have to see in the future, it's blossoming to notice it at its beginnings. And when this same thing has grown, we turn around to find the germ and the first sketches of it, it is these first stages in their turn which are hidden, destroyed, or forgotten" *([Te2] p.129).*

While **Teilhard de Chardin** affirms that Self-awareness is the prerogative of Man, making him capable of abstraction, logic, making choices, and innovating, which distinguishes him from mammals, Sir John **Eccles, Nobel** Prize of Physiology, deist, demonstrates it. Adhering to Sir Karl **Popper**'s "*3 Universe theory*"[548] they oppose the materialist theories, which the latter calls "*promising materialist theories*" (asserting that all our actions depend solely on the brain),

---

[548] K. Popper's Three Universe Theory *[Ec2]*. *The brain (superior primate or homo sapiens') is made up of two interacting universes:* the material *(1), which he calls* "Energy-Matter" *and the* spiritual *(2), that of the conscious experiences including 2 zones: that of the external senses (the 7 senses, Sight, Hearing, ...) and that of the internal senses (thought, emotion, memory, dream, imagination, intentions and perception).* In Homo sapiens *the universe 2 contains an additional zone:* "the central" *that the materialist philosophers call* "The Psyche," *the psychologists* "The Ego" *and the believers* "The Soul," *interacting with the external and internal senses. Moreover,* Homo sapiens *is endowed with a 3rd universe: that of* knowledge, *which behaves like a data bank, the cultural heritage of humanity, to which only modern man has access, which allows him to evolve from the state of a baby to that of a modern man, unlike the chimpanzee. The conscience of the self increases all the more that the man monopolizes the contents of the universe 3 which is itself in continuous evolution.*

with the theory of "*dualist interactionism*" stemming from the theory of "*common sense*" initiated by R. **Descartes** (ref.152, p.43). Drawing on the work of neurologists such as Dr. Hans **Kornhuber** and Dr. Luder **Decke**, who demonstrate that there are two areas in the brain, one of which is responsible for movement, while the other is responsible for the intention to do so, J. **Eccles** demonstrates, using the example of the baby who, like the chimpanzee, has only a primitive consciousness of self, that we live in two distinct universes: the material, including the brain, and the spiritual. But, what distinguishes "*the human person*" from the superior Primates is that in the first ones "*the evolution of the consciousness of the self is made thanks to bijective interactions between the universe 2 and a third universe: that of the knowledge*" which evolves permanently from where the qualifier of dualism.

Rejecting the "*genetic uniqueness*" - the materialist idea according to which our Self is the result of the infinity of connections existing between the tens of millions of cells of our brain - and the influence of the environment which he considers secondary, the neuroscientist demonstrates, by interpreting the results of experimental works, that each Soul "*is a new Divine creation which is attached to the fetus between conception and birth.*" Not being entirely separated from the body, and influencing the events in progress, the soul intervenes in the synapses of the brain, at a particular point of the cerebral cortex.[549] In 1991 he met the theoretical physicist Friedrich **Beck**[550] at a conference, with whom he developed a quantum model explaining how the Spirit (the Soul) interacts with the brain. If, for a while, this model has been put in default, it comes back in fashion after some updates.

-      **Today,** as we gaze into the vast expanse of the universe, we are captivated by its continuous expansion, a mesmerizing dance of cosmic proportions. It's awe-inspiring to think that the universe's temperature, a mere 2.7K, was prophetically foreseen back in 1930 by visionaries like G. **Lemaitre** (ref.342, p.102) and G. **Gamow** (ref.349, p.105). In 1965 A. **Penzias** and I. **Wilson** validated this theory through groundbreaking experiments (ref.364, p.108).

While the cosmos might seem uniform from a distance, the captivating images captured in 1992 (cf. p.125) by G. **Smoot** and J. **Mather** aboard the satellite COBE revealed intriguing patterns and fluctuations. Thanks to the European Space Agency

---

[549] *Their work finds justification in the behavioral study of Parkinson's patients who, although they have the will to make a movement, have the greatest difficulty in doing so, as well as in the study of patients who have undergone cerebral ablations or of children who have had a delayed development following parental mistreatment.*

[550] **Beck** Friedrich *(Wiesbaden, DEU 1927-2008) Theoretical physicist. Doctoral student of Max* **von Laue**, *he defended his thesis in 1952 on superconductivity. Worked at MIT, then in Munich where he obtained his habilitation in 1958 for his work in nuclear physics. Professor in theoretical physics in 1963 at the University of Darmstadt, then left to teach at the Lawrence Laboratory in Berkeley, US-CA, and in various universities.*

(ESA), since 2013 we've been blessed with a thermal mapping of the heavens, painting an extraordinary tapestry of celestial wonders (p.133).

Behold the fascinating story of cosmic evolution, an astonishing journey from an infinitesimally small and dense ball of energy to the complexification of matter, a process that, ultimately paved the way for the emergence of life and the rise of even more intricate beings, culminating in the existence of humankind.

The Big Bang model, born from the elegant principles of General Relativity, is according to A. **McGrath** (ref.739, p.271):

> *"a major scientific advance which **seems** to reinforce what Christians have always been saying: God created"* [CEA].

But be careful it proves nothing, it presents us with a compelling narrative: a universe created by God at a precise moment. The celestial mechanics, guided by a few fundamental constants, showcase a profound intricacy that cannot be mere chance. Numerous experiments have lent credibility to this remarkable theory, yet, humbly, it remains open to further exploration. Big Bang doesn't explain everything but it raises some unanswered questions that invigorate the minds of scientists, sparking an insatiable quest for clarity and understanding. Among the mysteries that beckon us are:

- **1**. The enigma of the initial singularity, a realm of infinite energy density and space-time deformation, as proposed by the theory of general relativity (cf. chap. 2, Penrose and Hawking's proposals years 1970 & 1975), remains an enduring source of wonder and inspiration, spurring us to unravel the majestic secrets that lie at the very heart of our universe.

- **2**. The end of our universe, if there is an end. The Hinduists think that **Sri Krsna**, creator at a given moment of the universe, will annihilate it when he decides to do so because His supreme Will is in the background of all cosmic events [KD- p.5].

Cosmologists predict it in about ten billion years. According to S. **Hawking** [HA2-p.121] the evolution of the universe as proposed by the Theory of General Relativity begins with the hypothesis of a singularity and should end either by a:

- <u>Big Crunch</u>. The force of attraction taking over, the expansion of galaxies ceases, producing a contraction of the universe. We end up with a singularity characteristic of the collapse of our Universe, and therefore of its end. What makes him, state: *"The universe having begun without man it will end without man."*

- <u>Big Rip</u>.[551] The expansion prevails, the galaxies continue to move away

---

[551] Big Rip: *Cosmological model describing the end of the Universe proposed by the cosmologist* Robert R. **Caldwell***, professor at Dartmouth College since 2005. He is also working on the Cold*

from each other, which requires a hypothetical *"phantom energy,"* producing a tearing of all the structures of the Universe; while inside each galaxy the contraction, due to the decrease of the *"fuel"* of the stars, leads to a collapse of these last ones and the creation of super black holes which will leave the place to radiation.

- **3**. The expansion at the time of the Bang. How could it happen when the force of gravitation - attractive - was preponderant? The Cosmological Inflation theory (chap. IV.1-p.177) provides an answer.

- **4**. The accelerating expansion of the universe observed by Perlmutter and Coll. refs. 443-445, p.132) in 1998. Gravitation, having only acted attractively since the beginning, how can we explain it? With the hypothetical Cold Dark Matter as some cosmologists suggest?

* Intriguing reflections.

1) Unveiling the subatomic realm. As we delve further into our journey, the realm of quantum physics emerges as a pivotal guide, with the Principle of Indeterminacy captivating our curiosity. Within this realm, the luminous legacy of Newtonian physics undergoes a third remarkable extension: the recognition of indeterminism in unstable systems according to **Prigogine** (ref. 387-p.115). It stands alongside the iconic pillars of General Relativity and quantum mechanics, collectively weaving a tapestry of cosmic understanding.

2) The Mystique of Entropy. Before the explosive dawn of the Big Bang, a perplexing enigma emerged: entropy. Following the choreography of statistical physics as orchestrated by Ludwig **Boltzmann**, the universe's initial state should have been one of exuberant entropy, a chaotic symphony of disorder. Yet, in the delicate dance of probabilities, a faint whisper emerges: the possibility of a brief interlude of low entropy, a symphony of order, though infinitesimally rare, isn't extinguished entirely. Across the cosmos, pockets of intricate elegance come into existence: stars, galaxies, and the symphony of life itself, including the intricate symphonies conducted by our brains. Amidst this cosmic ballet, the rise of entropy yields an intriguing paradox. As stars radiate their brightest (ref. 488-p.150), entropy ascends; yet, the birth of complexity in localized corners of space invites entropy to recede, by the tenets of the Second Law of Thermodynamics. This intricate dance culminates not only in a crescendo of disorder but also in a universal cooling, suggesting that all galaxies will collapse in billions of billions of years, in the form of supermassive black holes, each harboring the masse of billions of suns. This is without counting on the genius of the Creator who surely spares us some surprises. Lurking within this cosmic choreography is the unforeseen unpredictability brought forth by the burgeoning dawn of quantum indeterminacy, casting a beguiling allure upon deterministic narratives.

Dark Matter theory, initially proposed by J. **Peebles** *in 1982. Cold, because dark matter moves slowly compared to the velocity of light.*

3) A glimpse into the Cosmic Hourglass. As we cast our eyes upon the epochs of history, the annals of Earth's story reveal five profound mass extinctions, an irrefutable testament to nature's relentless dance. Gazing into the future, the silhouette of a sixth extinction takes shape on the horizon. Armed with the tools of linear regression we chart a course into this enigma. The graph's contours whisper secrets (fig. 12). If the fourth extinction claims its place in this mosaïc of existence, we have a dash of uncertainty of 30 million years, and the sixth's arrival aligns with a tentative harmony, a great symphony conducted over 15 million years. Yet, in a narrative where precision is a rarity, the sixth's echo could have resounded 15 million years ago or await its revelation within the tapestry of the next 45 million years. Amid these cosmic rhythms, a stage is set for our species' terrestrial finale: Stephen **Hawking**'s prediction of humanity's twilight within a thousand years. A resounding testament to the grandeur of the cosmos, a realm brimming with mysteries that defy the boundaries of our imagination.

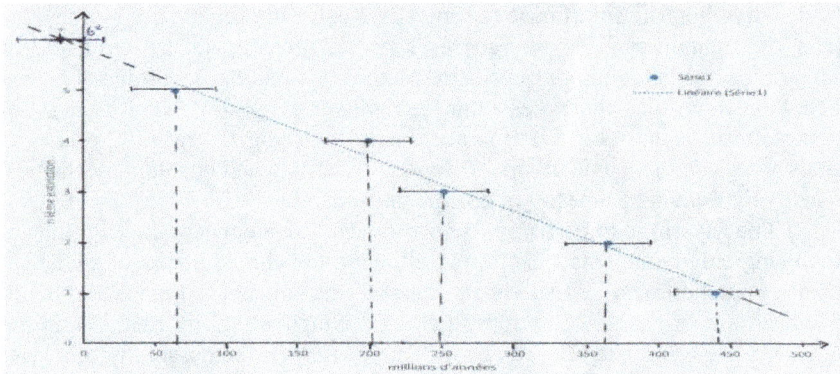

Fig. 12. Representation of mass extinctions as a function of time.

4) Exploring uncharted horizons. In the forthcoming chapter, our voyage through the cosmos continues, a tapestry woven with threads of intrigue and innovation. The Big Bang theory, though illustrious, shares the cosmic stage with other captivating theories that beckon us to embrace uncertainty, to explore the frontiers where Relativistic Physics and Quantum Physics coalesce into a harmonious melody. Possibilities burgeon: the universe's temporal vista isn't confined to mere infinity or isolated inception at singularity's embrace. The canvas of Quantum Physics unfurls a novel narrative, where space-time may be boundless,[552] akin to Earth's sprawling surface. Amidst such contemplation, there lies a realm where no corner is devoid of the laws we hold dear. While uncharted realms may stretch the limits of comprehension, the words of Robert **Boyle** (ref.196-p.59), echoing through time, reverberate:

---

[552] *It is a proposal of S. **Hawking** (but without proof) which deduces*: "It [the universe] would have been neither created nor destroyed. It would only be."

"The true man of science cannot force his way into the secrets of creation without perceiving the finger of God" (cf. [Bo] p.95),

a sentiment echoed in 1981 [553] by the wise pontiff, Pope Jean Paul II, reminding that the pursuit of understanding needs not to eclipse the awe of the divine creation:

"... not to look for what is at the origin of the Big Bang because it is the work of God."

As we embark on this voyage, the luminous beacon of faith remains undiminished, ever poised to illuminate the cosmos' unexplored corners. The quest for answers harmonizes with the hope for revelations, guided by an unwavering belief in the symbiosis of knowledge and spirituality.

I leave it to Ch. **Darwin** to conclude this chapter:

"There is grandeur in this view of life (that he expresses in [Dar]), with its several powers, having been originally breathed by the Creator into a few forms or one; and that, whilst this planet has gone cycling on according to the laws of gravity, from so simple a beginning endless forms most beautiful and most wonderful have been, and are being evolved" (cf. [Dn] p. 123).

Would an atheist have used the word *"Creator"*?

Note. As the mathematical physicist Paul **Davies** [554] a Christian, - who has developed several important contributions to theories concerning black holes, the beginning of the universe, the nature of time, and others - said during an interview with Phillip MacCarthy, Herald correspondent in 1995, when he received the prestigious **Templeton Prize**:

"If, as late as the 1960s the Big Bang theory was a mysterious and largely untested concept, today we have strong evidence from observational astronomy that the entire universe came into existence all at once about 15 billion years ago."

---

[553] Pope **John Paul II**, *at the end of a conference on cosmology held, at the Pontifical Academy of Sciences in the Vatican in 1981, thanked the participants - including S. **Hawking** - for their research on the evolution of the universe.*

[554] **Davies** Paul *(London, ENG 1945) Theoretical physicist renowned for his work on quantum gravity and quantum properties of black holes, cosmologist, and astrobiologist. He has held chairs of physics in major universities: Cambridge, College London, ... Director of BEYOND (Center for Fundamental Concepts in Science), SETI, department of the International Academy of Astronautics, Advisor to METI (Messaging Extraterrestrial Intelligence). Author of numerous scientific articles and books, his work has been rewarded with the Kelvin Medal, the Faraday Prize, ...*
Christian, his positions, and his works have attracted the criticism of atheists but have earned him the prestigious **1995 Templeton Prize**.

Appendix 3: Before the Bang.

Stepping outside the realms of space and time takes us beyond the confines of science. Nevertheless, I'll briefly discuss some symbolic and metaphoric concepts, open to diverse interpretations but not without intrigue, proposed by the French mystic **Martines de Pasqually.**[555] I don't doubt his sincerity when he identifies as a *Roman Christian*, independent of any church, though I don't personally endorse all his ideas. Nevertheless, some of them can be a basis to reflection.

Martines is renowned for his role in founding, among other endeavors, in **1767**, the "*Order of Knights Masons Elect Cohens of the Universe*," a mystical and esoteric movement. In **1775**, he published his midrash, a homily loosely based on the Bible, titled: "*Treatise on the Reintegration of Beings into their Original Estate, Virtues, and Powers both Spiritual and Divine*" [MP]. This work was the result of extensive research, reflection, and asceticism, originally intended for the "*Réaux-Croix*" (holders of the highest grade within the Order of the Elect Cohens) who came under the influence of **Pope Clement XII**'s 1738 papal bull "*In eminenti apostolatus specula*," which was issued against the Freemasons.

According to Martines, before the Big Bang, God had within His Divinity or Dominion (the Divine Immensity) an inexhaustible source of spiritual beings who could not act as they pleased. Instead, God emanated them for His glory according to His will, within a well-defined framework of immutable laws, precepts, and commandments. These spiritual beings, androgynous and eternal, possessed near complete knowledge of God: His Will, and His Actions. They were endowed with Free Will and lived in harmony with Nature to fulfill the tasks assigned to them. **Adam**, the first among them, was imbued with the power to command all spiritual beings and execute God's orders. His destiny was to become the God-Man, as originally, humankind was meant to be spiritual.

However, some angels, considering themselves equal to God, exceeded the boundaries set for them. The betrayal of these rebellious spirits led to the material creation of "*houses of correction*," the Hell managed by Satan, where God exiled them while, like Adam after his prevarication, granting them and his human descendants Free Will and immutable rights He had initially conferred on them.

Note. Martines de Pasqually's visionary concept of spiritual beings, fashioned by God before Adam's prevarication, resonates intriguingly with certain

---

[555] **Martines de Pasqually** *(Fr: Grenoble 1727-1774 St Domingue), a* Roman Christian *with a Spanish* Jewish *ancestry, was a remarkable figure in the rich tapestry of esoteric history. His father, a Marrano (Iberian Jew who, while outwardly practicing Catholicism, secretly clung to his Jewish heritage) infused Martines' life with a unique blend of JudeoChristian mysticism. Creating a foundation for his teachings from which later developments in Western esotericism would spring forth. His teachings emphasized spiritual illumination, divine communion, and restoration of the primordial harmony between humanity and the cosmos. De Pasqually's ideas laid the ground for later developments in Western esotericism, influencing various occult traditions and Masonic rituals.*

Eastern philosophies that posit God's allocation of a predetermined number of souls at the moment of our birth or conception. However, the resonance in Christ extends even further, finding a remarkable echo in the teaching of **St. Paul the Apostle**, as eloquently expressed in his Epistle to the Ephesians:

> *"Blessed be the God and Father of our Lord Jesus Christ, who has blessed us in Christ .... For he chose us in Him before the foundation of the world to be holy and blameless in His presence..." (1:3-6).*

This profound statement by St. Paul elucidates the timeless divine plan, akin to the notion of existing spiritual beings in Pasqually's vision.

Chapter 4.

Beyond the Bang:
Unveiling the Cosmic Tapestry

The world is an infinite sphere whose center is everywhere,
the circumference nowhere.
B. Pascal

In this new chapter, we journey beyond the confines of the Big Bang Theory, exploring horizons that stretch the fabric of our cosmos' understanding. This path takes us in the domain of cosmic inflation, the multiverse hypothesis, and the intriguing concept of conformal cyclic cosmology. Building upon the foundations laid by the Big Bang Theory, we delve into these theories, each offering unique insights into the universe's expansive history and potential. Through their lenses we seek to enrich our understanding of the universe's enigmatic tapestry, pushing the boundaries of our comprehension and sparking new questions about the nature of reality itself.

4.1. The Cosmic Inflationary Expansion.[556]

4.1.1. Description

As we have seen so far, the attractive force of gravitation, and more particularly the gravitational constant G (one of the 7 universal constants) introduced by Isaac **Newton,** has played a leading role in the history of the evolution of the Universe. It has nevertheless posed a problem for the scientists who support the Big Bang theory. If currently, many questions remain unanswered, the inflationary model provides us new perspectives on certain points. It explains in particular, how the brutal expansion of the initial ball of matter could occur while the force of gravity maintained the matter enclosed in an infinitely small and dense ball. It also induces new questions; to which, the unification of the theories of General Relativity and Quantum Mechanics should provide answers.

---

[556] *Read* "The Magic of the Cosmos" *by Greene which is very educational and can be read like a series of riddles to which scientific answers are proposed.*

Let's go back to 1917 (p.86), when A. **Einstein**, a supporter of a stationary universe, despite the opposition of some scientists, including A. **Friedmann**, reformulated the expression of G, the universal constant of gravitation proposed by I. **Newton**, and introduces in his equations the cosmological constant, generating a repulsive force balancing the attractive gravitational force. With the experimental proof brought by E. **Hubble** in 1929 that the universe is expanding, this constant was forgotten by most cosmologists. But A. Einstein's intuition was successfully revised in the 1970s by various scientists and in particular in 1979 by Alan **Guth** (ref.391-p.116), post-doc at SLAC, who proposed the inflationary model according to which the Bang is due to the existence, at a point, in the extremely dense universe:

- at a given time (that of the Bang[557]), of a supercooled Higgs' field (comparable to Einstein's cosmological constant[558] but with an intensity 10^{100} times greater!), called *"inflaton"* having very high excitation energy and producing a negative pressure sufficient to generate a colossal repulsive gravitational force causing, during 10^{-35}s (time at the end of which the energy of the inflaton will be minimal) the inflationary expansion of the universe at a very high increasing speed. During these 10^{-35}s its size was multiplied by a factor varying between 10^{30} (a DNA molecule would occupy all of the Milky Way) and 10^{100}, even more, inducing a weak curvature.[559] Some stars have moved at such a speed v > c (speed of light),* that the radiation they emit has not yet reached us, and should not reach us before the Sun dies (which it should do in about $4.5.10^9$ years) which gives us an idea of the size of the universe.

* The fact that v > c is not contrary to the theory of relativity because the speed limit only concerns the motion in space and not the motion due to the expansion of space itself.

- after 10^{-35}s the energy of the inflaton having sufficiently decreased the explosion ended, and the repulsive push ceased. The remaining energy was used to create particles of matter and ordinary radiation which filled, almost uniformly,[560] all the space in expansion. We join then the theory

---

[557] *This presupposes that we are in a "pre-existing" universe - fluctuating, of high entropy as predicted by L.* **Boltzmann**, *until the state of inflaton is in the favorable conditions that follow - still not explaining the creation but the evolution of the Universe.*

[558] B. **Greene** assimilates this cosmological constant to the dark energy entering the mass/energy composition of our universe.

[559] According to the theory of General Relativity, space-time can have 3 forms depending on $d_{M/E}$ (its density in mass or energy) relatively to d_C, its critical density = 10^{-6}kg/m3: **1.** flat (closing on itself) if $d_{M/E} = d_C$; **2.** spherical if $d_{M/E} > d_C$, ; **3.** bowl if $d_{M/E} < d_C$. (> and < meaning respectively larger and smaller than).

[560] *Almost uniformly because in some regions of space, the density was greater, which the inflationary model can explain.*

of the Big Bang: the space continues its expansion by cooling. Some particles located in the regions of greater density, gather under the effect of gravity, creating structures: stars, galaxies, planets ..., the universe. During 7 billion years these constituents will exert an ordinary gravitation (g > 0) slowing the speed of spatial expansion. But, as the universe grows and dilutes, its density decreases, so does the gravitational attraction and the expansion of the universe accelerates again without stopping as we could verify experimentally.

4.1.2. Contributions.

Cosmic inflation has breathed new life into the Big Bang theory, captivating the minds of researchers such as Andrei **Linde**,[561] Andreas **Albrecht**,[562] Paul **Steinhardt**,[563] and many others. Their contributions have intertwined this model with the Principle of Indeterminacy (ref.316-p.93), applying it not only to particles but also to fields, particularly the "*inflaton.*" In this fascinating realm, new interpretations have emerged, shedding light not only on the macroscopic world of stars and galaxies but also on the mysterious origins of our universe. These interpretations propose mathematical solutions to three intriguing enigmas:

a) The puzzle of inhomogeneities and the formation of structures like galaxies, in a nearly uniform universe. This inflationary hypothesis justifies the observed spatial, energy, and temperature, inhomogeneities by assuming that the expansion halted at different points in space. Quantum Physics, leaning on the Principle of Indeterminacy, further justifies the presence of these inhomogeneities in the initial microscopic state. As the extraordinary inflation begins, it carries these inhomogeneities isotropically in space, creating small, inhomogeneous regions in the expanding space: galactic clusters and their derivatives, the galaxies, the stars, ... in what seems to be a "*homogeneous*" universe. It also accounts for tiny variations in the temperature of the fossil radiation, hovering at T = 2.750 +/- 0.001 K.

The American Space Observatory WMAP (Wilkinson Microwave Anisotropy Probe), launched in 2001, and the European satellite Planck, launched

---

[561] **Linde** Andrei D. *(Moscow, RUS 1948). Trained in Moscow he works at Stanford, CA in the field of theoretical physics. He is a specialist in cosmic inflation. In 2002, together with A.* **Guth** *and P.* **Steinhardt**, *he received the Dirac Medal. He was awarded other prestigious prizes, as the most lucrative: the 2012 Prize of Fundamental Physics.*

[562] **Albrecht** Andreas *(Ithaca, US-NY 1957). Student at Cornell (University of Ithaca), he defended his thesis on the inflation of the universe in 1983 at the University of Pennsylvania under the direction of P.* **Steinhardt**. *Worked at the University of Texas at Austin, Fermilab, Imperial College London.*

[563] **Steinhardt** Paul (Washington-DC, USA 1952). Studied at Harvard (Cambridge near Boston, US-MA). Cosmologist at Princeton, US-NJ, he received in 2002 with A. **Guth** *and A.* **Linde** *the Dirac Medal, as well as other prestigious prizes. One of the promoters of cosmic inflation, he works on the* "Ekpyroptic" *model, a cyclic, branary model (§ 4.2).*

in 2009 under the aegis of the ESA (European Space Agency) not only confirmed these predictions[564] but also significantly advanced our understanding.

b) The conundrum of the immense initial energy required. The Big Bang theory suggests an astronomical amount of energy, surpassing human imagination. On the other hand, the inflationary model demonstrates mathematically that the energy density of the inflaton remained steady during the first 10^{-35} seconds of inflation, while the total energy increased exponentially. Assuming a conservative scenario where each dimension of the universe multiplies by a factor of 10^{30} during inflation, the volume and consequently the energy increase by a minimum factor of $(10^{30})^3 = 10^{90}$. Greene [Ge] suggests that an inflaton initially weighing just 10kg in a sphere of 10^{-28}m diameter could have generated all the energy of the universe. However, this hypothesis assumes an energy density $(10^{85}$kg/m$^3)$ approximately 10^{70} times greater than that of a proton.[565]

c) The mystery surrounding the arrow of time's origin. It hinges on the intriguing hypothesis that the inflaton giving birth to our universe was in a highly energetic chaotic state with very high entropy, as predicted by L. **Boltzmann**. A some point, it transitioned into a state of low entropy, a rare occurrence. This set the universe's initial gravitational entropy as low as possible. Subsequently, with the emergence of life due to solar activity, entropy began to grow following irreversible matter transfers, in line with the second principle of dynamics at the macroscopic level. According to the *Jewish atheist* Ilya **Prigogine**, 1977 **Nobel** Prize of Chemistry (ref.387, p.115) this simultaneous creation of entropy and time defines "*the arrow of time*," originating from the Big Bang, which he names the "*intrinsic arrow*." Although sensitive to the fact that Brownian motion is responsible for macromolecular micro-reversibility (cf. the work of Lars **Onsager**[566]), the agnostic biophysicist Dan **Vasilescu**[567] sharing the opinion of Ilya **Prigogine** but, pushing his reflections further, graphically demonstrates that the average mass of a

---

[564] *As early as 1992 the COBE (Cosmic Background Explorer) satellite had observed the existence of temperature anisotropies.*

[565] Proton: radius of the order of 10^{-15}m, mass = $1.67.10^{-27}$kg.

[566] **Onsager** Lars *(Oslo, NOR 1903-1976 Coral Gables, FL-US). Physico-chemist, 1968 **Nobel** Prize in Chemistry* "For the discovery of the reciprocal relations which bear his name and are fundamental in the thermodynamics of irreversible processes."

[567] **Vasilescu Dan** *(Paris 1937). Biophysicist, a graduate of the ENSET, D.Sc. Lyon in 1964. Appointed director of the Biophysics Laboratory, of Nice's university, he directed 25 theses. Invited member of various international scientific congresses, co-author of 148 scientific articles, and contributed to the writing of 14 books. Member of various international organizations (UNESCO, WHO, ...). Professor Emeritus in Biophysics at the Université Côte d'Azur, his main interest is Quantum Molecular Modeling. Agnostic.*

human (M_{Ho}) is the geometric mean[568] between the mass of the hydrogen atom (M_{Hy}), one of the first elements created, and the mass of the Sun (M_S).

It's worth noting that in the fourth century **St. Augustine**, though not a scientist, posited [Ag2] that:

"...time and the earth had been created simultaneously."

Despite the remarkable agreement between the mapping of the sky's temperature (revealing subtle numerical anisotropies, cf. p.135) provided by the ESA in 2013 and the theory, cosmologists remain intrigued. The theory, rooted in the mystery of low entropy, hasn't completely satiated their curiosity.

With the work of F. **Zucky**, V. **Rubin**, and others (cf. chap. 2) enhancing our knowledge of the universe's total mass, the idea of a flat universe has captivated minds. A variant of **Guth**'s inflationary model has emerged:

4.1.3. Flat Infinite Space.[569]

According to this theory, space was already infinite from the outset, as multiplying or dividing infinity by any number always results in infinity. At the moment of the *"explosion,"* energy density and temperature reached colossal values, but across the entire expanse of space. Multiple Big Bangs occurred at various points in space. Post-explosion, space expanded without increasing in size, given its infinite nature. **Hubble**'s experiments, however, affirm the expansion observed: distances between galaxies and stars are indeed growing. This expansion led to a decrease in temperature and, at certain temperatures, significant changes akin to phase transitions (referred to as *"Higgs' field"*).

4.2. Multiverse or Multiple Universes.

We dive into an intriguing cosmological concept that stretches the boundaries of our cosmos' understanding. The *"Foam"* model, birthed by the brilliant mind of the theoretical physicist A. **Linde,** proposes a fascinating notion: the "inflaton" at the origin of our universe may not have been alone and others, which don't have the same fundamental constants as ours and obey different laws, may have been created. In these conditions, **Our universe could be just one of many, in which no life comparable to us exists.**

Imagine a tapestry of universes, each with its distinct rules and possibilities, where life as we know it might not exist. This idea harks back to ancient philosophies, echoing **Anaximander** (6th century BC) or **Leibniz**, for whom: *"By*

---

[568] The geometric mean of X and Y is equal to the square root of the product X.Y. Generally speaking, the geometric mean of n numbers is equal to the nth root of the product of these n numbers.

[569] *We can make a representation of the expression* "flatness of the Universe" *by imagining a sphere of radius R increasingly large. When R tends towards infinity the cap that we can observe appears flat.*

the infinite multitude of simple substances, there are as many different universes, which are however only the perspectives of one, according to the different points of view of each monad."

Closer to us, drawing from the thoughts of notable thinkers like the *Protestant Theologian* Alvin **Plantinga**[570] or the *Anglican Theologian* John **Polkinghorne,**[571] [Po] **2002 Templeton** Prize, a renowned theoretical physicist, we begin to see our universe as intricately connected to the future, blurring the lines between the temporal and the eternal. This perspective allows for a harmonious blend of beliefs: a direct journey to eternal life or an anticipation of the Parousia, intertwined with an intermediate state resembling purgatory.

Reflecting on the words shared between Christ and the *"good thief"* (who is crucified like him but judges Christ's suffering undeserved):

"Truly, I say to you, today you will be with me in Paradise," [Lk:23-43]

we glimpse a broader understanding, a promise of not just redemption for one soul, but all the souls who follow a righteous path. In this vast expanse of possibilities, the essence of God's plan reveals itself in mysterious and awe-inspiring ways (cf. § 5.5.c).

4.2.1. The fascinating Branary Universe.

It is a marvel of M Theory, a fusion of five-string theory variants masterfully unified by E. **Witten**, (ref.404, p.119) presenting an awe-inspiring panorama. This theory not only harmonizes all matter particles and messengers into a singular string but also offers answers beyond the reach of the Standard Model (ref.382, p.113).

In this wonderful realm, we envisage a branar world ([Gr] pp.449-491), that could potentially encompass not only our universe but also p-dimensional marvels known as *"p-branes."* These spatial dimensions, extend beyond the familiar four

---

[570] **Plantinga** Alvin *(Ann Arbor, MI-US 1932)*. Protestant theologian and professor *of Philosophy trained at Harvard, and Yale, he teaches at the Catholic University of Jamestown in Michigan, and defends the concept of* "possible worlds" *due to* **Leibniz**.

[571] **Polkinghorne** John (Weston-Super-Mare, ENG 1930). Theoretical physicist, his thesis supervisor was the 1979 **Nobel** Prize in Physics Mohammad A. **Salam** (ref.394-p.117). After studying Mathematics at Trinity College, Cambridge, he left for the USA to *work at Caltech (California Institute of Technology) under the direction of Murray* **Gell-Mann,** *1969 Nobel Prize in Physics. He taught at Cambridge from 1958 to 1979. During all his years devoted to Science, he collaborated with the greatest universities (Princeton, Berkeley, Stanford, CERN) and his scientific work earned him many distinctions, including an appointment to The Royal Society.* Theologian. *In 1977, he opted for the priesthood and enrolled at Wescott House in Cambridge (Anglican theological university).*

He was ordained an Anglican *priest in 1982, then Dean of Trinity Hall Chapel (University of Cambridge), and President of Queens' College until 1996. He served as a* Canon Theologian *at Liverpool Cathedral until 2005. He received numerous awards including Knight Commander of the Order of the British Empire in 1997.* **2002 Templeton Prize.**

of Einstein's space-time [ref.268, p.81]), dwell at the scale of the Planck length $(1.616.10^{-35}$ m) eluding our direct perception.

Intriguingly, within the string model the Electromagnetic, the Weak, and the Strong Nuclear forces' messengers are trapped within our 3+1-dimensional space-time, veiling the mysteries that lie beyond. The gravitational force, however, allows gravitons to cross over the other seven spatial dimensions, establishing connections with distant universes. It's a force that potentially empowers us to influence or be influenced by these hidden dimensions.

Yet, a formidable challenge awaits resolution. The theories requiring spaces with ten spatial dimensions suggest a gradual decrease[572] in gravitational force, endangering the stability of our solar system. S. **Hawking** speculated the existence of regions where this holds, potentially void of life in forms we recognize.

The branar theory illuminates a path towards the "*cyclic model*," (below) aligning with the Hindu tradition of an eternal succession of cycles under the watchful eyes of Shiva, Brahma, and Vishnu. As a captivating tidbit, consider the humble jellyfish "*Turritopsis nutricula*" in the Caribbean Sea, measuring just a few mm. This marvel possesses the extraordinary potential for immortality, embarking on new life cycles after reaching reproductive maturity, embodying the cycles of life mirrored in the magnificent cosmos.

4.2.2. Cyclic universes.

4.2.2.1.

In the 1930s Richard **Tolman**[573] proposed the cyclic cosmological model. According to this, our universe would be one among others having preceded it. In expansion at present, it should, at a certain moment pass in the compression mode and set out again in a mode in expansion, and so on... without beginning or end. The reason for its abandonment is that according to the second principle of thermodynamics the system, being closed, the entropy S must increase continuously, lengthening the duration of each cycle according to the Theory of General Relativity (TGR), cf. Chapter II, the year 1917.

4.2.2.2.

P. **Steinhardt** et al. proposed the "*Big Splat*" model.[574] According to them,

---

[572] The gravitational force *decreases when the number of dimensions increases. It is always a function of the distance d between the bodies but, while in a 3-dimensional space, it is in 1/d2 (inversely proportional to the square of d), in a 4-dimensional space it would be in 1/d3 and in 1/dn-1 in an n-dimensional space.*

[573] **Tolman** Richard C. *(USA: West Newton, MA 1881-1948 Pasadena, CA). Physicist, chemist, and cosmologist, he studied at MIT, where he defended his thesis in 1910. Appointed at Caltech, he had L. **Pauling** as a student. Specialist in Statistical Mechanics and Quantum Mechanics, he published numerous articles with A. **Einstein**, including a cyclic model of the universe in 1930. Member of the American Academy of Arts and Science, recipient of numerous distinctions.*

[574] Big Splat *cf. [Gr] pp. 484 - 487.*

we live in a flat 3-brane universe which, with a periodicity of some 10^6 to 10^9 years, collides with another 3-branes causing a bang initiating a new cosmological cycle. There would be another 3-brane universe next to us, very close (at a fraction of a millimeter) whose gravitational force is so weak (no longer following Newton's law) that we have no way of seeing it, and the same in the opposite direction. We then find conditions similar to those of the inflationary theory, but with a longer duration, and join the observations that allowed us to establish the Big Bang model. This model is attractive but lacks experimental support for the existence of additional dimensions accessible only to gravitational forces.

4.2.2.3.

Everett's theory.[575] Based on Quantum Mechanics, he introduces the notion of a "*universal wave function*" ψ of the universe. Like any wave function (WF), representing the quantum state of a particle or a system, it is equal to the linear combination of the various eigenfunctions ψ_i associated with each allowed quantum state. The paradox with Quantum Mechanics is that, although the Schrödinger equation describes perfectly, in a deterministic way, the temporal evolution of the WF, the norm of the calculated eigenfunctions ψ_i defines the probability that the particle or the system is in a given state Ei. Thus, when a measurement is made, a result is obtained that depends on the experimental conditions and corresponds to a single eigenfunction ψ_i. We say that the "*wave function collapses.*" H. **Everett** does not share this idea. According to him, each of the possible results of a measurement exists in one universe or another. Hence his model, called a "*Multiverse*" or "*Parallel Universes*" all disconnected from each other. We dive into metaphysics.

The *Atheist* theoretical physicist David **Deutsch**,[576] a specialist in quantum computing, supports **Everett**'s theory by taking up mathematically the concept of the Omega point suggested by the Cosmologist Franck **Tipler,** a Theist (ref. 695, p.248).

4.3. CCC theory (Conformal Cyclic Cosmology) [Pen]

Published in 2005 by Sir R. **Penrose**, confirmed in 2010 from WMAP data, and in 2013 by Sir R. **Penrose** and Vahe G. **Gurzadyan;**[577] this idea did not generate much enthusiasm among the scientific community. Nevertheless, he will

---

[575] **Everett** Hugh *(Washington D.C. 1930-1982 McLean, US-VA). Studied at Princeton where he defended his thesis in 1956 under the direction of* J. **Wheeler.** *As a mathematician, he generalized the method of calculating* "Lagrange multipliers." *As a physicist, he specialized in quantum mechanics and cosmology.*

[576] **Deutsch** David (Haïfa, ISR 1953). *Professor of Theoretical Physics at Oxford, Paul Dirac Prize 1998, Dirac Medal in 2017.*

[577] **Gurzadyan** Vahe *(Yerevan, ARM 1955). Physicist, Ph.D. in theoretical physics and mathematics in 1988, post-doc in Japan, Sussex, Italy...he is interested in non-linear systems, stellar dynamics, cosmology, and ancient history ... Collaborated with geneticists at Duke University to establish a new method for detecting genetic mutations, ...*

publish on 2/03/2020 an article [An] together with the mathematician Daniel **An**, the *Christian* theoretical physicist Krzysztof **Meissner**[578] et al in memory of S. **Hawking** in which they experimentally confirm this theory. According to this, the universe is uniform before the Bang because it is part of a cycle of universes with a periodicity of several billion years. Each cycle begins infinitesimally small but ultra-smooth, then expands by generating clusters of matter that are held by supermassive black holes which over time, transform into Hawking radiation thus producing the uniformity from which the next Bang will come. But here again, the probability that the initial conditions and the universal constants are satisfactory is extremely low.

This theory will support Th. **Monod**'s[579] opinion:

"What does it matter if man disappears from the globe. ..... There will always be relays in nature ....

And the evolution can draw a circle, which will close again on the new origins, that is to say prehistoric" [Md].

Biologist, geologist, archeologist, scientist,... Th. Monod, one of the greatest world specialists on the Sahara in the 20th century, a *Protestant* and humanist who fought vehemently to defend Nature and pacifism,... - wrongly defined by some as a *"Christian anarchist"* [580] - defines himself as a *"Christian apprentice"* because according to him:

"The word Christian is so full of meaning and responsibility that one does not dare to call oneself a Christian, but one can say that one tries

---

[578] **Meissner** Krzysztof *(Warsaw, POL 1961), a theoretical physicist specializing in elementary particles. Studied at the University of Warsaw: Ph.D. in 1989, habilitation in 1997, Professor in 2006. Director of the Nuclear Research Institute in 2009. He collaborated with* R. **Penrose**. *Author of many scientific articles he puts science within everyone's reach.* Catholic.

[579] **Monod** Théodore *(Fr: Rouen 1902–2000 Versailles). Biologist, explorer, humanist, ...* Protestant, *he obtained a doctorate in Sciences in 1926 from Sorbonne University. After studying the monk seals on Mauritania's Cap Blanc Peninsula he searched vainly for sixty years meteorites in Sahara. Nevertheless, he discovered numerous plant species, Neolithic sites, and in particular in 1960 the remains of the first black person (9,500 to 7,000 years old): the* "Asselar Man".... *Professor at the French "Muséum national d'Histoire naturelle (from 1946 to 1973), founder of the "Institut Fondamental d'Afrique Noire" in Senegal, member of the* "Academie des sciences d'outre-mer" *in 1949. In 1954 he descended with* J. Y. **Cousteau** *to 2,100 m below the sea level in the Toulon (Fr) pit. Member of the French Academies of Marine in 1957 and Sciences in 1963. He was co-founder of the* World Academy of Art and Science *in 1960. Author of About 1,200 scientific articles. He was awarded 1960 the Patron's Medal of the Royal Geographical Society .....* Protestant *he was the founder and president of the Francophone Unitarian Association (the Fraternal Assembly of Christian Unitarian), he attended the Protestant Temple of* "l'Oratoire du Louvre" *where his father (Wilfrid) was a Pastor.*

[580] Christian anarchist: *political theological movement claiming that Christians are only answerable to God's authority, basing their ideas on the sermon on Mount Mt. 5-7).*

to advance in that direction, always on the side of the mountain, looking at the summit from far away."

It should be noted that current scientific knowledge does not support a Big Crunch, which requires a greater amount of matter that is known for positive gravity to take over the expansion.

The cyclic theory is compatible with the Buddhist theory which does not believe in the necessity of a Creator. For some Buddhists, the sensible world is illusory, hence the term *"Mâyâ"* (= magic), one of the three nodes; the other two being *"Ego"* or self-consciousness, and *"Karma":* the destiny of every living being is determined by the sum of all his past actions in his previous lives. Nodes, which each one must get rid of to stop the cycle of death and rebirth. This can be summarized by saying that Buddhism teaches the *"way of deliverance"* while Christianity teaches the *"way of salvation."*

At the end of this brief overview of some of the existing theories - whose relevance is questionable because they are far ahead of our technology and therefore difficult unless a miracle occurs, to confirm experimentally – we understand that science is far from answering all our questions. But it is characteristic of science to generate one or several new questions each time it brings an answer to a problem. Nevertheless, there are some, such as gravitational waves, which could be verified. The razor principle of the *Franciscan* **William of Ockham,** [581] according to which the simplest solutions are the best, is still applicable. In other words, why create an infinite number of universes (around a thousand) without life, so that there is only one in which life as we know it, finds its place?

No one can answer this question with certainty because God's design is inscrutable; nevertheless, I leave it to S. **Hawking** to conclude. At the interface between the twentieth and twenty-first centuries, relying on the scientific knowledge of the moment, he adapted the strong anthropic principle, which his colleague Brandon **Carter** had stated in 1973 (ref.648, p.229), and wrote ([Ha1]-p.151):

> *"Either several different universes exist, or several different regions coexist within a single universe: each of them would have its initial configuration and perhaps even its laws. In most of these universes, the conditions necessary for the development of complex organisms would not be fulfilled and only a small number of universes like ours would allow the development of intelligent beings...."*

---

[581] **William of Ockham** *(Ockham-Surrey, ENG 1285-1347 Munich, CHE)* Franciscan, *was called the* "invincible doctor" *or the* "Venerable Initiator." *After studying at Oxford and Paris, he taught at Oxford. Philosopher, logician, theologian, imminent representative of the nominalistic school (according to which the existence of God cannot be demonstrated), opposed to the Thomistic one.*

Continuing his reasoning he demonstrates that it is not different from the weak anthropic principle:

> "*If the universe is very large, even infinite, spatially and/or temporally, the conditions necessary for the development of intelligent life will be found only in limited regions of space-time. If intelligent beings exist somewhere, they should not be surprised that the corner of the universe they inhabit meets the conditions necessary for their existence. A bit like a rich person living in a rich neighborhood, who would not know what poverty looks like.*"

This idea of several worlds is not new for scientists. Two centuries earlier, the greatest of them all, Sir Isaac **Newton** (refs. 185&204, pp.59&61) an Anglican, wrote in his treatise on optics (cf. [Ne1]-p.146):

> "*As space is infinitely divisible, and matter is not necessarily in all parts of space (which we justify chap. 3); God can diversify the laws of Nature and make different worlds in different parts of the Universe.*"

These captivating strands of thought form an alluring constellation of ideas that continue to beckon humanity's most daring explorers. As we venture ever deeper into the cosmos believers can hope that there may be at least one other universe governed by laws different from ours that await us. Having been created in the image of God, it seems obvious to me that we will not be reincarnated with the same carnal envelope, it would not make sense, it would be terrible for the handicapped, the crippled, and the old, ... The second part of our life, which must be eternal, will be done under another form, which can only be more beautiful, and that we will discover when the time comes.

Chapter 5.

Scientists speak out

"The question is not whether God exists or not.
But rather: who is He, and what is He playing at?"
Hubert Reeves [Re1]

In the vast tapestry of our existence, questions arise about our origins, the nature of reality, and the existence of a divine power that has tantalized human intellect for millennia. As we peered (chapters 2-4) into the depths of the universe, we were confronted with a mesmerizing symphony of particles, energies, and laws, a carefully orchestrated ballet that brought the cosmos to life. Today, we stand at the threshold of understanding the universe in ways our ancestors could only dream of. This chapter delves into the fascinating interplay between science, philosophy, and the concept of God.

Throughout history, the greatest minds have grappled with these questions. Thinkers like Isaac **Newton**, Gottfried **Leibniz**, Max **Planck**, and Edwin **Schrodinger**,... pioneers of science and philosophy, not only expanded our understanding of the universe but also held a profound belief in a divine creator. Like Pythagoras, Plato, and others they saw the elegance of mathematics and the orderliness of the cosmos as a testament to a higher intelligence.

Hubert **Reeves** (ref.4, p.1), a modern astrophysicist, and Arno **Penzias** (ref.364, p.108) a Nobel in Physics,... echo this sentiment, stating that for many contemporary scientists, the existence of God is no longer a question. It's a certainty, a profound acknowledgment of a purposeful creation. In the quest to unravel the mysteries of the cosmos, belief in a higher power is not seen as irrational, but rather an avenue to deeper understanding. Yet, akin to the philosophical ideals of René **Descartes**, doubt is a crucial facet of intellectual growth. Every thinking being, regardless of belief, reserves the right to question and ponder the enigmas of existence. As we embark on this exploration of ideas, let us embrace the questions that spark curiosity, fueling the relentless pursuit of knowledge. The first question that beckons us:

5.1. Should the text of Genesis be taken literally?

"The Bible contains a hidden meaning, apart from the literal meaning. It is discovered by lending each letter an esoteric and divine meaning."
I. Newton [Br]

*"Have we any right to assume that the Creator
works by intellectual powers like those of man?"*
Ch. Darwin [Dar 91]

Any reasonable person has understood that Genesis shouldn't be taken literally, the God's word is timeless and addresses all peoples of different cultures. But let's leave it to the scholars.

In ancient times, when sacred texts like the Pentateuch were presented to the Hebrews, the Vedânta[582] to the Hindus, and the New Testament to the Christians, allegorical narratives, parables, and imagery were commonly employed. This was because it was impractical to convey complex concepts such as the creation of the Earth and natural phenomena to the largely uneducated and illiterate people of the time. Thus, figures like **St. Paul**, found it was necessary to tailor their language to their audience.

Even in the modern era, the tangible world we experience seems vastly different from the one described by physicists and cosmologists, who delve into its mystery using abstract mathematical tools (cf. chap. 2). Consequently, scientists often are reduced to creating visual representations to demystify complexes ideas and make them understandable to the general public. Similarly,

---

[582] Vedânta *(ref.6, p.2) Let us note that* **Sri Krsna** *(8th avatar of* **Vishnu***), God, the omnipresent Supreme Person, primordial Lord, first Cause of all causes, the central deity of Hinduism who appeared in the guise of* Caitanya **Mahāprabhu** - "the Great Apostle of God's love, the perfect preacher" - *in Sridhäma (Bengal) in 1486 and died in 1534; condemned any interpretation of the Vedanta:* "The principle is that no one, from his imperfect reason, can surpass the Vedas." **Buddha** *rejected the authority of the Vedas and developed his religion. As we saw (ref.309-p.92) Christian E.* **Schrödinger** *was a fan of Vedanta's. He said:* "Vedanta teaches that consciousness is singular, happenings are played out in one universal consciousness and there is no multiplicity of selves." *Closer to us, the Hindu* George **Sudarshan***, a theoretical physicist, who claimed vainly to share the 2005 Nobel Prize in Physics with* Roy J. **Glauber***, often lectured Vedanta.*
* **Sudarshan** George *(Pallam British India 1931–2018 Austin, TX-US). Theoretical physicist, raised in the Syrian Christian religion, left it for Hinduism. Awarding the Master's degree in 1952 from Madras Christian College he moved to the University of Rochester in New York where he received his Ph.D. in 1958. He taught physics in various universities: Rochester, Austin, Chennai, IND. His contributions in the fields of elementary particles, and quantum optics, are recognized by the greatest physicists of his time and have earned him various awards except for the 2005 Nobel Prize of Physics which many high-ranked scientists believe he deserved.*

to prevent deterring both young and older readers unfamiliar with the Bible, it is crucial to heed the advice of *Christian* biologist Jean Marie **Pelt:** [583]

> *"Genesis should not be read as a scientific text, a truth, but should, on the contrary, be considered like a founding myth where it is the moral that matters"* [Pe].

Other theologian scientists, such as B. **Warfield,** echo this sentiment, asserting that the Bible is not a compendium of scientific facts, but a source of spiritual and moral guidance. Using allegories, metaphors, anthropomorphisms (the female designating matter, the male form), and hyperboles, can sometimes complicate interpretation, leading readers astray. This issue is particularly pertinent today, as young individuals, steeped in rationalism and newly introduced to philosophy, may find a literal interpretation of certain biblical passages surrealist. [584]

This reflects a shortcoming in contemporary religious education, but it begs the question: where does the responsibility lie? The observations I share are not news:

-   St. **Augustine** (4th century, A.D.), one of the *Fathers and Doctors of the Church*, is a good example of this because, at first, his reading at the first degree turned him away from the *Christian* faith, although he had received a good education. Once he had assimilated the Neoplatonic method he understood the Bible's allegoric meaning, and returned to its study, devoting the rest of his life to making his reading intelligible.

---

[583] **Pelt** Jean-Marie *(FR. Rodemack-Moselle 1933-2015 Metz). Doctor of Pharmacy in 1958, he pursued scientific studies and then taught botany and plant biology at the Faculty of Pharmacy in Nancy, FR from 1962. From 1972 he taught botany and plant physiology at the Faculty of Sciences of Metz. He chaired the European Institute of Ecology:* "Economic growth can only continue at the cost of ecological degrowth...."
He was a great humanist, a practicing catholic, and he deplored the fact that atheists had seized Darwinism.

[584] *At the time the Bible was written, allegoric language was common. Other difficulties:* **1.** *when Genesis was given to the Hebrews they were in a polytheistic environment, it was necessary to use strong images to distinguish themselves from others -* **2.** *The punctuation system that we know did not exist; neither did the vowels, which complicates the translation. Let's take the example of three-letter words whose first and third consonants (f & r) are the same and let's introduce between them different vowels a, o, and u). We can obtain three different words: far, for, and fur with different meanings. The Bible was translated from Hebrew or Aramaic into Greek before being translated into a living language which, by definition, is constantly and rapidly evolving, a quotation cannot be separated from its context, and in a few millennia, many things have evolved. Hence the need for an updated and contextualized reading. The Hebrew language uses roughly 8,000 words, whereas the Grand Robert defines more than 100,000 words, the choice of which is not always easy. Let's take a simple example: the Hebrew word 'erets' can be translated as Earth, our planet, and land: a piece of land. Did the Hebrews, like the first Christians, know the solar system?*

- For Moses **Maimonides** (12th century A.D.), *the second Moses*, founder of *Jewish* rational theology, professing an allegorical reading of the Bible, it was necessary to use:

"... homonymous words, so that the vulgar men could take them in a certain sense, according to the measure of their intelligence and the weakness of their conception, and that the perfect man who received instruction could take them in the other sense,"

as well as metaphors and amphibological[585] expressions. What, in common language, is named the application of hermeneutics.[586] In the Book of the Lost [Ma1] **Maimonides** explains that: it does not matter whether the common man understands the true meaning or not, the ignorant will not find implausible what is impossible. The perfect and distinguished man (for him this was the rabbi) will understand that there are two meanings: an external meaning, which would give him a bad opinion of the author, and an esoteric meaning, which would give him a good one. Maimonides had read the gospels. Also, what he wrote brings me back to the questioning of the apostles and disciples of **Jesus.** Following his parable of the Sower for example, the apostles ask Jesus (Mt 13:10-17, Mk 4:10-12 & Lk 8: 11-15) *"Why do you speak to the people in parables;"* he replied: *"Because the knowledge of the secrets of the kingdom of heaven has been given to you but not to them....."* Meaning that only a small number of his followers could understand that he was proposing to the *Jews* a new covenant in addition to the old Jewish Law; while for the others, it was necessary to use pedagogy by returning to concrete things like the grain that can only bear fruit if it falls in the right place.

The great *Dominican theologian* **St. Thomas Aquinas** (ref.127, p.36), will take one century later, a similar position to that of M. **Maimonides**, adopting the Aristotelian philosophy which was abandoned for several centuries. Regrettably, it was not until 1893, and **Pope Leo XIII**, that the theory of the four senses of reading the Holy Scriptures, proposed by **Origen** in the third century (ref.70-p.22), was once again adopted by the Catholic Church [Le]:

"... the literal meaning itself conceals other meanings which serve either to illuminate the dogmas or to give rules for life,"

"... One who teaches the Scriptures will also be careful not to neglect the allegorical or anagogical meaning attached by the Holy Scriptures to certain words."

Facing an assault from Ludovico **delle Colombe,**[587] a staunch advocate of a

---

[585] Amphibological: *ambiguous, equivocal, giving rise to two interpretations.*
[586] Hermeneutic: *scientific interpretation of religious texts. There is a reference to Hermes, the name of the Greek god, the messenger of the gods.*
[587] **delle Colombe** Ludovico *(Florence, ITA: 1565 - 1616), nicknamed* "pipionne" *(pigeon, to see a mockery there), a member of the Florentine Academy (philosophical and literary) was an*

literal word-by-word interpretation of the Scriptures, who accused Galileo Galilei of manipulating the text of Genesis to align with theories that deviated from scriptural teachings, Galileo replied by stating:

> "... I admit that it is not possible that the Sacred Writings can deceive or be mistaken... nevertheless, some of its interpreters or presenters can sometimes be mistaken, in different ways ... mainly when they want to keep their true meaning to the words, which can lead to serious and blasphemous heresies. Leading to attributing to God: feet, hands, ... human feelings ... ignorance of the future ... many of the propositions of the Scriptures are written in such a way as to fit the ignorance of the common man."

E. **Schrödinger**, speaking [Sc] of the fundamental vision of the Vedânta, concludes (p. 40):

> "...reasoning allows us to arrive at least to this, that to grasp the substance of the phenomenon by reasoning is in all likelihood impossible, since reasoning itself belongs to the phenomenon, and is entirely included in it; we may ask ourselves whether we should give up using an imaginary, allegorical view of the situation, simply because its relevance cannot be rigorously established."

A. **Einstein**, too, warned us [Ei1 p. 171]:

> "...sense perceptions offer only indirect results about this world or about "physical reality"...Then only the speculative way can help us understand the world. But we must always be ready to transform these ideas...."

During an interview, the writer and journalist Edmond Blatten [Bl] asked the well-known scientist Théodore **Monod**, Protestant: "It is not necessary to take the Bible, the Gospel to the letter?" Th. **Monod** answered:

> "Of course not! We must put these texts in their chronological framework. These people knew some things, but they also knew very little ....."

Closer to home Freeman **Dyson**,[588] a Christian scientist said in an interview [Dy]:

Aristotelian philosopher defending **Ptolemy**'s system because it was following Genesis. He is mainly known for his attacks against **Galileo**. His brother Raphaelo, a Benedictine close to Archbishop **Marzi Medici**, is said to have been behind the attack on Galileo by the Roman Inquisition.

[588] **Dyson** Freeman (Crowthorne, ENG. 1923). Mathematician, physicist, holder of the **2000 Templeton Prize**, and many other scientific distinctions such as the Lorentz (1966), Max Planck (1969), Oersted (1991), ... Enrico-Fermi Prize (1993), Henri-Poincaré (2012) ... In 1945,

"Technology is a gift from God and it is perhaps after the gift of life the best gift from God. ......"

Awaring the difficulty for most of us to understand quantum mechanics he added:

"None of our contemporaries, gifted with reason, using everyday devices derived from the knowledge of quantum mechanics, can doubt its validity, nor its language, even if they do not understand it."

Sir Nevill F. **Mott**,[589] an *Anglican*, 1977 co-**Nobel** Prize in Physics - with Philip W. **Anderson** and John H. van **Vleck** *"for their fundamental theoretical investigations of the electronic structure of magnetic and disordered systems,"* went further. According to him, every practitioner has the right to pray in the church without accepting all its rites:

"It is possible to worship and yet interpret or even reject some of what it is said" [Mot1],

This aphorism reflects the contemplation of a talented scientist who embraced religion late in life and stands out from the average Christian while respecting his ideas. In the book he edited in collaboration with coauthors [Mot2], talented scientists, and religious too, he explains that the Supernatural Christian God in whom he believes does not break the laws of nature which he had created, to perform *"miracles,"* but stays by those who believe in him and helps them while respecting the laws of nature. He did, however, respect those who believe in miracles and apply orthodox doctrines dating back to the 4th century AD for some. John **Polkinghorne**, a contemporary scientist, and *Anglican priest*, wrote [Col2]:

Master of Cambridge in his pocket, he left to study in the USA at the universities of Cornell and Princeton where he intervened as a professor. He was elected a member of the most important scientific societies.

[589] **Sir Nevill F. MOTT** *(U.K.: Leeds 1905–1996 Milton Keynes), was awarded 1922 a major scholarship to enter St John College of Cambridge. From 1924 to 1928 he studied mathematics and theoretical physics at Cambridge, where he started research under R.H.* **Fowler**. *Then, in Copenhagen under* Niels **Bohr** *and in Göttingen under* Max **Born**. *Chair of theoretical physics at Bristol in 1932, began studies on semiconductors. His work has earned him a lot of prestigious medals (Copley, Hughes, Royal Society ...), including a knighthood in 1962 and the* **Nobel** *Prize in Physics in 1977. From 1954 to 1971 he was appointed Cavendish professor of physics.*
Mott's faith. *After their marriage in a Unitarian Church, his parents did not practice Religion. He was not baptized nor confirmed until he was 50 years old. In the meantime, he had attended a few meetings organized by Christians Scientifics at the college, but without taking any interest in them. Later in Cambridge, having been invited to integrate a group of senior Christian scientists, he joined the* Anglican Church *and took part in readings during which he explained his Christian vision of God. When he retired in Aspley Guise (Buckinghamshire, ENG) he and his wife were constant churchgoers.*

"The Bible cannot be treated as if it were uniformly inspired and authoritative ... its human writings contain both eternal truths and also writings about historical and cultural facts."

He does not share all N.F. **Mott**'s ideas.

5.1.1 The Creation of the Universe.

"God blessed the seventh day and sanctified it because on that day He rested from all the work of creation that He had accomplished." [Ge 2: 3]

Let's note that many **Greek philosophers**, such as **Aristotle, Plato,** ... and, then later, others like **Averroes,** having adopted the Aristotelian doctrine, did not believe in the creation - from scratch - of the universe by God, the Eternal, as taught by the *Abrahamic religions*, but in its formation by God, the Eternal, from the shapeless matter, also eternal, which was at his disposal and which he modeled as best as he could. The difference is important because, at that time, the formation did not imply, among other things, the obligation for God to create Man, whereas, in the theory of creation.

For **Plato** and his admirer **Philo of Alexandria**, for example, the sensible world is the image of the real world which is Idea [Th] (ref.65-p.21).

Shortly before his death (67-68) in a Roman prison, **St. Peter**, the *"Prince of Apostles,"* 1st bishop of Rome, wrote in his second epistle (3:8) addressed to the Christians of Asia Minor:

"... with the Lord, one day is like a thousand years, and a thousand years are like a day."

He had well understood and explained that for God, who is timeless (because He is the Creator of time itself) time does not exist, which should allow us to understand the true meaning of the above epitaph. Later: **Origen**, the *"Father of Christian exegetes,"* stated[590] at the beginning of the third century A.D:

"What intelligent person would consider it a sensible statement that the first three days, evening and morning, existed without sun, moon, and stars? .... No one doubts that these are related figuratively in Scripture and some mystical meaning may be indicated by them, through a semblance of historical narrative."

Since history is a discipline, events are classified chronologically, but this was not the case at the time, they were placed in order of importance; this is, it seems to me, what St. **Augustine** (4th century A.D.), *Doctor and Father of the Church*, who was not a scientist, understood:

---

[590] **Origen** in *[Or]* or online: "De Principiis, Book IV," *New Advent, § 16.*

"... the creation itself was accomplished in an instant and the work of the six days is only the picture of the development of the riches of the universe, by an evolution that may have lasted for many centuries and continues before our eyes."[591]

We find in this quotation not only a *"prediction"* of the Big Bang theory (chap. 3) but also that of the evolution theory (§ 5.3). His theological approach:

"It must therefore be admitted that the world at the moment of creation received with its being a causal virtue or power of evolution in which all future beings were contained, not formally, but in germ as the acorn contains the oak tree,"

although interesting, remains in the domain of metaphysics, but is not contradicted by **Teilhard de Chardin**. On the other hand, interpreting the sentence of Genesis (Ge: 1-16):

"God made two great lights [The Sun and the Moon] ... on the 4th day"

which obviously cannot be taken, because the notion of a day that we have adopted is that of a rotation of the Earth on itself, and before *"the lights"* were created, it made no sense to speak of day, the great rabbi Moses **Maimonides** gets around the difficulty. According to him, the lights were created on the first day but suspended on the fourth day.

For the proponents of an *"ancient earth,"* the Earth was indeed created 4.5 billion years ago, but the 6 days of Genesis must be assimilated into eras or geological periods. Based on the Theory of Relativity and taking into account the speed of expansion of the universe, the physicist Gerald **Schroeder,** [592] an *orthodox Jewish*, deduced that a cosmic day is worth approximately 10^{12} terrestrial days (time is relative), which, except for the uncertainty about the origin of the Bang, justifies, according to him, that life appeared at the end of the 6th day of Genesis. The translation *"end of the 6th day"* is moreover controversial, some rabbis speak: *"beginning of the 7th day."*

The **1966 Fields Prize**, Alexander **Grothendieck,**[593] the greatest mathematician of the 20th century, Atheist until the age of 16th, became a *Deist* following a

---

[591] *See [Th] p. 244.*
[592] **Schroeder** Gerald *(New York, USA 1938). Theoretical physicist, and* orthodox Jewish, *he defended his Ph.D. in Nuclear Physics and Earth and Planetary Sciences at MIT in 1965. He worked there for 5 years before emigrating to Israel in 1971 where he held research positions in various institutes including the Weismann Institute of Sciences. He currently teaches at Aish Ha Torah College of Jerusalem on the relationship between Science and Religion. I wrote* "would justify" *because it is quite controversial among* some Orthodox. *Cf. Wikipedia. Schroeder was a member of the American Atomic Energy Commission.*
[593] **Grothendieck** Alexander *(Berlin, DEU 1928-2014 St Lizier, FRA) Born in an anarchist family, was interned by the nazi from 1940 to 1942; then, he attended a Cévenol Collège (Fr), where he became fascinated by mathematics. After the war, he studied mathematics at the French*

course on evolution, given by Mr. Friedel, his professor of Natural History. In **1991**, after his withdrawal from the mathematical community, he devoted himself to politics and religions: *Buddhism*, then *Catholicism*. He was influenced by the mystic catholic Marthe **Robin** who, from 1930 until she died in 1981, ate only the consecrated host. Let's note that in 2014 **Pope Francis** recognized her *"heroic virtues"* and declared her *"venerable."* In 2002, Grothendieck had an original idea ([GK], p.61):

> *"As for the seven days of Creation, there is no doubt that these are the "days" when He drew out of nothingness the eternal laws (spiritual, physical, biological) that govern the Cosmos and the Universe - like a Master Painter who carefully prepares his canvas and frame for a painting he is about to brush."*

Everyone is free to think what he or she wants because the notion of time is complex, which is why many agree that God, being timeless, does not participate in time. Indeed, **Maimonides** says ([Ma1 Chap.1 p.199]):

> *"Time is an accident that competes[594] with movement, when, in the latter, one considers the idea of anteriority and posterity."*

It refers moreover to **Aristotle** for whom, time not being movement, nor change must necessarily be: *"something of movement."* It is recognized by what is before and after an intermediate point, which is the present. The explanation that we find in Genesis (1:14), *And God said*:

> *"Let there be lights in the vault of the sky to separate the day from the night, and let them serve as signs to mark sacred times, and days, and years...."*

This definition of time is at the origin of many calendars (ref.66-p.21) more or less satisfactory for earthlings but which would be different on other planets or exoplanets harboring Life. The probability that the rotation speed of another planet, on itself, is identical to that of the Earth is infinitely small. A change of paradigm would be necessary; the adoption of the atomic clock, based on the

---

*Universities of Montpellier, Paris, and Nancy (ENS). Specialist in algebraic geometry and functional analysis, one of the Bourbaki group (composed of French alumni of ENS: Ecole Normale Supérieure de Paris, who prepared new textbooks in Analysis), he defended his Ph.D. in 1956 and received the **Fields Medal in 1966** (the equivalent of the Nobel in mathematics). He applied for French citizenship in the early 1980s. He was a political radical and pacifist activist.*
[594] *Compete: "that which belongs to...." Time is thus for these authors a concept linked to the past, the present, and the progress of an action. God being timeless, the time scales cannot be the same for him as for us (Cf. chapter 3. Big Bang Theory). Moreover, who can, among us, give an exact definition of time? For E. **Schrödinger** (ref.309-p.92) time is also the notion of "before and after," whereas for L. **Boltzmann** (ref.237-p.72) it is the "natural tendency of any ordered state to go from itself to a less ordered state, but not the contrary."*

immutability of the frequency of an atomic transition of ^{133}Cs could be a solution (1 second corresponds to 9,192,631,770 energy transitions of an atomic electron).

The *Lutheran* astronomer Sir William **Herschel**, [595] founder of stellar astronomy - of whom Baron Joseph **Fourier**,[596] *Catholic,* perpetual secretary of the Academy of Sciences, said on 5/06/1824 [Fo] during a public session of the said Academy:

> *"William Herschel, a member of this Academy, is one of those extraordinary men who, destined to honor their homeland and their century....... opened new roads in a sublime science; he saw stars until then ignored, and pushed back all the limits of the spectacle of the skies... he devoted all his life to immortal works, and during forty years the brightness of his discoveries resounded in all Europe;"*

Herschel thought that:

> *"The more the field of science expands, the more numerous become the powerful and irrefutable arguments proving the existence of an eternal Creator with unlimited and infinite power. Geologists, mathematicians, astronomers, and naturalists have all collaborated to build the edifice of*

---

[595] Sir William **Herschel** *(Hanover 1738-1822 Slough, Eng.).* Lutheran, *British astronomer. Composer of music. He discovered Uranus with the first telescope that he built. Appointed astronomer to* **King George III***, he detected the movement of the Sun toward the* "apex," *a point in the constellation of Hercules. He discovered the atmosphere of Mars, highlighted the infrared radiation that he called* "heat rays," *and binary stars, and proposed a Universe model that could be a rough outline of the Milky Way. In 1784, he put in evidence many nebulae, and the cluster NGC 2264 or* "snowflake." *Located at 2,000 light-years, in the constellation of the Unicorn, this cluster contains less than 50 stars and is 9 million years old. He was laureate of the 1781 Copley Medal. His musical work is important: 24 symphonies, more than 14 concertos, many sonatas, and church music. He played, violin, harpsichord, and in 1766 became an organist in Bath*

[596] **Fourier** *Jean-Baptiste J. (Auxerre 1768-1830 Paris),* Catholic. *Orphaned by his father and mother at the age of 6 years, he studied at the military school of Auxerre run by the Benedictines. Following the dissolution of the congregation by the revolutionaries, his novitiate was interrupted just before he took his vows. He then devoted himself to the sciences. He was appointed lecturer at the Ecole Normale in 1795, then a professor at Polytechnique in 1797 where he taught* "Analysis." *He was awarded the Grand Prix of Mathematical Sciences in 1812. He was responsible for the theories on the* "Fourier series," "Fourier transforms," *indispensable tools for the study of vibrations, acoustics, and electricity...* "The analytical theory of heat," "Fourier's Optics." *The Italians have built, in his honor, the* "Planetary Fourier Spectrometer PFS" *used by the ESA for the exploration of Mars, and Venus, ... A lunar crater bears his name. He was elected to the French Académie des Sciences in 1817, the Royal Society of England in 1823, and the Académie Française in 1826. He participated in the French Revolution and Napoleon's Egyptian campaign, became a diplomat, and prefect of Isère, Fr in 1802, created the imperial faculty of Grenoble in 1810, and became its Rector.*

science, which is, in truth, the foundation of the Supreme Greatness of God the only one."

J. **Joule** (ref.224-p.68), a great experimenter in Physics, and a *sincere Anglican* recognized, without ambiguity, that God is the Creator:

"After the knowledge of and obedience to the will of God, the next aim must be to know something of His attributes of wisdom, power, and goodness as evidenced by His handiwork." [Lam]

Below, I reported quotations from 4 contemporary Nobel Prizes (3 in Physics and 1 in Chemistry):

- The *Christian* radio astronomer A. **Hewish** (ref.371-p.111), 1974 **Nobel** Prize said:

"God certainly seems to be a rational Creator. That entire terrestrial world is made from electrons, protons, and neutrons and that a vacuum is filled with virtual particles demands incredible rationality."

Let's remember that virtual particle pairs appear according to the principle of uncertainty, during a time too short to be detected directly. We enter the realm of theoretical physicists.

- His student, *Quaker* Jocelyn **Bell Burnell** (ref.370, p.110), the radio astronomer who discovered the first pulsars and deserved to share this Prize said in 2000:

"Any faith, any church...gets its authority from several roots. One is the holy writings, ... The second is the Church's history ... . And the third one is what is called "Continuing Revelation." God speaking to contemporary people, and Quakerism puts great emphasis on that third one... you are liberated from having to believe word-for-word what the book of Genesis says... You take responsibility on your shoulders for developing an understanding of God and God's will for this world and God's will for you and so on. And these developing and understanding are very like what a research scientist does."

To summarize, she confesses that it requires a lot of intuition, reflection, questioning, and self-confidence. This self-confidence that she has acquired over the years by seeing that her ideas in science often turned out to be right. This contextualization of the Scriptures is not specific to Quakers, cf. I. **Newton**, *Anglican,* and others among *Catholics,* but I gave her the floor because she is a very high-level scientist who speaks from the heart.

- For the theoretical physicist Abdus **Salam** (ref.394, p.117), 1979 **Nobel** Prize in Physics, an *Ahmadist:*

"The Holy Quran enjoins us to reflect on the truths of the laws of nature created by Allah; however, that our generation has been privileged to glimpse a part of His design is a generosity and a grace for which I give thanks with a humble heart."

- The physicist Arno **Penzias**, 1978 **Nobel** Prize (ref.364-108), a conservative *Jew* who also subscribes to God's creation of the Universe, had aptly answered the 12/03/**1972** to Malcolm W. Bröwne for the New York Times:

"How could a lambda individual take sides in this dispute between giants? On the one hand, the universe was created out of nothing, while on the other hand, the eternity of matter is proclaimed. Okay, let's consider the dogma of the eternity of matter. It comes from the intuitive feeling of people (including the majority of physicists) who do not want to accept the obvious observations that the universe was created, although the creation of the universe is supported by all the data that astronomy has produced so far."

Adding:

"... the most accurate data we have is exactly what I could have predicted from the 5 books of Moses, the Psalms, and the Bible."

"All reality, to a greater or lesser extent, reveals God's purpose. There is a certain relationship between the design and the order of the world in all aspects of human experience ...... This world is most consistent with the purpose of creation."

- The **1996 Nobel** Prize in Chemistry, Richard E. **Smalley** (ref.424, p.126), a *Christian* sharing his love for the Earth, said in 1960, during the 79th Annual Scholarship convocation at Tuskegee University (cf. Wikipedia Biography):

"Genesis was right, and there was creation, and that Creator is still involved... We are the only species that can destroy the Earth or take care of it and nurture all that live on this very special planet. I'm urging you to look into these things. For whatever reason, this planet was built specifically for us. Working on this planet is an absolute moral code... Let's go out and do what we were put on Earth to do."

We understand that he adhered to the evolution theory. In the late 1990s, he advocated for *"clean energy,"* the need for an alternative to fossil fuels, on the dangers of global warming, ...

I will conclude this paragraph with one of the most illustrious *Christian* scientists: Isaac **Newton**, an *Anglican*, who ends his treatise *"Optics"* [Ne1], an entirely scientific work, with a series of questions concerning some experiments that challenge him. It is interesting to note that the last one (XXXI) is dedicated to God:

> *"All this considered, it seems to me very probable that God formed in the beginning matter of solid, heavy, hard, impenetrable, mobile particles, of such sizes, figures, and other properties ..., so hard that they never wear out or break, nothing being able (according to the ordinary course of Nature) to divide what was primitively by God himself."*

> *"... it is through these principles [two or three general principles of motion] that matter was made, at the time of creation, of particles... variously combined by the will of an intelligent Being; for it is to him who created these particles that it belongs to put them in order. If he did it, it is not philosophical to look for another origin to the world, or to claim that the simple laws of Nature could have drawn it from chaos...."*

> *"It seems ... these particles are moved by certain active principles, such as that of gravity, of fermentation, ... I consider these principles, not only as occult qualities, which would result from the specific form of things; but as laws of Nature, by which the things themselves are formed."*

After these few quotations from exegetes, scientists, and often both, why should we put the one who has faith - on trial - even if he cannot rationally explain all the acts of God?

5.1.2. Man, the Image of God?

"... let us make man according to our image and likeness, ..." [Ge 1: 26]

"God, in his permanence, takes on different faces for different humanities"[Lr2].

Louis Leprince-Ringuet[597]

---

[597] **Leprince-Ringuet** Louis *(FRA: Alès 1901-2000 Paris). Graduated from "Polytechnique," Paris, in accordance with the family tradition. Telecommunications Engineer, and* Maurice **de Broglie**'s *student, whom he described as his* "spiritual father," *he contributed a great deal to the development of nuclear physics as an experimental physicist and designer of equipment and infrastructures. He is known for the early discovery of Kaon and coined the term "hyperon" in 1953. Commissioner for atomic energy from 1951 to 1971, he advocated strongly for the creation of CERN. He directed two large laboratories at Polytechnique where he taught physics from 1936 to 1969 and at the Collège de France from 1959 to 1972, was elected to the French Academy of Sciences in 1949, to the Pontifical Academy in 1961, and to the French Academy in 1966. Author of numerous scientific publications, he also wrote successful books such as* "Le Grand Merdier *or* l'Espoir pour demain" [LR] *in which, as a* practicing catholic, *he offers a thought of hope for us and for all peoples, a thought rooted in the* "message of Christ in which he finds a revolutionary wisdom in the love of others..." *(p. 245). He became president of the Catholic Union of French Scientists in 1949. His list of scientific and republican honors is impressive. He was vice chair (1956-69) and chair (1964-66) of CERN's scientific committee.*

The average person, conceiving God and His work from what he sees and understands, wrongly attributes to Him all the epithets relating to corporeality, forgetting that the attributes of God and ours are so different.

Also, how can we understand the epigraph of this paragraph attributed to God? Especially when **St. John the Evangelist** says [I 4:24]: "*God is spirit, ...,*" which at first sight is not our case. But the Apostle **Paul** in his 1st Epistle to the Thessalonians [5:23] gives us a clue when he pronounces a benediction of peace "*... let your whole being, spirit, soul, and body, be kept blameless until our Lord Jesus Christ comes ....*" He distinguishes in them, and therefore in us, three parts. Many of us separate the Soul from the Body, but few make a distinction between Spirit and Soul. So, in the Epistle to the Romans [Rom: 1:20], he encourages us to use the Intelligence and the Reason that makes us special:

> "*For since the creation of the world God's invisible qualities - His eternal power and divine nature – have been clearly seen, being understood from what has been made, so that people are without excuse.*"

Xenophanes (ref.41-p.13), the theologian par excellence among the pre-Socratics, rejected all anthropomorphism of the representations of the divinities of his time. And did not hesitate to declare immoral the gods of **Homer**.

Later, **Cicero** (ref.80-p.24), will say:

> "*...... God must have a form, but still the most beautiful of all, ...,*"

without defining it. It is a difficult question. Nevertheless, God does not have a bodily form identical to that of Man, notwithstanding the great Rabbi Abraham ben **David** de **Posquières**.[598]

Moses **Maimonides** explains[599] that we are in the presence of a homonymy: the Hebrew word "*Celem*" which has been translated, can be interpreted as "*bodily form*" - which according to him is an error and leads him to treat as heretics those who take this expression literally - and "*natural form*" which, on the other hand, is not. "*Natural form*" must do with our intelligence, differentiating us from animals. In fact, according to Maimonides, he endowed man with Reason (the

---

The above quote is from the lecture he gave on the occasion of his reception at the French Academy.

[598] Abraham ben **David** de **Posquières** *(FRA: Narbonne 1120-1197 Posquières, Gard). Philosopher, one of the greatest Provençal rabbis, directed the Kabbalistic school bearing his name. He was an opponent of Maimonides, whom he reproached for assimilating Aristotelian concepts to Scripture.*

[599] *First part of chapter 1 of* "The Guide to the Lost" [Ma1]. "Célem ...form ... designated the figure of a thing and its lineaments, which led to the pure embodiment of God...." *But* "this is a denomination that never applies to God ... Celem applies to the natural form ... which constitutes the substance of the thing ... in man, it is that from which comes the human understanding......intellectual" *that is to say, what characterizes human and distinguishes him from animals.*

ability to distinguish truth from falsehood), which constitutes his final perfection[600] and justifies the statement that he was made "*in the image of God and His likeness.*"

The word "*resemblance,*" a translation of the Hebrew "*demut,*" can refer to the physical image as well as to the image of an idea (for example, we say "*You are as beautiful as a star,*" which does not mean that one is physically comparable to a star but that he has his brilliance, his beauty, etc.).

In the "*Guide to the Lost*" ([Ma1] chapter LX), **Maimonides** proposes an apophatic approach:

> "*To get closer to the knowledge of God and His perception it is necessary to attribute all negative attributes to Him and to demonstrate that they cannot be attributed to Him.*"

What mathematicians call "*reasoning by the absurd.*" This method is fruitful in mathematics.

Saint Thomas Aquinas, a great *Dominican theologian*, in the tradition of **Maimonides**, while having his philosophy, explained it by what is called "*the analogy of being.*" This notion is rather difficult to explain, but some people summarize it by assimilating the word analogy to a relative resemblance between different realities, "*because of a certain relationship that He [God] has with the things from which our intelligence draws its concepts.*"[601]

This explanation is not in contradiction with "*Le grand Robert de la langue française*" [Ro] for which, until the middle of the 18th century, the words "*idea*" and "*image*" were used indistinctly because certain Hebrew and Greek words do not have their equivalent in French and can be interpreted in different ways, hence, some understandable variants in the translations.

The multiform scientist Ch. **Huygens**, *a Calvinist*, wrote [Mor] (in his time):

> "*The pagans and barbarians attributed to God a body similar to the human body, the philosophers attribute to Him a soul similar to the human soul and affections similar to ours, only different in perfection. They give Him a way of thinking, wanting, hearing, and loving. What else could they do? To admit that it far surpasses man to have an idea of God.*"

God's reason and power, are completely different from ours, it seems to me vain trying to understand His attributes.

The "Great Mr. **Newton**," a fervent *Anglican* [602] and follower of

---

[600] *Let us recall that in Psalm VIII, 6, speaking of Adam, it is written:* "Hardly did you make him less than a god (this refers to the angels forming the court of God), crowning him with glory and splendor."

[601] *The Twenty-Four Thomistic Theses, p. 22.*

[602] **I. Newton** *fought against the metaphysical theology which, according to him, transformed the apostolic message - which could be summed up as* "Love God and your neighbor" - *into a*

hermeneutics, rightly convinced that the prophetic texts, especially *"Daniel"* and *"Revelation,"* were written in coded language, gave them great importance, trying to decipher them and advising, in *"Interpretation of the Prophecies"* [Br], a method to achieve this. In this, he joined the Kabbalah.[603] He allowed himself - in his master scientific work: *"Principia Mathematica"* [Ne2 p. 177] - to make some remarks inciting his readers to undertake an allegorical reading of the Bible:

> *"... He (God) has no body nor corporeal form, so he cannot be seen, nor touched, nor heard, & he must not be worshipped in any sensible form... We have no idea of his substance .... We know him only by his properties & attributes, ... we admire him because of his perfections, ... He is present everywhere, not only virtually, but substantially, for one cannot act where one is not. Everything is moved and contained in him .... just as a blind man has no idea of colors, so we have no idea of how the Supreme Being feels & knows all things."*

He is very clear on this subject when he writes at the end of his mathematical description of the universe:[604]

> *"It is said allegorically that God sees, hears, speaks, rejoices, is angry ... because everything that is said about God is taken from some comparison with human things; but these comparisons, although they are very imperfect, nevertheless give some faint idea of him"* [Ne2 p.178].

According to the *Lutheran* polymath G. **Leibniz** (ref.182-p.53) the originator of the concept of the preestablished harmony between soul and body:

> *"... every substance* [he speaks here of the soul, entelechy or active principle] *is like a whole world and like a mirror of God or the whole universe..."* [Lz2 - 147].

He clarifies his thought:

> *"The only spirits are made in His [God's] image and almost of His race."*

Let's return to Martinès de **Pasqually**'s philosophy (Appendix 3, p.174). At its heart lays a profound emphasis on spiritual enlightenment, and divine communion. According to his perspective, before Adam's transgression, the

---

*list of mysteries erected by the clergy. He was a hard opponent of the Holy Trinity because he was convinced that the translations of certain biblical passages in **John** and **Timothy** for example, were erroneous and led to the introduction of the notion of the **Trinity** which he did not find in the Greek texts; leading him to accuse the Roman Church's translators of "pious frauds." That is why he always sought to return to primitive religion.*

[603] Kabbalah. *Collection of words transmitted (oral law) by God to Moses at Mount Sinai.*

[604] *[Ne2] Book 3 p.178.*

spiritual beings emanating from God were androgynous, immortal, and adorned with light garments, much like God. Endowed with *"free will,"* they inhabited the divine circumference known as the *"Divination,"* distinct from the supercelestial immensity that included both the celestial realm and the earthly domain, reserved for the animals they ruled over. Within them, resided *"a portion of the divine Domination; they could clearly and with certainty perceive all that unfolded within the Divinity, as well as everything contained within it."* They were fashioned in the image of God, at His resemblance, as they appear in the Genesis narrative.

But, when Adam, believing himself equal to God in every way, and influenced by the rebellious angels, transgressed the limits assigned by God; and sought to create spiritual beings as well, these beings assumed a material form! God expelled Adam from the Garden of Eden. Adam became the first mortal human, residing on Earth among animals. Merciful, God left Free Will to him and his descendants who were forced to work to survive. He nevertheless granted them the opportunity to redeem themselves to return to their spiritual form, regain their divine privileges, and benefit from immortality.

This concept of being equal to God finds parallels in the Bible, such as in Matthew 16:21, Where **Jesus** foretells his suffering, death, and resurrection to his disciples. Peter taking Jesus aside said: *"Far be it from you, Lord! This shall never happen to you."* This prompts Jesus' response: *"Get behind me, Satan! You are a stumbling block to me, you do not have in mind the concerns of God, but merely human concerns."* Nevertheless, this did not prevent Jesus from appointing **Peter** as the first Christian Pope.

For the 1927 **Nobel** Prize in Literature, Henri **Bergson,**[605] a *Jewish* with *Catholic* tendencies, since the middle of the 18th century the word *"image"*:

> *"... is opposed to reality or things that exist independently of the mind that thinks them."*

Here it must be taken as an allegory, metaphor, or comparison.

The **20th century** witnessed a profound transformation in the landscape of science, marked by the advent of advanced mathematical tools,[606] and a re-evaluation of fundamental concepts, such as substance, matter, energy, and

---

[605] **Bergson** Henri *(Paris, Fr: 1859 - 1941) philosopher elected to the French Academy in 1914, Nobel Prize for Literature in 1927.* Jewish *by birth, he wished in 1937 to convert to* catholicism *but gave it up to remain in solidarity with the* Jews *whose persecution was announced. He nevertheless obtained that* a catholic priest *officiates at his burial.*

[606] *The theories elaborated by mathematicians involve mathematical tools and spaces where the number of dimensions can be much higher than 3, contrary to the Euclidean space in which we evolve, daily, and which is however sufficient to solve many problems. On the other hand, when the theory of relativity (the infinitely great, speeds close to light) or quantum mechanics (the infinitely small, elementary particles) is used to interpret the universe and its laws, it is different. The images that we have had of threedimensional space and time seem very naive since the advent of the theory of relativity and its success in describing the universe.*

simultaneity, ... These foundational ideas, which trace their origins to thinkers like the Pythagoreans, underwent a metamorphosis in meaning and interpretation. In the realm of cosmology, as concepts evolved and abstract ideas took root, it became evident that describing subatomic particles necessitated a departure from strict mathematical symbols alone. To traverse the intricate domain of Quantum Mechanics, for example, and predict its outcomes, physicists turned to the power of imagery and parables, navigating the boundary between the conceivable and the tangible.

Erwin **Schrödinger**, a figure deeply routed in *Christian* beliefs and with a keen interest in *Hinduism*, profoundly observed the distinction between reality and perception, a philosophical concept echoed by many:

> "*Many philosophers have stated that a distinction must be made between a man's perception of a tree and the tree itself" in itself "... a good deal of consciousness is in this sense.*"

For *Hinduism*, reality transcends the material world, belonging to the realm of the spiritual, accessible only to the enlightened:

> "*The Absolute Truth belongs to the creative spiritual world and not to the Material Universe*" [Ka p- 6],

It's a reality that casts a mere shadow in the material world, and only the "*Great Souls*" can glimpse its true form. E. **Schrödinger**, employing a crystal metaphor with multifaceted perspectives, illustrates these multiple views of reality. There's the ineffable image of God, beyond human comprehension, and there are the images one can derive from interpreting ancient texts like Hindu scriptures or biblical prophecies (as **Newton** encourages us to do).

In the intricate tapestry of these perspectives, **Schrödinger** weaves the concept of Atman and Brahman, drawing parallels between Hindu and Christian spiritual beliefs. **Brahman**, the Supreme cosmic entity, embodies the essence of all energies and knowledge; an omnipresent, omniscient force. **Atman,** on the other hand, is the individual soul or consciousness residing within each living being, a fragment of **Brahman**. It's an eternal essence confined within mortal bodies, a glimpse of the greater whole. The core of yoga seeks a union wherein **Atman** merges seamlessly with **Brahman**, a state akin to ecstasy where the individual transcends the self and becomes one with the divine.

According to Viraj Kulkarni [Ku], at the end of his lectures, E. **Schrodinger** often wrote the following equation:

Atman = Brahman,

jokingly calling it: the "*second Schrödinger's equation.*" He also approved (cf. [Sc2]-p. 87) the poetic expressions through which they conveyed these profound truths:

> "*The one supreme divinity*

Being in all beings And living when they die, Who sees it is a seer.
For he who has found everywhere
The supreme God,
That man will no longer harm himself."

Niels **Bohr** (ref.297-p.89), the father of *"atomystic,"* [607] as Wolfgang **Pauli** (ref.310-p.92), a *Catholic*, respectfully called him in private, whose Copenhagen school was the rallying point of atomistic physicists, wrote [He]:

> *"Quantum theory thus constitutes an excellent example proving that one can have understood a certain state of affairs in all its clarity and that simultaneously one nevertheless knows that only images[608] and parables can evoke it... we can nevertheless, with the help of these images, come closer in a certain sense to the reality of things."*

In **1922**, during a walk on the sidelines of the Göttingen Congress, Niels **Bohr** expressed to *Christian* Werner **Heisenberg** (ref.312-p.93), a student of Arnold **Sommerfeld** (ref.299-p.89), his doubts about the atom planetary system as he had proposed it. Questioned by the latter on the images that he proposed, he answered:

> *"... these images (atoms) have been guessed from experimental facts ... we can only use language here in the manner of poets, who also do not seek to represent facts precisely, but only to create images in the minds of their audience, and to establish connections on the level of ideas."*

This led Werner **Heisenberg** to say[609] to his friend Wolfgang **Pauli**, who, working a lot at night, rested during the day when there was no communication:

> *"Bohr has acquired an intuitive idea of the structure of atoms, it is a picture that he tries to make understandable to other physicists by appealing to the electronic orbits inside the atom ... without perhaps believing in it."*

Pauli replied:

---

[607] Atomystic. N. **Bohr**, *one of the founders of quantum mechanics and leader of the Copenhagen school, improved in 1927 the atomic model of E. **Rutherford** (1911), by proposing the "stationary model" of which he had had the intuition, based on the stability of the matter, according to which the radiation emitted by the atoms is due to the passage of the electron of an orbital to another lower. His model justifies quite precisely the measured atomic radiation but defies the laws of electromagnetism. The quantum mechanics will confirm and improve it.*

[608] By image and parabola, *he means the complementary concepts of corpuscle and wave (which are two different, even contradictory, images associated with the same physical reality) as well as the notion of the electronic orbitals. We also speak of wave-corpuscle duality (ref. 304, p. 66).*

[609] *Cf. [He] pp. 59 to 65.*

"... Bohr's images must be correct, but how can we understand them, and what laws do they hide?"

The same W. **Heisenberg**, accustomed to manipulating abstraction, speaking of *"God the Creator,"* used the imaginary expression: *"the central order"* or the *"One"* of ancient terminology. What he justified, saying:

"... in atomic theory to formulate in an abstract mathematical language a unified order extending over vast domains; ... we see ... that when we want to describe in natural language the consequences of this order, we are reduced to use parables ... which contain paradoxes and apparent contradictions" ([He] Chap. XVII).

During the same conference, he explained to W. **Paul Dirac:**

"...the central order of things or phenomena, whose existence is not in doubt, ..." as *"one communicates eventually, with the soul of another human being."*

Answering one of his questions about the bijective relationship between God and himself, he said:

"I think that all these correlations can be better thought now since we have understood the quantum theory."

In effect, quantum theory, particularly the principle of complementarity, has fully changed our vision of material objects. It teaches us that our view is dependent on how we observe them. Here we have the germ of the ideas currently developed by certain theoretical physicists such as F. **Tipler** (ref. 695-p.247) for example.

Louis **Leprince-Ringuet**, an eminent *Catholic* Physicist, elected to the American Philosophical Society, told during his reception speech in the French *"Académie des Sciences"*:

"To summarize, the two aspects: the religious and the scientific, we can say "that Image" is not a carbon copy, but the fact that God gave man some of His attributes that distinguish him from animals, and make his specificity. It can even be admitted that Adam did not have all the qualities of God and could have had defects, although here again, it is possible to discuss the meaning to be attributed to Adam."[610]

Closer to home, Pope **Francis** wrote ([Fr] - p.177):

---

[610] *According to* Peter **Enns** *(Passaic, NJ-US 1961), an* evangelical theologian, *and Senior Fellow in Biblical Studies,* **Adam** *is the beginning of Israel, not of humanity. According to the same source ([Gi] Pp. 210 & 211),* **Adam** *represents a man, whereas* **adam** *represents humanity; hence the uncertainty about the exact role attributed to Adam in Genesis and especially about the meaning of* "the fall." *See YEC and Adam. 653 - p. 236.*

"... the son of God integrated into his person a part of the material universe,"

which for Christians is easily understood because God, the Creator, is present in all things.

I cannot imagine that this infinitely small man - placed at the center of this infinitely small Earth compared to the Milky Way, which is only a tiny part of the billions of galaxies that make up our universe, which some people think is only one among many - can have a clear vision of the infinite God, Creator of all this. Hubert **Reeves** seemed to have a similar view to mine on this point:

"One of the lessons of contemporary science is that the universe is infinitely more imaginative and inventive than we are."

It's hard to fathom how an infinitely tiny being, positioned at the center of a minuscule Earth within the vastness of the Milky Way – itself just a fraction of the countless galaxies constituting our universe – could grasp a clear understanding of the infinite God, the Creator of it all. This sentiment echoes that of Hubert **Reeves**, who shared a similar perspective:

"One of the lessons of contemporary science is that the universe is infinitely more imaginative and inventive than we are."

Regarding God, setting aside the intricacies of the Trinity, I view Him as dual. He took on human form through Jesus Christ (the Logos, the Word) and consistently revealed His presence in an immaterial guise as the Holy Spirit. The Greek translation of the Holy Spirit as breath, wind, or God's active force, hints at the possibility of assuming unknown forms or waves. [611] This allows communication at speeds surpassing[612] those of the observable electromagnetic spectrum while transmitting energy, akin to known waves. It also serves as a receiver, enabling dialogue with us. A scientific analogy could be drawn to the **Higgs'** field (ref. 449-p.133) or any other field, as suggested by Hubert **Reeves** [Re2] who emphasized:

"Who would be so foolhardy as to assert that we know and perceive all forces, all waves, and all means of communication?"

Similarly, the duality observed in entities like the electron (wave or corpuscle) depending on experimental conditions, highlights our limited sensory perception of reality. Theoretical physicists bypass this limitation by employing

---

[611] *If in 1865* **J. C. Maxwell** *imagined the concept of Electromagnetic Waves (E, H) it is only in 1888 that H.* **Hertz** *(1888) proved experimentally their existence. Until then, scientists had to rely on the ether: an invisible, immaterial gas (without mass and smell), energetic, filling the whole universe, a substance imagined by* **Aristotle** *in the 4th century BC.*

[612] *Let's recall that, according to the theory of relativity, only material bodies, including in a galaxy, have a speed limit: the celerity of light in a vacuum; but galaxies can move at speeds higher than c.*

mathematics, a realm of abstraction that may uncover the laws of nature, though **Gödel**'s theorem[613] reminds us of its inherent incompleteness.

In my perspective, describing God is ineffable. We should not seek a corporeal resemblance, for God and Man differ in their very nature. However, through essence (the intrinsic characteristics of beings or things), man mirrors God, endowed with an immortal soul. Yet, man, marred by imperfections, can only imperfectly reflect God's image. As we distance ourselves from this ideal, as noted in the teachings of **Plotinus** and the **Upanishads** (ref.6-p.2) our image further deteriorates.[614] Notably, Isaac **Newton**'s beliefs diverge from mine, as he did not adhere to the concept of the Trinity. But he nevertheless wrote in [Ne2 - p. 177]:

> *"God is one & the same God everywhere & always. He is present everywhere, not only virtually, but substantially, for one cannot act where one is not."*

Let's conclude this reflection with Antoine de **Saint-Exupéry**'s[615] wisdom:

> *"One can only see well with the heart, the essential is invisible to our eyes."*

5.2. Is the universe the result of chance or a divine plan? Insights from specialists.

> *"What is chance with regard to men, is design with regard to God."*
> **Bossuet**

> *"This Being governs all things, not as the soul of the world, but as Lord over all... For God is a relative word and has respect to servants, and Deity is the Dominion of God not over his own body as those who fancy God to be*

---

[613] **Gödel**'s theorem *or* **incompleteness theorem***: ([TXT] p. 62):* "... there is always a limit to our knowledge of a given arithmetic system because we are part of this system. To go beyond this limit, we must leave the system. .... The power of rational thought is not without limits. ... A computer could never solve some mathematical problems...." (**Gödel:** *ref. 184–p.56).*

[614] For **Plotinus** (ref.78-p.23) *there is a hierarchy: God (the ONE) above all, then intelligence (knowledge: the vision of the ONE, of oneself, and of the intelligible world), the Soul, and Matter ([Th] pp.169-179).*

[615] de **Saint-Exupéry** Antoine *(Lyon, FR 1900-1944 Mediterranean Sea) Christian. Pilot (Aéropostale from 1932, during the war, then as resistant in a squadron of recognition; the wreck of his plane was found recently off Marseille). A great reporter and a recognized writer, he was awarded the Prix Fémina in 1931 for* "Vol de nuit," *and the Académie française in 1939 for* "Terre des hommes." *An amateur scientist, he nevertheless holds a dozen technical patents. Brought up in the* catholic religion, *he affirms in* "Ecrits de Guerre" *p. 283:* "I simply and totally commit my conscience as a Christian" *and p. 118:* "It is necessary to serve the *Christian* idea of the Word who became flesh."

the soul of the world, but over servants.... but a being, however perfect,
without dominion, cannot be said to be Lord God."
I. Newton[616]

In the grand tapestry of existence, the question looms large: is the creation of the universe a product of mere chance, or does it align with a divine plan?

The hypothesis of *"chance"* at the origin of our universe, put forward by atheists, is not new but it is necessary to agree on the meaning we attribute to this name. 2.500 years ago, **Leucippus** and **Democritus** said:

"... necessity and chance are not opposed to each other, but they are both opposed to the action of an intelligence or a providence" [Pu].

An opinion shared closer to us by Jacques **Brunschwig**,[617] a 20th-century philosopher:

"... chance seems to be only the name we give to the cause when it is unknown or unrecognizable."

This definition of chance is that of the atheists. Facing a phenomenon inexplicable rationally, seeming to result from a combination of unexpected circumstances, they do not want to hear the non-scientific explanations of believing scientists. This is their right, but are they right? I don't think so, because there are many scientific facts that even the greatest scientists cannot explain, and nevertheless, the same people admit it without any proof. We have seen in Chapter 3 that the cause of the Big Bang, for example, is not yet known to scientists, who are still only at the stage of hypotheses, which does not prevent them from admitting the validity of this theory as long as no experiment will invalidate it. After this parenthesis, let's resume chronologically the ideas of some recognized thinkers and scientists.

In the philosophical lineage of **Anaxagoras** and **Socrates**, **Plato** [Th] contends that the Soul transcends the physical realm. According to him, the soul possesses the remarkable ability to comprehend both the ideal world, characterized by stability, and the sensible world, marked by perpetual motion. For Plato, the cosmic order is not a result of chance, it is the:

"handiwork of a divine Craftsman (dêmiourgos) the Royal Soul ... imposing mathematical order on preexistent chaos to generate the ordered cosmos."

---

[616] For **Bossuet** *(ref. 183-p.56) Cf. Le Robert Vol. 5 p. 115 and for Newton: [Ne2] pp. 175 & 176.*
[617] **Brunschwig** Jacques *(FR: Paris 1929-2010). Graduated (1st in 1948), from the Ecole Normale Supérieure, Paris, he passed the agrégation in philosophy in 1952. He taught ancient philosophy in Paris, Princeton, Austin, Cambridge, Berlin, Venice, Padua, ...*

Timaeus, a *Pythagorean philosopher* presents [Pla], on scientific knowledge bases of the time, Plato's vision of the formation of the universe based on the **Necessity** to reach his goals. Consequently, our universe would be:

"the product of rational, purposive, and beneficial agency."

Aristotle established that the eternal world and all the beings that populate it are not the results of chance. When he speaks of necessity we need to see it as emanating from a superior intelligence: God, acting in full consciousness, is at the origin of movement and time.

Later, **Avicenna**, a *Muslim*, and then *Rabbi* Moses **Maimonides**, believing in a world created by God, would endeavor to demonstrate that God wanted this world freely and did not act by necessity:

"God made all things by willing this universe;"[618]

Among the distinguished scientists who did not believe in chance, we selected the *Anglican* sir Isaac **Newton** who speaking of gravity established that:

"... gravity explains the movement of the planets but cannot explain who set the planets in motion. God governs all things and knows all that is or is to be done." [Ne1]

At the same time, his famed rival, Gottfried W. **Leibniz**, a *Lutheran* polymath stated:

"... God does nothing at random ... the effect must answer to its cause."[Lz2]

Chapter 2 (like the others), is intended to be factual; nevertheless, believers will be able to see that if God created our universe, it is based on fundamental laws and constants, allowing Man to apprehend them, what G. **Leibniz** calls the *"ordinary wills."* He also established *"extraordinary wills,"* that no created mind, however, enlightened he may be, can foresee by reasoning, because their understanding, distinct from the general order, surpasses his capacities. They are the cause of what is called *"miracles"* but are entirely consistent with the original universal order He intended. When they occur, therefore, it does not mean that God intervenes punctually like a mechanic to repair or adjust a faulty machine. According to G. **Leibniz**, what seems bad today will be good tomorrow because God wants the best for us, he created the most perfect universe, that is to say: *"the one that is at the same time the simplest in hypothesis and the richest in phenomena."*

The *Jesuit* mathematician Roger **Boscovich**,[619] sharing the opinion of I. **Newton** and G. **Leibniz**, stated:

---

[618] *Cf. [Ma1] chap. XX p. 172.*
[619] **Boscovich** Roger, J. *(Dubrovnik, Croatia 1711-1787 Milano, ITA). Jesuit, theologian, mathematician, astronomer, diplomat. He studied at the Jesuit College of Dubrovnik and then,*

"... the world could not be the result of chance... by a fortunate meeting of an imposing number of determined conditions, each of which would have arrived at reality among an infinite number of possibilities. It could not exist by itself, by its force... It was created, that is to say, sorted out by an intelligent and free choice from the infinity of possibilities...."

He also stated, alluding to the conclusions so obvious to him that anyone studying the universe can make:

"Non-pious astronomers must be crazy."

André-Marie **Ampère,**[620] a polymathe. For those who were not enthusiastic about physics in their final year, I will point out that he laid the foundation for the mathematical formulation of classical electromagnetism.

His Ampère's Circuit Law describes the relationship between electric current and magnetic fields; he also introduced the concept of Ampere, a unit of electric current. It is well known that J.C. **Maxwell** admired his work and gave him *"Newton of Electricity"* to highlight Ampère's significant contributions to the field of electromagnetism, similar to how Sir Isaac Newton made groundbreaking contributions to classical mechanics. We can also think that it is because he sought a quantitative law of interaction between two conducting wires traversed by currents I and I,' similar to that discovered by I. **Newton** to describe the force of gravitation between two masses M and M,' cf. (ref. 185-p.59). A-M. Ampère a devout *Catholic*, was a follower of the *"minor brothers."*[621] Interested in music and poetry he wrote:

"The true scientist Happy, the one who during his learned vigils,
Of this vast universe contemplating the wonders,

---

in 1920, entered the Society of Jesus (Jesuits, in Roma), and was ordained priest in 1740. In the meantime, he had been noticed for his gifts in Mathematics and Physics. Traveling through Europe he was recognized by his peers who admitted him to their scientific academies (St Petersburg, London (Royal Society), Moscow, ...). Built an astronomical observatory in Milano in 1762, of which he became director. At the same time, he taught mathematics in Pavia. Invited to France in 1773 by Louis XV he was, naturalized, and appointed director of the optical center of the Navy in Paris in 1778. He propagated **Newton**'s philosophical ideas in Italy.

[620] **Ampère** André-Marie (FR: Lyon 1775-1836 Marseille), Catholic. He was self-taught and learned from his father's library. At the age of 13, he wrote a treatise on the sections of conics and became fascinated by algebra. In 1799, married, he settled in Lyon where he gave private lessons and set up a small laboratory in his apartment. Noticed in 1804 by Joseph J. **Lefrançois de Lalande,** Free Mason, and his pupil J-B J. **Delambre** (both renowned astronomers and members of the Academy of Sciences) he taught analysis at Polytechnique and was appointed professor in 1809. He became a member of the French Académie des Sciences in 1814 and was appointed the Physics chair of the Collège de France in 1824.

[621] Friars Minor: is the first of the three Franciscan orders. Created in 1209 by **St. Francis of Assisi** who was joined in 1212 by **St. Clare** creating the Order of Poor Clares and then the Third Order for secular priests (not subject to monastic vows but subject to papal authority).

Before so much beauty, before so much greatness,
Bows and recognizes a divine creator.
I do not share the mad inconsistency
Of the scientist who, of a God disputes the existence,
Who closes his ear to the announcement of the heavens, And refuses to
see what shines in his eyes.
To know God, to love him, to pay him pure homage,
This is the true knowledge and study of the wise."

Albert **Einstein**, a *Jewish* pantheist with a Spinozian tendency (*"cosmotheist"*
according to R. **Millikan**):

"*A contemporary has said, not without reason, that in our age, generally
devoted to materialism, serious scholars are the only men who are deeply
religious. The scientist is imbued with a sense of the causality of all that
happens... His religiosity lies in ecstatic astonishment at the harmony of
the laws of nature, in which is revealed a reason so superior that all the
ingenious thoughts of men and their arrangement are, in comparison,
but a quite useless reflection.*"

"*This conviction, linked to a deep feeling of a superior reason, revealing
itself in the world of experience, translates for me the idea of God... I
affirm that the cosmic religious feeling is the most powerful and noble
motive of scientific research*" [Ei1].

According to Louis **de Broglie**, *Catholic*, 1929 **Nobel** Prize in Physics:

"*The apparent indeterminism of microphysics must be attributed to
experimental conditions.*"

For the 1966 Nobel Prize in Physics Alfred **Kastler,**[622] humanist, *Evangelical
Protestant*:

"*The idea that the world, the material universe, created itself seems to
me absurd; I do not conceive of the world without a creator, hence a*

---

[622] **Kastler** Alfred *(FRA: Guebwiller 1902-1984 Bandol).* **Nobel** *Prize in Physics in 1966* "For
the discovery and development of optical methods for the study of Hertzian resonances in
atoms." *His discovery will have applications in atomic physics particularly for lasers. Born in
the German zone, at the time, he became French with the attachment of Alsace and Lorraine to
France. After graduating from the Ecole Normale Supérieure, he was awarded the Agrégation
de Physique in 1926 and taught in high schools and various French universities as a professor.
Director of research at the CNRS in 1968. Recipient of various awards including the gold medal
of the CNRS, he was also Director of Higher Education.* Evangelical Protestant, he was involved
in many humanist struggles.

God. For a physicist, a single atom is so complicated, so rich in intelligence, that the materialist universe makes no sense."[623]

Théodore **Monod** naturalist, geologist,.... *Protestant* stated:

"Life is a considerable mystery ... I find it difficult to imagine that chance alone can be sufficient to explain the world" [Mdc].

During an interview in September 1971 on American television [Wi] it was asked to Wernher **Von Braun,**[624] *Lutheran*, rocket scientist at NASA: *"Can you tell me, you specialist in interplanetary physics, what is your thought about God? In your opinion, does God preside over this world of sidereal infinity, this cosmos, or is it appropriate to declare the sky empty?."*

He answered: *"For me, I don't have to prove God to myself, I can say that I see him, that I experience his presence in some way, for the only reason that it would seem inconceivable to me to think that one could make scientific calculations and forecasts so precise and so complex as those we have had to make, if this cosmic universe were not subject to precise and constant laws which, alone, can allow them. As far as I know, there have never been laws without a legislator... God is for me this supreme Legislator to whom the worlds obey."*

Let's note that he was talking about worlds and not universes.

For the 1933 Nobel Prize in Physics E. **Schrödinger**, *Christian*, one of the fathers of quantum physics (Chapter. 2):

"...the cogwheel in question (the world of life) is not of crude human manufacture but constitutes the most delicate masterpiece ever made according to the principles of the Lord's quantum mechanics." [Sc2]

Hubert **Reeves**, answering Patrick Poivre d'Arvor, presenter of French TV (cf. interview with Claude Lelouch [Le]):

---

[623] *Express of 12/08/1968.*
[624] **Von Braun** Wernher *(Wirsitz, Posnania [Prussia] 1912 - 1977 Alexandria, Virginia). Lutheran, a naturalized American in 1955, was the rocket specialist at NASA. In 1932 he was the leader of a research team on missiles, a project supported by Hitler, of whom he was adulated, which led in 1942 to the realization of the V2, several thousand of which were launched on Europeans. Enrolled in the SS, he received numerous medals, which raised doubts about the integrity of his character and the reality of his faith. Arrested in 1944 by the Gestapo for defeatist remarks, he was taken over by the Americans at the end of the war, along with the main German engineers specializing in the field, to develop the first American ballistic missiles. It was his Juno 1 rocket that put into orbit the first artificial satellite "Explorer 1." Created NASA to develop the "Saturn" rockets necessary for the Apollo project. Born into a family of wealthy aristocrats, he showed a music talent. After a mediocre start in mathematics and physics, he proved to be brilliant in these fields.*

"When I am at a concert, and I hear Mozart, Beethoven, I cannot believe that this is the expression of chance." (...) "There is something... what it is, I don't know. ... I have this idea that there is something that means something."

For this observer of the sky and the stars:

"...the beauty of the great works, the harmony of the universe that surrounds us, make us sense in the depths of ourselves this mysterious sense of the divine that, as Pythagoras said, great mathematician and philosopher of antiquity, allows us to recognize that Nature is wise."

Also, can we believe that Stephen **Hawking**, an *agnostic*, thought that the universe is the fruit of chance when he wrote:

"If the gravitational attraction of a star like the Sun varied with distance more or less than the inverse square of the distance, the orbits of the planets would no longer be ellipses but spirals hurtling towards or away from the Sun"?

The order of Nature shook the *agnostic* astrophysicist Robert **Jastrow** [Ja1]:

"Although an agnostic I must admit that nature has a beginning and do not see how she could have created it on her own. I must admit that a supernatural force, which we can call God, outside space and time, has done this."

As did theoretical physicist and *Anglican* priest John **Polkinghorne** who wrote [Po] in 1996:

"...the priceless features of the world, its order, its beauty, its ethical imperative, its worshipful experience, reflect the personal traits of the Creator who is rational, joyful, good and holy."

In drawing the curtain of this chapter, let's revisit its essence through the perspectives of two distinguished Nobel Laureates. Anatole **France,**[625] *a freethinking* recipient of the 1921 Nobel Prize in Literature, once mused:

"In life, one must take into account the element of chance. Chance is ultimately God"

On the other hand, we encounter Sir Roger **Penrose** (ref.368, p.109), a *Jewish agnostic* who clinched the 2020 Nobel Prize in Physics. Despite his agnostic

---

[625] **France** Anatole *born* François Anatole Thibault *(FR: Paris 1844 - 1924 St Cyr-sur-Loire). Writer and literary critic, recipient in 1921 of the **Nobel** Prize for Literature* "for the whole of his work." *The quotation is taken from* "Le jardin d'Epicure" *Coda - 2004, p. 102.* Freethinker.

stance (I think that he does not recognize himself in any religion, and I would classify him as a Deist) he expressed in the film "*A History of Time*" that:

"he believes that the universe has a reason for being, that Man is not the result of Chance; and that there is something very profound in the reason for the universe"

These laureates while acknowledging the existence of a creative force use different lenses. Anatole France sees it as "*chance*" equivalent to "*luck*," while Sir Penrose leans toward "*probability*." How can we reconcile these seemingly disparate viewpoints? In the context of the knowledge amassed over centuries, let's propose that God has established eternal laws of physics, enabling truly spontaneous events, a notion supported by **Darwin**'s evolutionary insights (§ II.9 & V.4) and justified by the principles of quantum physics (§ II.10). This perspective justifies the existence of an organized, self-complexifying universe. Over time, the physical states of this universe can change, bringing about renewal and complexity. Whatever the reason, life and consciousness, have emerged, allowing humanity to trace its origin back to the cosmic event known as the Big Bang. In contemplating our existence, we are propelled to explore the profound interplay between eternal laws and spontaneous events, paving the way for a deeper understanding of our place in this intricately woven comic tapestry.

5.3. Is life on Earth the result of chance?

"The more modern cosmology has progressed, the more it has discovered that the universe has been regulated in an extremely precise way for the appearance of life and consciousness, that the existence of life is inscribed in the properties of each atom, of each star and galaxy of the universe, and of each of the laws that govern the universe ...
From the beginning, the universe contains in germ the conditions required for the emergence of a living and conscious being."
Trinh Xuan Thuan[Tn]

Let's remember that the 1st terrestrial life form (prokaryotes) goes back $3,85.10^9$ years, that it was followed by 5 mass extinctions, the last one 65 million years ago, and that the appearance of the 1st hominids dates only from 7 to 10 million years ago, that is to say, a long time after the creation of the universe.

It seems clear that the answer to question 5.3 is yes for those who don't believe in God Creator of the universe and that the likelihood of believers saying no is high. While various beliefs and hypotheses exist regarding the origin of life, it's interesting to note a few contrasting points and also the evolution with the time of some scientist's views.

It is the case for Paul **Dirac**, 1933 **Nobel** Prize in Physics. In his youth, Paul **Dirac** was an *atheist* and did not consider the existence of God necessary, as evidenced by a discussion he had in Solvay (a meeting place for leading scientists) in 1927 with Werner **Heisenberg** *Christian,* and Wolfgang **Pauli** *Catholic.* At the end of P. **Dirac**'s tirade, W. **Pauli**, said:

> *"Well, our friend Dirac has a religion and his creed is: there is no God and Dirac is his prophet."*

But, in 1963 P. **Dirac** changed his opinion and wrote in the magazine *"Scientific American"*:

> *"It seems to be one of the fundamental characteristics of nature that the fundamental physical laws are described as a mathematical theory of great beauty and power, ... why is nature built along these lines? We just have to accept it. We could eventually describe the situation by saying that God is a mathematician of the highest order and that he used very advanced mathematics to construct the universe. ...."*

In a 1971 conference, he drove the point home a little further:

> *"... The chances of life emerging when we have the right physical conditions are 10^{-100}... I feel that under these conditions it would be necessary to assume the existence of a god to trigger life. .... If the physical laws are such... It would be unreasonable to assume that life would have begun only by blind chance, then there must be a god...."*

Nevertheless, among the believers, there is also a debate around the contingency[626] of man. Because, if you have non-scientists and scientists too, who can answer *"What would have been the use of the universe if not for the appearance of life"*? you encounter also believers like R. **Boscovich**, who think that God had the choice and he decided to create Man free of his choices with as a consequence: *"the evils,"* with which we feel burdened - often rightly - but that we must assume because we are responsible for them. With this digression out of the way, let's resume the discussion chronologically.

In Genesis (2:7), it's mentioned that:

> *"Yahweh-God fashioned man from dust taken from the ground,"*

a concept that prompted the poet **Ovid**[627] to express in his Metamorphoses [Ov] the idea that:

> *"A being [Man] was still missing, more marked with the divine seal, a more qualified repository of a Penetrating intelligence, and one who*

---

[626] Contingent: *That which can happen or not; that which can be or not be: God was not obliged to create it.*

[627] **Ovid** *(Sulmo, ITA -43 B.C. to 17 A.D.) whose real name was* **Püblius Ovidius Näsö***. Studied law in Rome and became a lawyer. Poet, he published the Metamorphoses, Amores (The Loves), Heroides (The Heroines), ... then Metamorphoses (230 fables). In the year 8, he was exiled to Tomis (Black Sea) by the emperor* **Augustus***, where he died.*

could exercise dominion over the rest of the creation" [Ov].

According to Ovid, man's creation is a pivotal moment; he further pondered whether:

> "... *that man was born, either that the creator god, author of a better world, had formed him from the divine seed, or that the earth in its newness, recently freed from the deep layers of the ether, had preserved some germ of its brother the sky, and that this earth, the son of Japhet,* [628] *by mixing it with the rain waters, had shaped it in the image of the gods, moderators of all things."*

This thought has intrigued various thinkers throughout history. This concept was embraced by the *Muslim* physician, Professor Ismaa Sadaf **Farooqi**,[629] who took a literal interpretation of the scriptures:

> "*He (Allah) initiated the creation of your father Adam from dust, then he created his offspring from semen or worthless water"* Quran [40:67], and decided that:

> "*There is a time set for every people: they cannot hasten it, nor, when it comes, will they be able to delay it for a single moment"* Quran [34:7].

The enigma of biogenesis reverberates through the realms of both science and philosophy, shaping our quest for the origins of life; an exploration that is, fundamentally, a search for understanding ourselves.

While some captivating theories, like **Cairns-Smith**'s[630] proposition of life emerging from clay crystals with the ability to capture organic molecules in the early 1980s, have piqued interest, they remain on the fringes of scientific consensus, as Biophysicist D. **Vasilescu** [Vu2] emphasizes. Consequently, we won't delve further into these less mainstream ideas. On the contrary, the more ancient concept of panspermia (ref.504, p.154) championed by **Anaxagoras of**

---

[628] *Cf. [Ov] p. 43. According to Greek mythology* **Japhet**, *one of the Titans, is the son of* **Ouranos** *(the sky,) named Uranus by the Romans, and of* **Gaia** *(the earth). Japhet is "the father of man" via his son* **Prometheus** *who created Man from clay. Other legends exist about him.*

[629] **Farooqi** *Sadaf (1969) Physician, surgeon, geneticist, Bachelor of Medicine, a bachelor of surgery in 1993, Ph.D. in 2001 from the University of Cambridge* "on the genetics of severe childhood obesity." *Professor of Metabolism and Medicine at the University of Cambridge,* Muslim. The quotations are drawn from the Internet, during an interview conducted on February 03, 2012.

[630] **Cairns-Smith** *(Kilmarnock, Sc 1931-2016 Uplawmoore, ENG). Professor of Molecular Biology and Organic Chemistry at the University of Glasgow. Proposed to the general public in:* "The enigma of life" *(cf. [Cs]) the hypothesis that the self-replication of clay crystals would have been the intermediary between inert matter and life. He also pointed out the role of quantum mechanics in human thought. Was a member of the Royal Society of Edinburgh.*

Clazomenes, [631] *"the physicist par excellence"* has resurfaced and is gaining traction among eminent scientists: biologists, astrobiologists, and physicists alike, regardless of their belief systems.

As we progress through exploration, we will turn to the perspectives of prominent scientists who defend this theory. For now, let's embark on a journey into the insights of distinguished biologists.

Alexander **Oparin,** [632] an *Orthodox Christian* biochemist, amplifies this sentiment by debunking in **1920** the idea of chance playing a significant role:

"... to allude to the stroke of luck that, among trillions [10^{12}] and quadrillions [10^{24}] of combinations, was able to form by chance precisely that indispensable sequence required for the synthesis of proteins is exactly as irrational as when one previously alluded to the "chance" formation of proteins (enzymes). The structure of these proteins is not only very complicated, but it is also extremely well adapted to the accomplishment of the defined catalytic functions that play an important role in the life of the whole organism; this structure is strictly designed "for this purpose," for that. Such an adaptation to its biological function, such a structure per its purpose, characterizes also the nucleic acids of the present organisms, and, that it appeared by chance, is as impossible as the assembly by chance, from its elements, of a factory capable of producing any particular product" [Op].

Oparin's perspective underscores the irrationality of attributing life's intricacies solely to chance.

In his **1921** book, *"The Direction of Human Evolution,"* [Cn] Edwin **Conklin,** [633] a prominent biologist known for his studies in embryology,

---

[631] **Anaxagoras** *(Clazomenae, TUR -500 to -428 Lampsacus (Asia Minor)). He was the 1st Greek Philosopher to introduce this hypothesis. He also brought the concept of "Intelligence," to the origin of the creation of nature. Advocating purely mechanistic astronomy, he predicted ([Bt] p. 42) the fall of a giant meteorite on Aegospotami (Turkish peninsula of Gallipoli) and interpreted the eclipses of the Moon. Not interested in religion, he was condemned for impiety and exiled to Lampsacus.*

[632] **Oparin** *Alexandre I. (RUS: Uglich 1894-1980 Moscow). Biochemist, author in 1920 of a theory on the origins of life according to which there would have been, before the appearance of life, a prebiotic* synthesis of nucleic acids, sugars, ... various elements entering the constitution of proteins. This theory has been taken up and is not contradicted for example by that of* John D. **Bernal** *(Nenagh, IRL 1901-1971 London, ENG). Physicists have made interesting discoveries) according to which life would have appeared at the bottom of the oceans. *Constitutes the passage from minerals to primary organic compounds. According to Oparin, for example, hydrocyanic acid (formed from methane and ammonia) could have dissolved in water (sea, lakes, etc.) and thus provided the building blocks of life.*

[633] **Conklin** *Edwin G. (USA: Waldo, OH 1863-1952 Princeton, NJ). A* Methodist Protestant, *he considered the priesthood but turned to Science and Biology in particular. He is one of the fathers of the evolutionary biology of development. He prepared his thesis on* "The Development

genetics, and evolution, and a *Methodist*, expresses his views on religion and philosophy:

"I have no quarrel with any Religion or with any philosophy which does not deny or belittle the facts of science or human experience. I believe that there is a great deal more in religion that can be proved by science or logic, but I also believe that there is nothing in true religion that is contrary to science or common sense. Religion deals with values, ideals, sentiments, and aspirations; science deals with facts. Religion is concerned with what ought to be; science is with what is. Religion is an expression of the spirit; science is an expression of intellect. Religion gives meaning and value to life; science gives power over nature. Religion makes life worth living; science makes living easier and more comfortable."

He demonstrates that religion and science are not antagonists but complementary aspects of human nature. A proponent of evolution through natural selection, he paints a vivid metaphor:

"The probability that life is the result of an accident is comparable to the probability that a dictionary is the result of the explosion of a printing press" [Cn].

Highlighting the staggering improbability of life, arising by chance, he leads to a perspective that steers toward the notion of a divine plan.

The 1932 Nobel Prize of Physic, Werner **Heisenberg** [He] *Christian*, one of the fathers of Quantum Physics, wrote in **1965**:

"Perhaps the chance that plays such an important role in Darwin's theory is, precisely because it obeys the laws of quantum mechanics, something more subtle than we might think."

In effect, since we know that the fundamental physical laws can be expressed statistically we are not limited to a determinist conception; as the Uncertainty Principle (proposed by W. Heisenberg) teaches us (ref.312-p.93). Furthermore, we must stop limiting God's actions by referring to our human possibilities which are infinitesimal compared to His. Let us not forget that God is outside of space-time, that He could at the time of the Bang, and still now, have a vision of the

---

*of Gastropods" at John Hopkins (attended by 37 Nobel Prize winners) in Baltimore, MD. He then left to teach at various universities, including Princeton (from 1908 to 1933). Then he became an independent researcher and lecturer. He was a leading critic of eugenics racism, and war, and advocated for international cooperation and peace. He defended **Darwin**'s theory of evolution. He held many honorary positions (President or VP of various scientific associations) and received various awards including the John J. Carty Gold Medal awarded by the National Academy of Sciences in 1943. A scholarship bears his name.*

future, allowing Him to proceed to the adjustments of the various stages of His creation by having a perception of the result as well as of the continuation that we will not know down here. As an adherent to the theory of evolution (cf. below), I share the opinion of A. **Penzias** (ref.364, p.108), an orthodox *Jewish*, 1978 Nobel in Physics:

> *"Nature, blind nature, is driven by chance, by the laws of probability which revolve around what one would call "the second law of thermodynamics"…… "Nature through random chance seeks disorder. Life, on the contrary, works against that, it organizes, makes things orderly."*

But we must remember that, for the believers, this second law was initially established by God.

Anthony **Flew,**[634] once a prominent *Atheist* philosopher underwent, in 2004 (he was then 81 years old) a profound transformation. Acknowledging the intricate complexity revealed by biologists' research on DNA:

> *"The research of biologists on DNA has shown, by the almost inconceivable complexity of the arrangements necessary to produce [life], that an intelligence must necessarily be involved,"*

Flew revealed *"with regret,"* much to the dismay of his readers, that he had shifted toward *Aristotelian deism*:

> *"Since people, have certainly been influenced by me, I want to try to correct the enormous damage I must have caused."*

Pierre **Grassé,**[635] a devout *Catholic* zoologist, interviewed by R. **Decout,**[636] rejects the notion of chance as a driving force in evolution. He aligns with the belief that the word *"chance"* was introduced by materialists, to supplant the idea of a divine creator. His perspective challenges the role of randomness in the genesis of complex biological structures like the eye.

---

[634] **Flew** Anthony Garrard *(ENG: London 1923 - 2010 Reading). Declared himself an atheist at the age of 15 but a deist at the end of his life. Philosopher, specialist of D.* **Hume,** *a university professor in Keele (Staffordshire), Aberdeen (N-Y), Oxford, Reading (Berkshire-England). He says [F1] that his experience of God is a pilgrimage of Reason and not of Faith.*

[635] **Grassé** Pierre-Paul *(FR: Périgueux 1895 - 1985 Paris), Military doctor, at the end of the 1st world war he resumed studies in Biology in Paris. Professor of Zoology in 1929, he was president of the Zoological and Entomological Societies of France. Member of the FrenchAcadémie des Sciences in 1948 and president in 1967. Commander of La Légion d'Honneur, Doctor Honoris Causa of many foreign universities. From* Catholic *culture, he supported* **Lamarck**'s *thesis (according to which living beings have the faculty to transmit to their descendants a faculty acquired during their life) against* **Darwin**'s *thesis on evolution.*

[636] **Decout** R. *(1918 - 2015) Editor in chief of the "Voix du Nord." Interview on: "How they see the world," in ŒIL, 1984, p.57.*

Jacques-Yves **Cousteau,**[637] a pioneer of modern oceanography, *Catholic,* stated:

"I have reached, at the end of my road, to a certainty: the living universe is not the result of chance; there is an obvious unity in all forms of life......"

Amid these perspectives, Jacques **Monod's** views[638] stand prominently. If I can agree with him when he claims that natural evolution is based on the interplay between chance (what does he exactly mean when he uses this word?) and necessity, bringing about adaptive evolutionary change, I cannot, when he asserts (1970) the universe's indifferent nature and mankind's emergence by chance:

"The old alliance is broken; man finally knows that he is alone in the indifferent immensity of the Universe, from which he emerged by chance. Neither his destiny nor his duty is written anywhere. It is up to him to choose between the Kingdom and the darkness" [Md].

He backed up his words:

"Finally, it must be added, and this point is of very great importance, that the mechanism of transcription is strictly irreversible. It is neither observed nor conceivable, that information is ever transferred in the opposite direction, that is, from protein to DNA. This notion is based on a set of observations that are so complete and certain today, and its consequences in evolution, in particular, are so important that it must be considered as one of the fundamental principles of modern biology."

---

[637] **Cousteau** Jacques-Yves (FR: St André de Cubzac 1910-1997 Paris) called "Le commandant Cousteau" was an officer of the French Navy. He was a member of the Navy's intelligence services, which facilitated his diving experiences. He was awarded the Croix de Guerre with two mentions and palms, Commander of "la Legion d'Honneur." In 1943 he perfected, with Emile **Gagnan** (engineer at Air Liquide), the autonomous diving suit. They invented the regulator. In the 1950s he had a "diving saucer" built, that allowed two people to descend to -350m. They could reach -500m. As an oceanographic explorer, his films have made us discover the ocean depths and the resources they hold. Elected in 1957 to the direction of the Oceanographic Museum of Monaco, he made *many expeditions aboard* "La Calypso" *in all the oceans. Author of successful books including* "The World of Silence" *he entered the Académie française in 1988. Contrary to some writings he did not convert to Islam, his family buried him according to the* Catholic *rite.*

[638] **Monod** Jacques *(FR: Paris 1910-1976 Cannes). Famous French biochemist, he obtained the 1956 Nobel Prize in Physiology or Medicine in the company of* François **Jacob** *and* André **Lwoff** "for their discoveries concerning the genetic control of the enzymatic and viral syntheses." *Their work on the biochemical mechanisms of the transmission of genetic information highlighted that the DNA is the starting point of the biochemical reactions which, by the intermediary of the messenger RNA, produce the proteins necessary for the life of the cells. But RNA is probably not the only factor responsible for life.*

Although co-laureate with François **Jacob** and André **Lwoff** of the 1956 **Nobel** Prize in Physiology or Medicine *"for their discoveries concerning the genetic control of the enzymatic and viral syntheses,"* the evolving field of biology, including discoveries like reverse transcriptase in 1970,[639] challenges the absolute certainty of his assertions. It's intriguing to consider how J. Monod might have responded to subsequent scientific findings and Nobel Prize-winning research that shed new light on the complexities of life's origins.

Iliya **Prigogine**, the laureate of the 1977 **Nobel** Prize in Chemistry, and a professed *atheist*, presents a unique perspective on humanity's place in the cosmos. He discusses the evolving alliance between humanity and nature:

> *"Our science is no longer this classical knowledge; we can decipher the narrative of a new alliance. Far from excluding him from the world it describes, science finds as a problem the belonging of man to this world"* [Pr].

In **1986**, I. Prigogine clarified his thoughts on *"this alliance"* in the company of the philosopher Isabelle **Strengers** [PS]:

> *"... our presence in the world is natural and our work of creation prolongs that of nature. Man was not expected in the cosmos, but a new alliance between nature and us becomes possible,"*

portraying man as a co-creator in this grand narrative, potentially offering new avenues for exploration.

This remark is interesting because for some *Christians*, one of the reasons God gave us *"Free Will"* is to allow us to continue his work of creation, something he could have done himself. I wonder what he must think when he sees what we do with our Earth and space.

Christian **Duve**,[640] a laureate (with Albert **Claude** and George Emil **Palad**) of the 1974 **Nobel** Prize in Physiology or Medicine *"for their discoveries*

---

[639] In *the same year (1970) several researchers highlighted an enzyme: the reverse transcriptase allowing to pass from RNA to DNA, that does not contradict Monod's assertion, since it is not a question of reverse translation but of reverse transcription. Nevertheless, in* **1975** *(3 years after their announcement at the VIth Symposium of Molecular Biology (Baltimore - USA 06/1972) the Americans* Howard M. **Temin** *(USA: Philadelphia, PA 1934 - 1994 Madison, WI), a* Jewish, *and* David **Baltimore** *(New-York 1938), together with* Renato **Dulbecco***, were awarded the Nobel Prize in Physiology or Medicine for* "their discoveries concerning carcinogenic viruses and the genetic material of the cell" *and I wonder if J.* **Monod** *would have been so affirmative if he had known about this work.*

[640] **Duve** Christian *(Thames-Ditton, ENG 1917-2013 Nethen, BEL). Belgian, was born in England where his Catholic parents had taken refuge during the war. He graduated from the Catholic University of Louvain (UCL), in Medicine in 1941, and Chemistry in 1946; he was a research professor at UCL and Rockefeller University in New York. He was a precursor in biochemistry and genetics, and co-founder of cell biology. In 1974, he created the International Institute of Molecular and Cellular Pathology, which was to be called the "Duve's Institute." A music lover and writer, he has invested much time and energy in the future of humanity and the*

concerning the structural and functional organization of the cell" with strong *Catholic* beliefs - of whom *Father* Christian **Ringlet**, vice-rector of the Catholic University of Louvain, said of him in his eulogy: "*this great humanist with a great spirituality*" - has a different view from that of Jacques **Monod**:

> "*I have opted for a meaningful and not meaningless universe. Not because I want it to be so, but because that is how I interpret the scientific data we have.*"

The well-known Astrophysicist Hubert **Reeves**, taking up the idea of a school of thought from the beginning of the 20th century according to which we would be "*the children of chance*," wrote:

> "*This point of view seems a bit simplistic today. The achievements of contemporary sciences, in particular astronomy, invite us to rethink the problem. The notion of evolution, initially introduced by biology, now invades all scientific discourse. For fifteen billion years, the matter has been evolving towards ever higher states of organization, complexity, and performance. From the primordial chaos, it has generated successively: nucleons, atoms, molecules, cells, and living organisms. Everything happens as if, to use a colorfull expression, the universe was animated by a life drive.*"

His perspective highlights the delicate balance between randomness and design in the cosmos. [Re]

Alexander **Grothendieck**, (ref.593, p.196) a renowned mathematician, **1966 Fields Prize** (regarded as one of the highest honors in Mathematics. It is attributed every four years to a mathematician under 40 years of age), and a *Catholic* sympathizer, with *deistic* inclinations comments[641] on J. Monod's essay [Md]:

> "*The book Chance and Necessity by Jacques Monod, if not a complete dogma of scientism, is certainly a particularly striking illustration.*"

Adding that it is a striking representation of scientism:

> "*... for someone aware of the simple raw facts, and in particular for the biologist... to invoke the "endless" chance that would have created such a cascade of wonders, all coming to contribute to a concerted harmony of such an unheard-of breadth and depth, is there a blindness that for me, from that moment on, came close on madness.*" Cf [Gk-p.81]

---

planet. Belgium's King **Baudoin** *named him Viscount in 1989. He has received numerous awards, including the 1974* **Nobel** *Prize in Physiology and Medicine, and was a member of various international academies.* Catholic.
[641] Alexandre **Grothendieck** "*la nouvelle église universelle*" Survivre 1971 N° 9 internet.

In this exploration, we have traversed a broad spectrum of beliefs and perspectives, offering a glimpse into an intricate debate surrounding the universe's genesis. From chance to divine design, the discourse among specialists continues to shape our understanding of existence and our place within this vast enigmatic cosmos.

Jérôme **Lejeune**[642] was a *committed Catholic*. In 1964 he was awarded the chair of genetics at the Sorbonne (he had refused the one at Columbia University in 1961), thus becoming the youngest French university professor. Co-discoverer in 1958 of the chromosomal anomaly responsible for trisomy 21 (presence of three chromosomes 21 instead of two) winner of prestigious scientific awards: Kennedy Prize in 1962, William Allan in 1969, Leopold Griffuel in 1992, ... Following his faith, he deliberately renounced competing for the Nobel Prize in Medicine, which was *"promised"* to him for his numerous works, when he decided in 1970 to publicly commit himself, in all conscience, against abortion, that we call in French *"Interruption Volontaire de Grossesse"* with the Acronym IVG that he defined as *"Interruption de Vie Gênante"* becoming in English *"Interruption of Embarrassing Life."* Nevertheless, aware of the problems posed by the birth of a child with Down's syndrome, he did his utmost to help the families and children concerned, and his foundation took over after his death. For him, Catholicism is the only religion that loves the child.[643] In my opinion, he was a *"Humanist*

---

[642] **Lejeune** Jérôme *(FR: Montrouge 1926 - 1994 Paris). As a young man, he wanted to become a doctor to devote himself to the poor and the sick. He was elected to many academies, including the Pontifical Academy of Sciences in Rome in 1974 by* **Pope Paul VI**, *the French Academy of Moral and Political Sciences (1982), and the Academy of Medicine (1983)... In 1994, Pope* **John Paul II** *named him the first president of the Pontifical Academy for Life. He began his research with Professor* Raymond **Turpin** *(FRA: Pontoise 1895-1988 Paris) on mongolism, and in 1961 he defended a thesis on the subject, the scientific repercussions of which led him to hope for the Nobel Prize. He received that year various prizes, including the silver medal of the CNRS, and that of L'Académie des Sciences. Lejeune was appointed by the French government, to the UN as an expert on the effects of ionizing radiation. In 1958, he unveiled the identification of the chromosomal anomaly, responsible for trisomy 21. In 1962 he discovered trisomy 16. Head of the cytogenetic unit of the Necker Hospital, he worked with the hope of curing children of this disease. He fought all his life against what he called* "chromosomal racism" *which went against the Hippocratic oath. In the 1970s, many pro-abortion scientists turned away from him. In 1982, the French State withdrew his credits, and he received numerous proposals for chairs in the USA but remained in France and could continue his work thanks to US donations. In 1975, he was invited by* Karol **Wojtyla**, *then archbishop of Krakow, to give a conference. In 1981, in the middle of the Cold War, K.* **Wojtyla**, *who had been elected Pope* **John Paul II** *in 1978, sent him to* Leonid **Brezhnev**, *General Secretary of the Communist Party of the USSR, to warn him of the dangers of atomic fallout on the human race. In 1997, Pope* **John Paul II** *opened a process for his beatification. The fact that he was declared* "Venerable" *in 2021 means that the Catholic Church has recognized the* "heroic virtues" *necessary to be beatified. The recognition of a miracle would make him a saint.*

[643] Christianity *is the first religion that placed the child [Jesus] above all else. According to tradition, the Magi came from the East to venerate Him, bringing Him rich gifts and making Him the* "king of the Jews," *a title he never claimed, having come to save men, and whose main theme was* "love one another."

Christian" in the true sense of the word, as J-P **Sartre**[644] (whose ideas I am far from sharing) defines it: "*... the humanist who loves men as they are ... who wants to save them ... in their life ... in their death ...*" and not like many dare to define themselves without knowing what that implies. He proved by his actions that for him life is contingent on creation. So it is with pleasure that I learned on 21/01/2021 that **Pope Francis** had just recognized Professor J. **Lejeune**'s "*heroic virtues*" and declared him "*Venerable,*" opening the way to the holiness that is worth a Nobel Prize. Let's note that recent studies[645] conducted by teams from the neuroscience and cognition laboratory (Inserm/University and CHU of Lille, Fr) and the CHUV of Lausanne, (CH) have proven that the hormone GnRH (Gonadotropin gonadotropin-releasing hormone) would have a potential role on various brain systems, including that of cognition (out of 7 patients with Down's Syndrome, 6 have seen their cognitive performance increase), J. Lejeune's Institute has approached them.

During the **2009** summer session of Does God Exist? *Christian* A. **Hewish**, 1974 **Nobel** Prize for Physics (ref.371, p.111) answered the question at the head of this paragraph:

> "*I believe in God. It makes no sense for me to assume that the Universe and our existence are just a cosmic accident, that life emerged due to random physical processes, in an environment that simply happened to have the right properties.*"

In the field of ethology and environmental conservation, Dr. Jane **Goodall**[646] stands out as a distinguished figure. Her incredible journey began with a profound love for animals and nature, associated with a deeprooted curiosity. Contrary to the notion of life on Earth being the product of chance, Dr Goodall argues that it is, in fact, the result of a kind of "*anti-chance*" *phenomenon* [Go]. She elaborated on this perspective in an article titled "*Goodall Thoughts on the Fire of Notre Dame*" published on the internet on 15/04/**2019**. In this article, she reflected on a pivotal moment in 1974, when her personal life was difficult to live, her visit to

---

[644] **Sartre** Jean-Paul *(Paris 1905-1980). Philosopher, he refused the 1964 Nobel Prize for Literature. The quote is from* "La nausée" *p. 149.*
[645] "*Une thérapie pour améliorer les fonctions cognitives*" Science 2022/09/02.
[646] **Goodall** Jane *(London, ENG 1936). Already in her childhood, she loved animals, nature, and reading. 19 years old she began to work as a secretary and then sailed to Kenya where Louis* **Leakey** *hired her. Turning to London on the advice of Louis Leakey, who had seen that she had untapped potential, she obtained a BSc in 1964 and a Ph.D. in 1966. Back in Africa, she could undertake her research, raise funds, and create institutions, for young people's promotion.*
She authored about 30 books and was the subject of 40 films from 1965 to 2023. She held a Stanford University professorship from 1975 to 1979. Her conferences and videos reached millions of people around the world. She received a lot of medals and awards, including the Tyler Prize for environmental achievements in 1997, the French Legion of Honor, ... and the **2021 Templeton Prize.**

Notre Dame cathedral in Paris *"marked an epiphany in my thinking about my place on Planet Earth and the meaning of my life."* She remarked:

> *"That moment, a suddenly captured moment of eternity, was perhaps the closest I have ever come to experiencing ecstasy, the ecstasy of the mystic. How could I believe it was the chance gyrations of bits of primitive dust, that had led up to that moment in time: the cathedral soaring to the sky; the collective inspiration and faith of those who caused it to be built; the advent of Bach himself; the brain, his brain, that translated truth into music; and the mind that could, as mine did then, comprehend the whole inexorable progression of evolution? Since I cannot believe, this was the result of chance, I must admit **anti-chance.** And so I must believe in a guiding power in the universe: in other words, I must believe in God."*

Dr. **Goodall**'s remarkable career took flight in 1957 in Kenya when she crossed paths with Louis S.B. **Leakey,** a prominent paleoanthropologist and archeologist, who was also a *Christian* and a member of the Church of England. Despite her young age, she was just 19 years old, and although she had formal college training, Leakey recognized her untapped potential. In 1958, he advised her to travel to London and study with Drs William Ch. **Hills** and John R. **Napier**, leading authorities in the field of primate behavior, and anatomy. Jane's enduring dedication to the great apes' cause led her back to Africa in 1960. At that time, while the scientific community recognized that humans shared approximately 98% of their genetic code with chimpanzees, it was not widely accepted in ethological circles to attribute mind or personality to animals. However, Leakey believed that studying great apes in their habitat could offer insights into the behavior of early hominids. He sent the *"trimates"* - Jane with two other young women - to the Gombe site in Tanzania, a National Park, with a relatively large and undisturbed population of chimpanzees.

With patience and gradual effort, she established trust with the chimpanzee groups by observing their movements, mimicking their behaviors, and offering them bananas. This led to her being accepted as a friend, and she formed relationships of mutual trust that have lasted six decades. She could observe social behaviors, emotional moments, and even ritual activities, such as the burial of chimpanzees and elephants. Living in such proximity, she noticed similarities in their behavior with that of human, such as tool-making to reach their objectives, such as hunting piglets, and monkeys for meat. In 1960, Leaky recognizing her brilliance, encouraged Jane to return to London, where she defended her Ph.D. in Natural Sciences, in 1968, focusing on *"the behavior of free-living chimpanzees."* This achievement marked the beginning of her journey as an independent researcher and the establishment of foundations, including the Jane Goodall Institute in 1977. These organizations, allowed her to continue her work in studying and protecting chimpanzees, while also improving the lives of local communities through education and training. Through her organizations, she has saved thousands of animals and inspired millions of people in all parts of the world, to take action in their communities. As she stated in 2007:

"the nonverbal body language is the same for chimpanzees as it is for us. They use the same gestures and postures in the same context."

Dr. Goodall's groundbreaking research on chimpanzee society and their natural habitats significantly transformed our understanding of these creatures. Her work, spanning six decades, has left an indelible mark on the world, inspiring countless individuals and prompting us to reevaluate our role in the interconnected tapestry of life on Earth. She was awarded the **2021 Templeton Prize**.

Sir Roger **Penrose** (ref.368, p.109) 2020 **Nobel** Prize in Physics, during an interview with the BBC, declared himself an atheist in the sense that he does not adhere to any religion, but said in the film "*A History of Time*," based on the book by S. **Hawking**, with whom he worked, that he:

"... believes that the universe has a reason to exist, that Man is not the fruit of chance, and that there is something very deep, in the Reason for the universe."

Trinh Xuan **Thuan**,[647] *Buddhist*. Astrophysicist, a follower of the: "*Complexity Principle*," substitutes it to the "*Strong Anthropic Principle*" proposed by Brandon **Carter**[648] because, according to him:

"...the universe is set for the emergence of any life and consciousness whether terrestrial or extraterrestrial,"[Tn] and consequently, the goal of fundamental laws governing the universe is Life, in any form whatsoever. Like many scientists, he justifies his assertion: "*I do not believe in randomness in the universe*" by relying on probability calculations. Taking as a hypothesis that if the value of only one of

---

[647] **Thuan** Trinh Xuan *(Hanoi VNM, 1948). Astrophysicist, he defended his Ph.D. at Princeton, US-NJ. Researcher at the Institute of Astrophysics of Paris and at the University of Virginia. In 2004, using the Hubble Space Telescope, he discovered with* Y. **Isotov** *that the galaxy I Zwicky is only 500 million years old, which makes it the youngest known to date. It is also only 45 million light years away from Earth.* Buddhist.
The Complexity Principle *or* Pyramid of complexity *concept was initially proposed by* P. **Teilhard de Chardin** *and then taken up by* H. **Reeves** *to explain how, from the initial elementary particles, we arrived at Life in all its forms (cf. chapter III p. 145-177).*
[648] **Carter** Brandon *(AUT 1942). Specialist in black holes, having worked with* S. **Hawking** *in particular. He works now in the laboratory* "Universe and Theories" *of Meudon. Cosmologist, he proposed in 1973, during the symposium on the 500th anniversary of* **Copernicus**, *the* Weak Anthropic Principle (WAP): "We must be prepared to take into account that our position (in time and place) in the universe is necessarily privileged to the extent of being compatible with our existence as observers. This is, in fact, a tautology*: we see the universe as it is because we exist." *Later, he proposed the* Strong Anthropic Principle (SAP): "The universe (and hence the fundamental parameters on which it depends) must be such as to admit the creation of observers within it at some stage." *From this point on, we enter the field of metaphysics, hence the fact that it has given rise to many different interpretations.* * Tautology: a formulation that can only be true, synonymous with a truism.

the fundamental constants was modified by a factor equal to 10^{-19}, the chances for life to have appeared would have been 10^{-43}! However, there are about fifteen fundamental constants, which leave no chance to chance.

For the *Christian* geneticist Francis **Collins** who, also, has a solid background in quantum physics:

> "*... there is intrinsically no conflict between the idea of a creator God and what science has revealed*" [Co1 p.81].

Adding that the hypothesis of God allows us to answer the questions about the pre-Big Bang. Starting from the "*Principle of Moral Law*" or "*Law of Good Conduct*" (cf. Kant ref.2, p.xi) admitted by theists and from the hypothesis that God exists, he develops a simple argumentation, leading him to conclude that not only God is eternal, but that the Universe He conceived and created is perfectly adapted to the sentient creatures that, because of natural selection, we are.

Note. At the beginning of this paragraph, I have occulted the interesting biological paradigm of panspermia, which connects life on Earth with the wider cosmos because it is still a speculative idea and lacks substantial empirical evidence. Nevertheless, panspermia is supported by prominent scientists because it holds the potential to address pivotal questions, such as:

-   Why Earth's life is not the first choice? It opens the door to extraterrestrial life,

-   Where did Earth's life originate? Was it on the seabed of oceans, on the ocean floor, near volcanic events, on the Earth's surface, or within the Earth's subsurface; a hypothesis supported by circumstantial evidence, but lacking certainty. Hypotheses associated with the numerous meteorite impacts that Earth has suffered since the Bang (cf. Chap. II).

So, I can't overlook the fact that Swante **Arrhenius**, the visionary scientist (ref. 411, p.122), published an article in 1903, the year he was awarded the **Nobel** Prize, in which he defended the hypothesis that certain microscopic forms of life could have arrived on our planet and others under the impulse of radiation emitted by the Sun or various stars, a process known as "*radio panspermia*." Given the high probability that these microorganisms would be destroyed by the same radiation, it received little attention from scientists. Research into panspermia languished for seven decades, until eminent researchers such as F. **Hoyle** (ref.351, p.105) and his pupil Chandra **Wickramasinghe** brought it back into the limelight, though without being able to impose it for lack of evidence, by calling on "*litho panspermia*," a mode of transport reinforced since the discovery of exoplanets. According to litho panspermia, microorganisms are transported well away from radiation by various debris resulting from collisions between meteorites and planets, or even by meteoroids and asteroids, which, as we have seen in Chapter I, have repeatedly targeted the Earth. Supporters of panspermia saw this as a way of rejecting the Darwinian theory of the evolution of life, as well as justifying certain viral epidemics such as COVID. Among the various origins of this dust, Mars is the most talked-about target, as it could at one time have hosted life, as Earth did later.[649] Hence the interest of the research undertaken on samples taken on Mars. If this proves to be the case, Life on Earth would then be a second choice, which might not be the only one, reviving the public's attention; for, under these conditions, we would not be the only living beings, which is conceivable.

Even the atheist Francis **Crick,**[650] 1962 **Nobel** Prize in Physiology or Medicine, with J. Watson *"for the discovery of the structure of DNA"* acknowledged the seemingly miraculous nature of life's origin. In his own words, he mused:

"The origin of life currently, seems to be a miracle because so many conditions must be met to bring it about."

Crick's contemplation delved into the intricate complexity of life's genesis, hinting at a profound understanding of the universe. Amidst the myriad of speculations, one theory stands out: the captivating notion of direct panspermia. In 1973, **Crick**, in collaboration with the chemist Leslie **Orgel** - who had made important experiments on life origin - proposed the presence of a *"cosmic genetic code"* embedded in the fundamental laws of physics, aptly named the *"Direct panspermia"* theory. According to this intriguing concept, life itself might have been dispatched to Earth billions of years ago by extraterrestrial intelligence, driven by a purpose that remains shrouded in mystery. Could it have been a quest to perpetuate the common genetic heritage of all terrestrial life?

Yet, as with any hypothesis, direct panspermia raises as many questions as it answers. Who were these scientifically advanced aliens, able to craft and launch life across the cosmos? The enigma deepens, compelling us to question the very nature and origin of the cosmic architects.

In the vast landscape of scientific thought, direct panspermia finds allies among other visionaries, such as S. **Hawking** (ref.369, p.110), the renowned theoretical physicist. Hawking envisioned the possibility of microorganisms traversing the cosmos, jumping from planet to planet, via meteorites. Similarly, the astrophysicist Paul **Davies,** a *Christian* (ref.554, p.173), 1995 Laureate of the

---

[649] Life on Mars. *According to more recent data Mars could have been habitable 500 million years before the earliest known Earth lifeform, when the planet had liquid water and a warmer climate. However, 3.5 billion years ago Mars lost most of its atmosphere and became much colder and drier, making life improbable. I say improbable rather impossible because some scientists suggest that life could have survived in the subsurface where water and heat may still exist.*

[650] **Crick** Francis Harry C. *(Northampton, ENG 1916-2004 San Diego, US-CA) won in 1962,* with James Dewey **Watson** *(Chicago, IL-US 1928) and* Maurice H.F. **Wilkins** *(Pongaroa, NZL 1916-2004 London, ENG), the* **Nobel** *Prize in Medicine* "for the discovery of the double helix structure of Deoxyribonucleic Acid (DNA) and of the mechanism of its duplication, which allowed us to understand the transmission of genetic information." *He has written that he rejects religion cf.* "What a mad pursuit. A personal view of Scientific" *Discovery 1968.*

prestigious Templeton Prize, expressed during the ceremony a profound sense of wonder:

"The contrived nature of physical existence is just too fantastic for me to take on board as simply "given." It points forcefully to a deeper underlying meaning to existence."

The Panspermia hypothesis, finds support not just in the musings of brilliant minds but also in the words of Harvard geneticist George **Church**, for example. During a conference in December 2018, G. Church hinted at our imminent capability to mimic the actions of those hypothetical extraterrestrial beings.

While panspermia remains speculative and awaits experimental verification, it harmoniously coexists with our beliefs. Far from contradicting our understanding, it provides a unique lens through which scientists may unravel the mysteries surrounding the origin and evolution of life on Earth. Moreover, it opens profound avenues for contemplating our place in the cosmos and the tantalizing possibility of encountering extraterrestrial life; echoing S. **Hawking**'s cautionary words: *"If aliens landed on Earth, it could have the same consequences as the landing of the conquistadors had on the Amerindians."*

As we journey through the uncharted territories of scientific exploration, the allure of direct panspermia persists, an enticing hypothesis that beckons us to reconsider our cosmic origins and the potential tapestry of life woven across the vast expense of the universe.

5.4. Does the theory of evolution (1859) contradict Genesis?

"We see a world in a continuously capable process, through its one inherent property and natural character, of producing new living forms, matter is now seen to be self-organizing."
A. Peacocke[651] [Po]

"The World appears as a mass in the process of transformation...
Evolution of Matter
Resulting in an increasing complexity until Life can appear.
Life is, and can only be, a magnitude of evolutionary nature or dimensions."
P. Teilhard de Chardin [Te2].

In the year **1850**, Ch. Darwin, a former devout *Catholic* unveiled his magnum opus *"The Origin of Species"*[Dn]. This groundbreaking work, a result of exhaustive research, introduced the revolutionary *"theory of evolution."* According to Darwin, evolution, understood as the gradual modification of species over time, reveals the inherent genetic differences within populations: be they humans, animals, or plants. These differences equip individuals with unique abilities to adapt to their environment.

"Natural Selection [652] *acts exclusively for the preservation and accumulation of diversities, which are favorable to the organic and inorganic conditions to which each creature is exposed throughout the different periods of life. The ultimate result is that each creature tends to become more and more capable of responding to these conditions."*

---

[651] **Peacocke** Arthur *(Watford, ENG 1924-2006),* Anglican theologian, *and biochemist, studied at Exeter College, Oxford: BA, MA, D. Phil (1948), D. Sc (1962) as well as a doctorate in theology at the University of Birmingham (1960) where he taught from 1948 to 1959 before teaching Biochemistry at Oxford. Ordained as a priest in 1971, he was appointed Director of Theological Studies at Cambridge. President of the Science and Religion Forum from 1995 until his death. Recipient of numerous awards, he was made a member of the Most Excellent Order of the British Empire by the Queen and received the* **Templeton Prize** *in 2001. A supporter of Darwinism, which according to him is not an enemy of religion, he published** "The Evolution: The disguised Friend of Faith," *an essay in which he compiled 13 of his articles on the subject. The book has been adapted for the stage. The note is taken from [Po] p. 16. * Templeton Press, Nov. 28, 2004.Cf. § 5.5.b.*
[652] **Darwin** *specifies ([Dn] p. 45) that the expression* "Natural Selection" *should be taken in the sense of* "Survival of the fittest" *and p. 61 that it acts very slowly on few people because of interbreeding, crossbreeding, hybridization of diverse populations. The quote is from p. 75. He never spoke of the law of the strongest what is current with eugenic tendencies that have seized upon Darwin's theory to propagate this* "Fake news."

It's important to note that Ch. **Darwin** stated, both in his introduction and on page 115:

> "*I am convinced that natural selection has been the main but **not the exclusive means** of modification.*"

Beyond Darwin's assertion lies an intriguing perspective.

Catholic botanist J-M. **Pelt** (ref.583, p.191), an advocate of evolution, posits alternative selection mechanisms such as mimicry, suggesting a complexity that extends beyond the survival of the "*fittest*":

> "*... I am confronted with breathtaking phenomena of mimicry, when an orchid disguises itself as an insect, takes on the insect's scent, arranges its hairs like those of the insect's female to attract the male, who finds himself irresistibly attracted, ... copulates and embarks on the pollen to deposit it on another orchid, which is thus fertilized. To suppose that this lure appeared by the play of chance even cleverly baptized "coevolution" questions me*" (cf. Pp. 156 & 157 of [Pe2]).

Darwin himself, while emphasizing the gradual nature of natural selection and the extinction of species, also alludes to a "*plan of creation,*" and acknowledges the existence of "*laws imposed on the matter by the Creator.*" He concludes:

> "*There is grandeur in this view of life, with its several powers, having been originally breathed by the Creator into a few forms or one.*"

This nuanced stance challenges the notion that evolution and faith are irreconcilable. Far from denying the Creator, Darwin's worldview aligns with a belief in a guiding force. This suggests a *Deistic* inclinational, though some interpretations hint at occasional *Theistic* tendencies.

Theological echoes of evolution stretch back centuries, as exemplified by St **Augustine**'s indirect prediction in the fourth century (§V.1.1, p.195). Even John **Calvin**, a prominent *Protestant reformer,* entertained evolutionary ideas, reflecting a historical undercurrent where faith and evolution found potential harmony.

However, **Darwin**'s discoveries ignited divergences reminiscent of **Galileo**'s era. Critics emerged, overlooking the fact that many religious scientists see science and religion as partners in the pursuit of the truth. Their approach, championed by figures like Maimonides, Galileo, Pascal, Newton, and Einstein, emphasizes an allegorical reinterpretation when scientific discoveries seemingly contradict scriptures. Darwin, a dedicated *Theist* and evolutionist, aimed not to challenge the divine narrative but to contribute to scientific inquiry. He asserted:

> "*It seems to me absurd to doubt that a man can be an ardent theist and an evolutionist.*"

Despite these reconciling perspectives, as soon as 1865, at the initiative of an influential London *philanthropists' team* - including among others Anthony A. **Cooper** the 7ᵗʰ Earl of **Shaftesbury**, an *Evangelist Anglican* - founded the **Victoria Institute or Philosophical Society of Great Britain** to defend *"revealed truth"* against scientific challenges. In effect, the cosmological content of the Bible was challenged not only by **Darwin**'s theory; but also, by new natural science treatises and new ideas coming from Germany.

At the same time, as a reaction to the Darwin theory and the geological influence of an ancient Earth, has appeared the Young Earth Creationism Movement (YECM),[653] supporting a *"young earth"* created in the 4th millennium B.C. in 6 days of 24 hours (which makes no sense) following the flood of which **Noah** was the main protagonist. Firstly influenced by the writings of the *Seventh-Day Adventist*[654] George **McCready Price**,[655] this movement gained influence with the adhesion of two evangelical *Christians*, the hydraulic engineer Henry Morris, and the theologian John Whitcomb, who co-authored in 1961 *"The Genesis Flood"* a best-seller. The YECM influence has grown in the 1970s and 1980s by the creation of Institutions and associations promoting YEC through publications, debates,.... Jean Marie **Pelt**, says of them [Pe2, p. 152]: *"I am the opposite of creationists, who are adults who have remained in childhood."* The same is true for biology professor Darrell R. **Falk**,[656] *Evangelist*, who defends theistic evolution, cf. [Fa].

---

[653] Young Earth Creationists: *Despite all the clues that can be found in Genesis, they have not understood that* **Adam**, *who is about 6,000 years old according to the scriptures, is not the first man, nor the only one created by* **God**, *but the* "legal man," *i.e. the one with whom* God *made a contract but who did not respect it, believing himself capable of accessing the absolute knowledge alone. Hence his rejection of Eden as a* "human laboratory" *by* God. Adam *makes Man enter history, and with him is born the questioning that characterizes Man.*
Among contemporary Muslims, the Turkish Harun **Yahya** born Adnan **Oktar** *(Ankara, TR 1956), an interior designer is the most charismatic representative creationist. He is a negationist and anti-evolutionist whose reference work is forbidden in most Western countries because it has no scientific basis. Involved since 1979 in religious activism, he was arrested in 1986 on the grounds of promoting theocratic revolution and interned for 10 months in a mental hospital. Arrested in 2018 for the creation of a criminal enterprise, sexual intercourse with minors, kidnapping, and financial fraud,... he was sentenced in 2022 to 8,658 years in prison*
[654] The Adventists *(a reference to the revival and therefore to the return of Christ) are Protestant Christians, constituting the twelfth most important religious organization in the world. Let's note that it's not an obligation for the members of the Seventh-Day Adventist to adhere to YEC.*
[655] **McCready Price** John *(Havelock, CAN 1870-1963 Lorna Lind, US-CA),* Adventist, *Creationist Author.*
[656] **Falk** Darrell R. *(Vancouver, CAN 1946), Ph.D. in Biology in 1968 (University of Alberta) Postdoc at the University of British Columbia in Vancouver, then at the University of California in Irvine, CA. Professor of Biology at Syracuse University, NY, then at OHIO. Professor Emeritus in Biology at Point Loma, San Diego, CA.*

It's noteworthy that in 1865, Gregor **Mendel**,[657] the founder of modern genetics, read a paper to the Natural History Society of Brno, that he published in German in 1866: "*Experiments in Plant Hybridization*" [Me] laying down the theoretical foundations of genetics and heredity, known as **Mendel**'s "*laws of heredity*." As a result of a decade of methodical experiments on the hybridization of two types of peas G and Y. He demonstrated that Y, which he called the "*dominant allele*," appeared 3 times more than G (the "*recessive allele*") and, attributed this to the existence of invisible factors, which he called: "*heritable factors*" now called as: "*genes*," determining the specific characters of the offspring. We would have arrived earlier at the current conception of evolution: a series of unpredictable mutations of the genetic code of an organism X giving organism Y slightly different from the strain X. At present, science has not discovered a law that pushes or encourages these mutations in one direction or another. They appear to be genuinely random, but researchers have not said their last word. Unfortunately, the synthesis of **Darwin**'s theory with **Mendel**'s work had to wait more than thirty years before being achieved. An acceleration of the process could have avoided some clashes. Nevertheless, some prominent scholars defended the harmony thesis between evolution and religion, unveiling the beauty of creation. In **1865**, prompted by the prevailing sentiment, the Victoria Institute rejected the theory of evolution. However, by **1870**, a notable shift occurred in the opinions of its members. This transformation was catalyzed by George **Stokes** (ref.226, p.68), a prominent devout Anglican, and the Lucasian Professor of Mathematics at Cambridge, who also served as the secretary of the Royal Society. After several refusals, he eventually agreed to address a lecture to the members of the Victoria Institute, marking a significant turnaround in the views of these distinguished institution members. **Stokes** explained in a nuanced form that there was a misunderstood between the two parts and that "*there is no discrepancy between the book of Nature and the book of Revelation if rightly interpreted.*" With the arrivals in 1875 until 1883 at the head of the Victoria Institute of the *Anglican* Botanist and Surgeon Joseph D. **Hooker**,[658] director of the Royal Botanical Gardens, the *Anglican* Mathematician and Physicist William

---

[657] **Mendel** Gregor *(CZE: Heinzendorf bei Odrau, Silesia 1822-1884 Brno), the* "Father of genetics," *a lifelong learner, scientist, and* man of faith. *Botanist,* Augustinian Catholic friar. *Raised on the family farm, he became interested in crops, beekeeping, etc.*
He had to interrupt his studies in philosophy when his father died, and in 1843 he entered the St Thomas' Abbey in Brünn (Brno) as a novice, a place renowned at the time for its scientific work and teaching. He took his vows in 1845 and was ordained a priest in 1847. *He studied Philosophy and Physics at the Philosophical Institute of the University of Louts and, devoted his free time to meteorology which made him known in the scientific world, and to botany, a field in which he was too far ahead to be recognized.*
[658] **Hooker** Joseph *(UK; Halesworth 1817-1911 Berkshire). He graduated in Medicine in 1839, an assistant surgeon on the HMS Terror an exploration vessel for Antarctic studies (botany, ornithology, magnetism…), he was one the closest friends of* **Ch. Darwin**. *He was Chairman of the Royal Society (1873-1878).* Anglican.

Spottiswoode,[659] and the *agnostic* Biologist Thomas H. **Huxley**[660] one of the harshest voice for pre-eminence of science over religion - three outspoken scientists members of the X-club (Social Network of Science) who supported a clear separation between religion and science - the level of discussion rose. Nevertheless, the relevance of science was strongly debated within this institution, where defenders of the Bible against science were active. **Stokes'** presidency (1883-1903) of the Victoria Institute played a determining role in its rehabilitation. Lord **Kelvin** (ref.227, p.68), *Presbyterian* succeeded him comforting its authority.

In the meantime, the leading Botanist Asa **Gray**[661] *faithful Presbyterian* whom Darwin quotes on several occasions in *"The Origin of Species,"* because Gray did not hesitate to communicate information to him (compiled in a book called *"The Darwiniana"*). **Gray** was also a *devout Christian* and supporter of a **Theistic Evolution** (under the Creator's rules). In Darwinania he wrote (cf. SMF, ref. A. Gray):

---

[659] **Spottiswoode** William *(ENG, London 1825-1883). He studied Mathematics and Physics at Balliol College, Oxford, Businessman, he was co-editor and co-publisher after his father died. He also worked on physical and mathematical problems such as symmetry and determinants. He authored over 100 papers. Was Elected Chair of the London Mathematical Society in 1870 and of the Royal Society (1878-1883). Anglican, he was buried in Westminster Abbey.*

[660] **Huxley** Thomas H. (GB: Ealing 1825-1895 Eastbourne). *Biologist and Anthropologist autodidact, he supported* **Darwin**. *In 1845 he was an assistant surgeon on the HMS Rattlesnake and sailed toward the southern hemisphere where he studied marine invertebrates. Elected Fellow of the Royal Society in 1850, professor of Natural History at the Royal School of Mines in 1854 President of the Geological Society (1868-1870) of the Royal Society (1883-1885), and of the Marine Biological Association (18841890). He edited a lot of books, and received many important awards: the Royal Medal (1852), the Copley Medal (1888), the Darwin Medal (1894)...Although* agnostic *(because for him only what is demonstrable is certain), he campaigned for reading the Bible in school. He coined the word* "agnostic" *derived from the Greek* "agnostos" *meaning* "unknowable" *to describe his philosophical position.*

[661] **Gray** Asa (US: New York 1810-1888 Cambridge, MA). Graduating from the Fairfield Medical College in 1831, NY, he began his career as a physician but turned to Botany, he was a leading botanist in the US. In 1838 he was appointed professor of Botany at the University of Michigan, then professor of Natural History at Harvard in 1842. He authored numerous botanical works that he communicated to Ch. **Darwin.** He was a great advocate of this later. Member of the Presbyterian Church, he served as a Deacon *at the First Church in Cambridge (cf. Peter E. Hodgson* "ASA Gray: Bridging Faith and Evolution" *Science Meet Faith 17/11/2020, Internet). His close relationship with* John **Torrey's*** *family led him back to the religion he had abandoned in his youth.*

***Torrey** John (USA: New York 1796 - 1873), Physician, botanist, and chemist. At 16, Amos Eaton (1776 – 1842), a pioneer in natural science, taught him the structure of flowers and the basic principles of botany. In 1815 he began the study of medicine and took his Medical degree in 1818 but he was more interested in botany, he became the leader of American botanists. Appointed professor of Chemistry, Mineralogy, and Geology at West Point, he takes a post of professor in Chemistry and Botany at New York. He wrote many reference books. Faithfully protestant he said that the legitimate results of scientific inquiry would not be inimical to the Christian religion*

"Agreeing that plants and animals were produced by Omnipotent fiat does not exclude the idea of order and what we call secondary causes. The record of the fiat - "Let the land produce vegetation: seedbearing plants and trees[Ge1:11] ... let the land produce living creatures according to their kinds: the livestock, .. the wild animals,...after his kind.....[Ge1:24]" *- seems even to imply them."*

The Rev. Dr William H. **Dallinger,**[662] an *Anglican Methodist* proto-zoologist, for his part, experimentally verified the theory of evolution on unicellular organisms and was an early supporter of Darwinism. Likewise, the *Christian* paleontologist Charles D. **Walcott**[663] who made a fundamental contribution to the study of trilobites, as an assistant of James Hall.

In **1901,** the *Anabaptist Evangelist* botanist, Hugo Marie **de Vries,**[664] seduced by the ideas of Ch. **Darwin**, made a groundbreaking announcement, species change by sudden and discontinuous variations, variations that he named *"mutations."* These mutations, without intermediate form between two species, as if they resulted from a spontaneous cut in the genome, are due for the **1933** Nobel Prize for Physics E. **Schrödinger** (ref.309, p.91), a *Christian*:

"... to quantum jumps in the gene molecule."

This led Schrödinger in **1944** to call this theory *"the quantum theory of biology"* (cf. [Sc2] p. 34).

Benjamin **Warfield,**[665] a conservative *Protestant* bishop, in favor of a Young Earth (less than 20,000 years old), who studied Mathematics and Physics at Princeton, saw no contradiction between Genesis and the natural laws governing

---

[662] **Dallinger** *William H. (England 1839-1890). Protozoologist (the science of unicellular living organisms),* *English* Methodist Minister.
[663] **Walcott** Charles D. *(USA: New York Mills, NY 1850-1927 Washington, D.C). Although he interrupted his studies at the age of 18, he was one of the greatest Paleontologists in the USA. He was interested in minerals and fossils from an early age and traded in them. In 1872 he made one of the most important discoveries of Trilobites (a marine species with 3 lobes that lived from 521.10^6 years before our era to the 3rd mass extinction (-252.10^6 years)). Since 1876 he has been interested in paleontology, working from 1879 for the* "US Geological Survey" *he made important discoveries. Elected to the US Academy of Sciences in 1896, President of the American Geological name to various sites, a warship, ...* Fervent Christian *([Col] p. 166).*
[664] **de Vries** Hugo Marie *(NDL: Haarlem 1848-1935 Ede). Botanist, geneticist, Professor then director of the Institute of Botany of Amsterdam. Discovered in 1900 the laws of heredity without knowing Mendel's work. In 1901, his work on a grass that he named Œnothera allowed him to observe brutal variations leading to new species. He called these variations* "mutations." *A foreign member of the Royal Society, he received the Darwin Medal in 1906... He was an* Anabaptist: an evangelical Christian *who believed in* "adult baptism," *which is performed by immersion at the request of the impetus, and which can renew the baptism performed as a baby.*
[665] **Warfield** Benjamin *(USA: Lexington, KY 1851-1921 Princeton, NJ),* presbyterian protestant. *Graduated from Princeton in 1871 with honors in Mathematics and Physics, entered Princeton Theological Seminary in 1873, and graduated in 1876. Professor of Theology at Princeton from 1887.*

evolution, operating under God's control who uses the process of evolution to create the diversity of life on Earth:

"I am free to say that, for myself, I can discover no irreconcilable antagonism between the general statement of Scripture that God created all things, and their subsequent development according to natural laws, and any scientific facts in the realm of nature, whether about the unity or the diversity of species" [Wa1].

"We must not, then, as Christians, assume an attitude of antagonistic toward the truths of reason, or the truths of philosophy, or the truths of science, or the truths of history, or the truths of criticism...... we must be careful to keep ourselves open to every ray of light. None should be more zealous in them than we, none should be quicker to discern truth in every field, ..." [Wa2].

Teilhard de Chardin, a visionary champion of evolutionary theory, boldly asserted the shared ancestry of Man and Great Apes as documented in references (Chapter III). An advocate of orthogenesis,[666] Teilhard delved into the profound concept of *Noogenesis,* [667] embraced by psychologists, and psychiatrists. His anthropological perspective on evolution unfolded, portraying a journey culminating in the emergence of Man. In Teilhard's perspective:

"the change[668] *of biological state leading to the awakening of the Thought [the Reflection] affects the Life itself in its organic totality... it marks a transformation affecting the state of the whole planet."*

"Man [the] rising arrow of the biological synthesis. The Man constituting to him only, the last-born, the freshest, the most complicated of the Nappes of the Life."

---

[666] Orthogenesis: *A suite of variations which, in the course of the evolution of living beings, occur in the same direction, through several species or several kinds. We have noted on p. 88 of* "The Origin of Species" *this sentence by Ch.* **Darwin***:* "The one who believes in separate and innumerable acts of creation can say that, in these cases, it has pleased the Creator to make that a being of a certain type takes the place of another, belonging to another type." *For the proponents of theistic evolution, this choice was made from the beginning of Creation, God not being a tinkerer. If he intervenes over time, it is only exceptionally.*

[667] Noogenesis: *the period that extends from the appearance of Man and during which thought was progressively constituted.*

[668] **Teilhard** *uses simple examples so that we can understand these state changes. We have all learned that any heated body sees its temperature increase until it changes state. Pure water for example, at atmospheric pressure, changes from a solid state (ice) to a liquid state at 0°C and to a vapor state at 100°C.*

His ultimate destination, the point Omega[669] underscores Teilhard's belief in an evolving consciousness.

Teilhard's narrative endows Man with Reflection and freedom, placing him "*in charge*" of Evolution and entrusting him with the responsibility for the collective Life of Humanity. Within the grand scale of the Universe, we are merely in the infancy of Humanity, and Teilhard urges his Ultra-Human[670] to guide this journey toward adulthood, fostering a collective human consciousness in the Noosphere,[671] the Earth's thinking envelope, and the dwelling place of Christ at the point Omega.

It's noteworthy (see [Sc]-p.43) that W. **Schrödinger** aligns with Teilhard, adhering to the Vedânta idea that one is "*a whole within the whole.*" **Teilhard** concludes ([Te2] p. 279) that:

> "*... Evolution is an ascent to Consciousness "which" must culminate in a supreme Consciousness.*"

He:

> "*... supposes everywhere the presence of a personal and creative God, who provokes and directs the Evolution of the World, ... Christ is the axis and the end of the whole event of the world, the mysterious Omega point towards which all the ascending forces converge, so that the whole creation appears to him according to the Incarnate Word.*"[672]

As a biologist, **Teilhard** observes that the arrow of time[673] points towards complexity, emphasizing the evolution from monocellular to multicellular organisms:

---

[669] Point Omega: "*By structure, Omega, considered in its last principle, can only be a distinct Center radiating in the heart of a system of centers.*"

[670] *In his* cosmogenesis, **Teilhard** *prophesized a new stage of evolution that should give birth to an* "ultra-human" *resulting from the Thought of Man that God created free.*

[671] Noosphere*: the sphere of the spirit (Noûs in Greek meaning Spirit), surrounding the Biosphere. We have been taught that the Earth is made up of a metallic core (containing 9/10 of iron) called the:* "Barysphere" *and of 4 concentric spheres: the* Lithosphere *(rocky), the* liquid Hydrosphere, *the* Biosphere or Ecosphere *(sum of all ecosystems), and the* atmosphere *(gaseous); the* Noosphere *becoming the 5th.*

[672] *Cf.* [Te2] pp. 23 and 22. "The Human Phenomenon" *was forbidden to be published by the Roman Curia and only appeared a few months after Teilhard de Chardin's death because he did not want to disobey, going so far as to refuse a professorship at the Collège de France.*

[673] *This does not contradict the entropic arrow of time of physicists (working on* "closed systems"*) which goes towards disorder because the arrow of biological time that* Paul **Davies** *(ref. 554, p.175) calls the* "optimistic arrow of time" *applies to* "open systems" *exporting entropy towards the outside.*

"Life is, and can only be, a magnitude of nature or evolutionary dimensions ... It is getting increasingly sophisticated."[674]

He introduces the *"biological law of complexity"* supporting the idea that:

"...The nervous system is constantly developing and concentrating ... the differentiation of the nervous substance stands out, ..., as a significant transformation. It proves that there is a meaning to evolution," A view difficult to dispute.

In his geological inquiries based on the Paleolithic,[675] **Teilhard** questions:

"... can we dispute that the continents go on granitizing? ..." (cf. [Te2] Pp. 159 -).

An inquiry, with an obvious answer that aligns inexorably with the march of evolution.

Teilhard's multifaceted approach underscores his unwavering support for evolution, evident in his coherent perspective that Life presupposes Pre-life:

"Original Matter is something more than the particulate swarming so wonderfully analyzed by modern Physics. .... In a coherent perspective of the World, Life supposes inevitably, and as far as the eye can see before it, of the Pre-life" ([Te2] p. 56):

Teilhard's holistic view opens us to new perspectives, such as:

"is an Inside of the things, a Consciousness (p.157), a Spontaneity. ... the existence of a conscious internal face that necessarily doubles, everywhere, the material external face, only considered usually by the science."

A perspective shared with the British Biologist John B.S. **Haldane,**[676] an *atheist*. This conscious internal face, often overlooked by conventional science,

---

[674] *Some opponents of the theory of evolution have invoked the fact that if we accept evolution, going towards more and more complex systems (thus more ordered), we violate the second principle of thermodynamics, according to which the entropy of an isolated system and the disorder increase. This is erroneous because the system in which we place ourselves is that of the Earth, which is far from an isolated system, it is the receiver of solar energy responsible for photosynthesis for example.*

[675] Paleolithic: -3.3 million years to -12,000 BC, an era during which all humans were hunter-gatherers.

[676] **Haldane** John B.S. *(Oxford, ENG 1892-1964 Bhubaneswar, IND). Taught at Oxford, then at Cambridge, as a professor of Genetics then of Biometry at* "University College London." *In 1957, he left to teach in India and created a laboratory of genetics and biometry. Member of various scientific academies, recipient of numerous awards: Baly, Darwin, ... atheist.*

aligns with Haldane's anticipation of finding traces of thought or life throughout the universe:

> *"We find no obvious trace of thought or life in the Matter... But if the modern perspectives of Science are correct, we must expect to find them eventually, at least in rudimentary form, throughout the universe"* ([Hal]-p.153).

We found this idea with the pan-experientialist Ch. **Birch** (ref.541, p.165) and his mentor A. **Whitehead** (ref.542, p. 166) author of the *"Process though"* (ref. 543, p.168).

Teilhard's intellectual prowess, recognized by his membership of the French Académie des Sciences (1951), not only embraces evolution but, seamlessly integrates it into *Catholicism*. Despite facing intellectual opposition, Teilhard aligns his thoughts on Matter with **Leibniz**'s notion of monads (p.53), drawing parallels that resonate with the explorations in *"Spirit and Matter"* by E. **Schrödinger** ([Sc2] part 2).

In this rich tapestry of thought, the parallels between Teilhard's ideas and those of the polymath G. **Leibniz**, and the insights of E. Schrödinger with the *Protestant* Physician and Psychiatrist Carl G. **Jung**,[677] for whom the soul is the reservoir of Science, associating [Ju] God the Father with the Self,[678] create a compelling narrative that intertwines the source of energy in the psyche (a self-regulating system), Jesus with an emergent structure of the consciousness that he assimilates to the Ego, and the Holy Spirit the mediator between the two.

In the contemporary era, the endorsement of Theistic Evolution by religious authorities, such as Cardinal Henri **Sonier de Lubac** (ref.73, p.22) a *Jesuit, Catholic theologian* who worked hard to bring the various religions closer together, and stated [Lu] in **1944**:

> *"The law of evolution does not tend to evacuate theology any more than it does Science, but to "purify" both, by differentiating them,"*

---

[677] **Jung** Carl G. *(CHE: Kesswil 1875-1961 Küsnacht). Born into a family of* Protestant *pastors (Swiss Reformed Church), his religiosity and his work as a psychotherapist are strongly intertwined. He is the founder of analytical psychology; Freud saw him as his successor. Author of numerous works linking religion and psyche.*

[678] *According to* |Rob] S. **Freud** *distinguishes three instances in the psychic apparatus: the* Ca, *a reservoir of psychic energy whose contents (primitive tendencies, elementary instincts) are unconscious and partly hereditary - the* Ego, *the conscious part of the psyche, in charge of the defense of the personality and mediation between the impulses of the Ca and the prohibitions and imperatives of the* Superego, *3rd instance at the origin of moral conscience, of the formation of ideals; playing the role of censor with regard to the ego, by unconsciously opposing the accomplishment of desires and emergence of drives.*

comforted in **1950**, through the encyclical *"Humani Generis"* by **Pope Pius XII,**[679] and subsequent statements by **Pope John Paul II,** on 22/10/**1996**; who, in front of the Pontifical Academy of Sciences, recognized the existence of various theories of evolution compatible with the Catholic dogma.

The *Catholic* biologist and entomologist Rémy **Chauvin,**[680] who specialized in the study of insects, particularly bees and ants, had unconventional and controversial views. He argued for example, that evolution is not a random and blind process, but rather a creative and purposeful one, guided by a *"cosmic intelligence"* transcending the material world (cf. F. Crick and L. Orgel authors of *"Directed Panspermia"* §V.3). A view different from *"theistic evolution"* but compatible.

As it appears in the epigraph of §V.3 the *Anglican* theologian Arthur **Peacocke** advocated for a view of evolution compatible with *Christian* faith. He maintained that God had created the world through a process of self-organization and natural selection according to His initial laws of nature. If he did not believe in miraculous interventions he supposed that God, having created the man in His image, had a special relationship with him. Although an ordained *Anglican priest*, he identified himself as a *panentheist*: everything in God.

In **1997,** the paleontologist Stephen J. **Gould**[681] coined the acronym NOMA in his essay titled: *"Non-Overlapping MagisteriA,"* an acronym, endorsed in 2007 by the *Catholic* biologist Pereda Francisco **Alaya,**[682] a defender of the theory of evolution. Alaya believes that purely materialistic interpretations are the work of naive people.

---

[679] **Pius XII**, *pope from 1939 to his death, real name: Eugenio, Maria, G. G.* **Pavelli** *(Rome, ITA 1876-1958 Castel Gandolfo VATII). Appointed in 1929 Cardinal and* **Pope Pius XI**'s *Secretary of State he worked with him on the drafting of* "Mit Brennender Sorge," *an encyclical written in German (whereas they are usually in Latin) condemning Nazism and distributed in 1937 in German churches. Although the Jewish authorities thanked him after the war for having helped save many Jews, some condemned his positions during the war, but he wanted to preserve the unity of the Church.*

[680] **Chauvin** Rémy *(FR: Toulon 1913-2009 Sainte-Croix-aux-Mines). Biologist, Entomologist, and Essayist. Professor emeritus at the Sorbonne. Schenk Prize* Elected at the French Academy of Sciences in 1956, Holder of the Outstanding Career Awards of the Parapsychology Association in 2002. He joined the CNRS in 1942 and was director of research at the INRA (National Institute for Agricultural Research) from 1948 to 1964.*

* *attributed in honor of the great German naturalist (*Alexander G. **von Schrenk** *(Tula-Russia 1818-1876 Livonia)).*

[681] **Gould** Stephen J. *(NY-US 1941-2002) Paleontologist, biologist, theorist of the evolution of life, and historian of Science he teaches at Harvard since 1974. Author of numerous books for the general public.* Agnostic Jewish.

[682] **Alaya** Pereda Francisco J. *(Madrid, ESP 1934). Ordained Dominican in 1960, he renounced his vows and got married. Emigrated to New York in 1961 where he defended his thesis in Biology in 1964 under the direction of* Th. **Dobjanski** *(539-167). He became an American citizen in 1971. Professor in Biology at the University of Irvine, CA, he is also a professor in Philosophy, Ecology, Evolutionary Biology, ... His work on genetics and evolution has allowed a new approach to certain diseases which earned him the title of* "Renaissance Man of Evolutionary Biology." *He defends the theory of evolution versus creationism.* Catholic.

On 16/02/**2004**, the *Christian* geneticist and Academician Jacques **Ruffié,**[683] was interviewed by Alia David for the special issue of La Recherche: *"Religion and Science. The genetic code fruit of chance?"* On the question: "Does accepting the theory of evolution deny God and the creative act?" he answered:

> *"No. In my opinion, which is the opinion of a scientist and not of a believer, if there is an intervention of God, it is not in the appearance of man, but in the appearance of the genetic code which is the most mysterious and decisive moment in the history of evolution. I think that the day when the genetic code appeared, everything was set. Was it by chance? Can we imagine writing a book, even with a billion years ahead of us, by pulling the letters out of a bag with our eyes closed?"*

According to Biologist Francisco **Varela,**[684] a Buddhist, we are far from having understood everything about evolution. There are [Va]: *"several paths of change, all of which are viable if there is an uninterrupted line of organisms. It is not a question of survival of the fittest, it is a question of survival of adaptation."* This is what, he and his teacher Humberto **Maturana,**[685] a *Theist* Biologist, call *"natural derivation"* [Mt]

But, even with the support of institutions like the American National Academy of Sciences (NAS),[686] the conflict has seen a resurgence at the end of

---

[683] **Ruffié** Jacques *(Limoux, FR 1921-2004), Doctor of Medicine, Geneticist, member of the Académie des Sciences, of the Académie nationale de Médecine. Professor of Hematology in Toulouse, Professor of Anthropology at the Collège de France since 1972, Professor in New York. Founder of the Hematotypology Center at the CNRS (discipline allowing to find the history of populations from blood characteristics). He demonstrated that if the notion of race had a meaning at the beginning of humanity, it disappeared 6,000 years ago, following the displacement of populations. Resistance fighter during WWII, he is a holder of the Médaille Militaire and Grand Officier de la Legion d'Honneur. He has written many books and publications. Le Monde of 02/07/2004 describes him as a* Christian.

[684] **Varela** Francisco *(Santiago, CHL 1946-2001 Paris). Ph.D. in Biology at Harvard, Teacher at the Pontifical Catholic University of Chile from 1964 to 66, then at the University of Chile from 1965 to 67, Director of Research at the CNRS ("La Salpêtrière," a Hospital in Paris), Author of numerous books in Theoretical Biology and Cognitive Sciences. He is a pioneer in bringing together Buddhism and Cognitive Sciences, and in 1990 he founded the* "Mind and Life Institute" *in the USA for this purpose.*

[685] **Maturana** Humberto *(Santiago, CHL 1928). After studying medicine in Chile, he left in 1954 to study anatomy and neurophysiology in London; In 1958 he obtained a Ph.D. in Biology at Harvard. Professor in Biology at the Faculty of Santiago from 1965. Author with his student and collaborator F.* **Varela** *of the concept of* "autopoiesis": *property that a system or a structure in interaction with the environment has to produce or maintain its organization despite a change of components. Author of numerous books, he received various awards including the Chilean National Science Prize in 1994. Maturana, like Varela was raised catholic but he turned to other religions like Buddhism and Hinduism and made a mix of them.*

[686] NAS: *was established in 1863 by an act of Congress, to advise the federal government and the public.* It is composed of eminent scientists. The NAS supports the theory of evolution since its inception in 1850.

the 20th century. A conflict, that has been fueled not only by the voices of *agnostics* such as Jean **Rostand**,[687] materialists, and anticlericals, whose religion was a "*combat atheism*" that relied on what they understood of this theory to deny the existence of God. Discordant voices within the highest echelons of the Roman Curia were evident. Thus, Cardinal Georg **Gänswein**, a proponent of a conformist perspective within the *Catholic Church*, and the secretary of **Pope Benoit XVI**, adopted 2018 an intriguing stance. He went out of his way to emphasize that the theory of evolution was not a doctrine of the Church but rather a scientific hypothesis open to verification. This emphasis seemed to oppose the 1996 statement of St Jean-Paul II and align with an effort to bolster the Intelligent Design Movement (IDM) position.

Founded in the **mid-1990s**, the IDM originated from a philosophical idea proposed by lawyer Phillip **Johnsson**.[688] Over time, charismatic religious scientists like the *Catholic* Biochemist Michael **Behe**[689] propelled it into a powerful movement in the USA. While IDM acknowledges the undeniable aspects of Darwin's theory of evolution, it emphasizes that certain complex biological systems remain unexplained. IDM suggests an "***Intelligent Design***" or creator, carefully avoiding the term "God." Notably, ID[690] is presented as a

---

[687] **Rostand** Jean *(FR: Paris 1894-1977 Ville d'Avray). Biologist (he had his laboratory), writer, and philosopher. In 1959 he was elected to the French Academy and obtained the Kalinga Prize, an international prize for popularization. He would have said of the theory of evolution:* "It is a tale;" *which can be summarized as follows:* **Darwin** *draws on certain facts to come up with a plausible theory but ignores the facts that bother him.* Agnostic, *he wondered a lot:* "Those who believe in God, do they think as often of Him as we, who do not believe in Him, think of His absence" *(J. Rostand* "Pensées d'un Biologiste, *J'ai Lu n° D5" p. 75).*

[688] **Johnson** Phillip E. *(USA: Aurora IL 1940-2019 Berkeley, CA). Law clerk then Law Professor at Berkeley.* Presbyterian.

[689] **Behe** Michael *(Altoona, US-PA 1952). BSc in Chemistry in 1974, Ph.D. in Biochemistry in 1978, post-doc on DNA structure, Professor of Biochemistry at Lehigh University, US-PA since 1985.* A Catholic *biologist, he recognizes the Big Bang theory and validates some of Darwin's demonstrations, notably the fact that Man and chimpanzee have a common ancestor, but defends Intelligent Design as an idea, and the principle of irreducible complexity, the only ones able, according to him, to explain the evolution of certain complex biological systems that* **Darwin**'s *theory of evolution cannot explain; but which we are entitled to think that, given the progress made in these fields, we are on the verge of achieving. The eye, whose complexity was a problem for* **Darwin**, *is no longer a problem nowadays because many simple organisms have been found to have a very high sensitivity to light and complexity that is justified by evolution. Dr.* **Behe** *suggests that all the genes and necessary materials were* "pre-loaded" *a long time ago and* "woke up" *several hundred million years later. But no organism containing such genes has been found to date. Moreover, given the knowledge of genetics on the mutation rate of genes, for example, such storage seems highly unlikely, and these genes could not have survived. It should be noted that although he intervenes as an expert in trials relating to I.D., his university colleagues let him full responsibility for his ideas which they do not endorse.*

[690] Intelligent Design (ID). *Starting from the observation that certain biological effects could not find any more explanations in the theory of evolution than in the scientific theories of the time, the* Presbyterian *lawyer* Phillip **Johnson** *published in 1989* "Of Pandas and People" *a bestseller in which he qualified the theory of evolution as atheistic, and proposed, on purely philosophical*

scientific theory although it is only based on philosophical arguments. and above all, it is not talking about a theory but a philosophical idea.

Key figures within IDM, such as an *Evangelical theologian* and mathematician William **Dembski,**[691] and *Protestant* Chemist Henri **Schaefer**[692] a self-proclaimed *"Jesus supporter,"* approach the concept with a nuanced perspective. **Dembski,** adopting a rational stance, acknowledges the potential fragility of ID as a scientific theory, if alternative explanations for intricate systems emerge.[693] It's noteworthy that recent research from the Department of Biology at Oxford, led by Professor T.A. **Richards** [RG], challenges ID's notion of irreducible complexity. Studies on microorganisms like warnowiid dinoflagellates reveal the roles of organelles like plastids, found in plant and algal cells, and mitochondria present in eukaryotic cells possessing a nucleus, providing alternative insights into the complexity of biological systems.

A broadened view of ID is that of the *Catholic theologist and biologist* Denis **Lamoureux**[694] who, after a parenthesis in the YECM, returned to his religion of origin. He proposes an intriguing perspective, suggesting that Intelligent Design can be viewed as an acknowledgment of nature's beauty, complexity, and functionality. Instead of a formal scientific theory, it becomes a religious

---

bases, a theory which called itself "theistic realist." *According to him, complex biological systems can only be explained by appealing to an* "intelligent" *cause, which he did not call God.*

[691] **Dembski** William *(Chicago, IL-US 1960) Mathematician, Computer Scientist, studied at the University of Chicago, Princetontrained* Evangelical theologian, *Philosopher, and Writer. In* "The Act of Creation: Bridging Transcendence and Immanence" *Global Journal of Classical Theology Vol 01: 3/07/1999 he writes: "The important thing about the act of creation is that it reveals theCreator. The act of creation always bears the signature of the Creator."*

[692] **Schaefer** Henry F. *(Grand Rapids, MI-US 1944). Studied at MIT and received his Ph.D. in physical chemistry from Stanford University, CA in 1969. In 1979 and 1980, he was a Professor of Chemistry and the first Director of the Institute of Theoretical Chemistry at the University of Texas in Austin. Theoretical Chemist and Director since 1987 of the Center for Computational Chemistry at the University of Georgia. An active member of various academies, and president of some, he is since 2004 professor emeritus at Berkeley, CA.. Author or co-author of 1,200 publications in prestigious journals. President from 1996 to 2005 of the World Association of Theoretical and Computational Chemists. Holder of numerous awards (the last two being the Gold Medal of the American Institute of Chemistry in 2019 and the P. Debye Prize in 2014), he is a potential Nobelist. He was a very* active protestant.

[693] *For W.* **Dembski** *if it were indeed established that certain intricate systems such as the bacterial "flagellum" or the eye, which he attributes to irreducible complexity, could be adequately explained through alternative interpretations, then the theory of Intelligent Design (I.D.) would inevitably crumble. Let's note that our understanding of the gradual formation of the eye, despite its high level of complexity, has made significant advancements.*

[694] **Lamoureux** Denis *(Alberta, CAN 1954). Ph.D. in Dentistry, Theology, and Biology. Chair of Science and Religion at St Joseph College, University of Alberta, Edmonton, Canada. Raised in the* Catholic *faith he lost it when he began his studies in Biology at the University. After completing his studies, he joined the* Young Earth Creationist *movement. While preparing his Biology and Theology thesis he adhered to the theory of God-guided evolution and returned to* Catholicism.

sentiment, recognizing the presence of an *"Intelligent Designer"* without blurring the lines between scientific inquiry and matters of faith.

Some scientists within IDM, such as cosmologists Franck **Tipler,** [695] a specialist in quantum mechanics, and the *Protestant* John **Barrow,** [696] present a *"quantum perspective"* involving an Omega point[697] (different from Teilhard de Chardin's one) and a paradisiacal universe. These views, distinct from traditional *Christian* beliefs, showcase the diversity of thought within IDM.

Facing them are high-level scientists, geneticists, biochemists, believers of all confessions, and even atheists like Richard **Dawkins,** [698] *"The Darwin's Rottweiler."* Dr in zoology, Dawkins stands as a staunch defender of evolution and a vocal critic of *Creationism and ID*. Dawkins' engagement in debates and controversies reflects the ongoing discourse between scientific inquiry and religious perspectives. Among the defenders of the *"Theistic Evolution of Life,"* we find mainly biologists. In effect, it is necessary to master this science, to debate with creationists and Intelligent Design advocates, and to propose compelling perspectives. They are spearheaded by eminent figures like the *Evangelist* Physicist Karl W. **Giberson,** [699] and F. **Collins** (refs.433 & 435, p.129)

---

[695] **Tipler** Franck *(Andalusia, AL-US 1953). Professor of Mathematics at Tulane University, NW, Cosmologist. Although a supporter of* Intelligent Design, *he does not defend creationism, which is opposed to the idea of evolution.* Theist.

[696] **Barrow** John, D. *(London, ENG 1952). Physicist, Mathematician, Astrophysicist,* Protestant. *BSc Math & Phys at Durham in 1979, Ph.D. Astrophysics at Oxford. Postdoc at the University of California at Berkeley, he became Professor and Director of the Astronomy Center at the University of Sussex in 1981; then, Professor in the Department of Math. and Theoretical Physics at Cambridge and Director of the Department of Astronomy at the University of Sussex in 1999. He received numerous awards such as the Templeton & Faraday Prizes in 2006, the Zeeman Medal in 2011, the Dirac Medal in 2015, and the Gold Medal of the Royal Astronomical Society in 2016. Member of the Royal Society of London since 2008, and the Academia Europea since 2019.* Intelligent Design supporter.

[697] Their Omega Point *[Ti]&[Bw] (not referring to Teilhard de Chardin) is a cosmic god, assimilated to a quantum hypercomputer, endowed with immeasurable capacity and speed, evolving in a multi-universe. He would be able to resurrect - physically or virtually, a little before the Big Crunch - the dead, thanks to all the data he would have collected on them, by making avatars, which is contrary to the teaching of* Jesus. *This is a bit like what Artificial Intelligence is doing today by creating holograms reproducing Johnny Halliday for example on the stage of the Olympia in front of his fans* "meeting" *their idol. It brings us back to the notion of image.*

[698] **Dawkins** Richard *(Nairobi, British Kenya 1941). Graduated in zoology at Balliol College, Oxford in 1962, then Research student of the ethologist Nikolas Tinbergen, 1973 Nobel Prize in Physiology or Medicine, he received his Ph.D. in 1966. He founded 2006 the* "Richard Dawkins Foundation for Reason and Science (RDFRS)." *He was awarded many medals & prizes for his good communication of science to public audiences, such as the Michael Faraday Prize in 1990 or the Fellowship of the Royal Society in 2001.*
He identifies himself as an **atheist,** *a* "cultural Christian" *and a* "cultural Anglican" *about the fact that he was raised as an Anglican whose reading of Ch.* **Darwin** *and B.* **Russel** *led to refute God's existence while adhering to Christian value.*

[699] **Giberson** Karl *(Bath, CAN 1957). Physicist and* Evangelist *committed to the defense of the theory of evolution, which is why he gave up in 2011 the teaching positions in Mathematics and in Relations between Science and Religion, which he had held since 1984 at Eastern Nazarene*

evolutionary *geneticist*, and director of NIH who dismisses ID theory as the "*god of the gaps*" theory. This expression meaning that IDM uses God to explain what science cannot is also used by another distinguished scientist, Professor Michal **Heller,**[700] a thinker who bridges Science and theology. Being parallelly a *Catholic priest*, he championed a worldview combining Mathematics, Cosmology, Philosophy, and Theology pointing the way to a "*new theology*." On receiving the 2008 Templeton prize, Heller said:

> "*Science is but a collective effort of the human mind to read the mind of God from question marks out of which we and the world around us seem to be made.*"

Having developed original concepts on the origin and cause of the universe he could answer a usual question: Does the universe have a cause?:

> "*Does the universe need to have a cause? It is clear that causal explanations are a vital part of the scientific method. Various processes in the universe can be displayed as a succession of states, in such a way that the preceding state is a cause of the succeeding one. If we look deeper at such processes we see that there is always a dynamic law prescribing how one state should generate another. However dynamical laws are expressed in the form of mathematical laws. By doing so we are back in the Great Blueprint of God's thinking of the universe. The question on ultimate causality is translated into **Leibniz**'s question....when asking this question, we are not asking about a cause like all other causes. We are asking about the root of all possible causes*"

One can find in the pivotal work of **Collins,** responsible for the international public team that participated in the sequencing of the human genome, reasons to become a supporter of Theistic Evolution. It reveals to us[Col] that at the DNA level, 99.9% of humans are identical, confirming that we all belong to the same

---

College in Quincy, US-MA. He earned two bachelor's degrees: one in Philosophy and the other in Mathematics/Physics. He holds a master's degree and a Ph.D. in Physics from Rice University in Houston, TX. Since 2012 has been a course leader for "Science and Religion" at "Stone Hill" University in Easton, MA. Elected a member in 1913 of ISSR (International Society for Science & Religion), a field in which he is very involved, both as a lecturer and prolific writer. His book "Save Darwin" was voted the "Best of the Year in 2008" by the Washington Post Book World. He has been a guest speaker at various universities and the Vatican.

[700] **Heller** Michal (Tarnow, Pol 1936) Roman Catholic priest and cosmologist. In 1959 he obtained a Master in Theology and was ordained priest. He followed Mathematics studies, obtained a Ph.D. in cosmology in 1966, and graduated from Dr ès-Sciences in 1969. Professor at the Pontifical Academy of Theology in Krakow since 1990, he has been an associate member of the Vatican Astronomical Observatory since 1981. Author of 35 books, and co-author of various papers in theoretical physics and cosmology, he is a laureate of many prestigious awards and honors, among which the prestigious 2008 Templeton Prize that he devoted to the Copernicus Center in Poland.

species. Geneticists agree that we descend from a group of 10,000 genera that lived 100 to 150 years ago in East Africa, confirming the fossils discovered by geologists. This means that humans share a common ancestor with other living species such as chimpanzees, with whom we share 98-99% of DNA, which the *Jesuit* Paleontologist **Teilhard de Chardin** (ref.505, p.155) contributed to demonstrate. Let's note that the chimpanzee has 24 pairs of chromosomes, so we suppose that at a certain time, two of them merged to arrive at the 23[701] of man. We can admit that since there was a period before man, there will be one after, which will arrive with the Parousia[702] according to some Christians, and perhaps even before if we do not manage to slow down the climatic effects observed by everyone and confirmed by meteorologists. We are indeed in the presence of an evolutionary theory. K. **Giberson**, like these prominent scientists believes [Gi] that everything in the laws of the universe must allow Evolution to choose the path leading to Man. Similarly, for one of the greatest contemporary world paleontologists, the *Christian apologist* Simon C. **Morris,**[703] the human species is an inevitable consequence of evolution. Based on the concept of convergence of the different possible paths introduced by the theory of evolution, he "*demonstrates*" that it is wrong to think that Life has no predictable direction or goal. He *postulates* that just like physics and chemistry, biology is governed by laws that make everything possible and that Life was inevitable [Mo] on our Earth which, according to him, must be a rare planet in the whole universe. Taking the opposite view of another famous paleontologist Stephen J. **Gould** (ref.681-p.243) a *Jewish agnostic*, for whom, it is not said that another species or Intelligence would have appeared if an asteroid had not hit the Earth causing, 65 million years before our era, the 5th mass extinction and leading to the disappearance of dinosaurs.

---

[701] *22 pairs are called* "autosomes" *(body chromosomes) while the 23rd* "allosome" *defines the sex. Men have one X and one Y chromosomes while women have two X chromosomes.*

[702] Parousia: *Greek term meaning* "arrival" *or* "presence." *According to* **Matthew** *25-31, it is the return of Jesus Christ, victorious over the devil, to establish the kingdom of God. When Ed. Blattchen asked* [Md3] *Th.* **Monod** *(ref. 681-p.187):* "Do you believe in the Parousia, the return of Christ on the cross, victorious"? *the latter said:* " ... *I can hope for it of course. I hope, especially for the prophecies of* **Isaiah** *(Jerusalem: -766 to -701):* the holy mountain on which no harm or damage will come, the time...). *For the Muslims, though he is not referenced in the Koran - only in some hadith literature - he is* **Mahdi** *(the Guided), a descendant of* **Muhammad,** *the messianic figure appearing instead of* **Jesus** *who, is only a prophet for Muslims.*

[703] **Morris** Simon Conway *(Carshalton, ENG. 1951). He studied Geology at Bristol and received his Ph.D. at Cambridge. One of the world's greatest paleontologists, a specialist in the* "Cambrian explosion" *(541 to 530 million years B.C.) which corresponds to the appearance of most multicellular animals (metazoans) whose first fossils were discovered in Burgess, Canada. Since 1995, he has held the Chair of Paleobiology and Evolution at Cambridge. Elected Fellow of the Royal Society at the age of 39! Holder of various medals and prizes: Walcott, of the Nat. Acad. of Sc. of USA; Lydell 1998; W. B. Hardy in 2010. A convinced Christian, he participates in many scientific and religious debates. According to him, evolution is compatible with the existence of a creative God. He authored numerous books on evolution.*

Combining their theistic conceptions and their adherence to scientific discoveries, such as the theories of the

Big Bang, evolution, and natural selection but planned by God, Giberson, and Collins argue [Gi] more insistently that:

> "...*God's creative work can be done through the laws of nature, and not merely by breaking or suspending those laws.*" .... "*We suggest that once life arose, the process of evolution and natural selection permitted the development of biological diversity and complexity, and humans are part of this process.*" "*Moreover, once evolution got underway, no supernatural intervention was required*" (cf. [Gi] p. 115).

In his essay: "*Is Evolution a Chance Process?*" [Ar], the *Baptist* apologist Denis **Alexander,** [704] a prominent biologist, well-known for his criticism of creationists and I.D. supporters, a Fellow of the Royal Society of Edinburgh, argues the case that:

> "*The evolutionary process is far from being a theory of "chance" from biological, mathematical, or indeed, philosophical and theological perspectives.*"

He backs up his claims with advances in biochemistry, to which he has broadly contributed. If chance plays a role in the occurrence of random mutations - as quantum mechanics do in various fields - natural selection is not a random process it plays a key mechanism. It acts as a filter in the evolution process and, depends on environmental conditions, genetic resources, and others. Natural selection favors mutations that provide advantages for survival and reproduction in a given environment.

It is important to note that the synthesis of **Darwin**'s theory with the knowledge of the evolution of the Universe (cf. chap. III) offers an interesting paradigm for some theistic scientists such as the mathematician Brian

---

[704] **Alexander** Denis (East Goscote, ENG: 1945), Emeritus Fellow of St Edmund's College where he graduated in biochemistry under A. Peacocke in 1965; was awarded in 1968 a Ph.D. in neurochemistry at the Institute of Psychiatry, at Cambridge. Previously Chair of the Programme of Molecular Immunology at The Babraham Institute, Cambridge. From 1971 to 1986 he developed new *research programs outside the UK (Turkey, Lebanon). Returning to London he was a Project Leader at the Imperial College (from 1986 to 1989) and then at Cambridge (1989-2008). He co-authored many scientific publications.*
*Christian (Baptist) apologist, wrote his first book on Religion and Sciences: "Beyond Science" in 1972; since then, he has written, lectured, and broadcasted widely in this field. He is the Emeritus Director of the Faraday Institute for Science and Religion which he co-founded with Bob **White** in 2006. In 2012 he was awarded the Templeton Prize, and in 2017 was appointed Member of the Order of British Empire (MBE) "for services to science and to the public understanding of science and religion"...*

Swimme,[705] an evolutionary theist, who introduced him to the work of **Teilhard de Chardin**. In effect, adopting the idea that everything in the universe has a physical and a spiritual component he does not see evolution only as a mechanistic and random process but, attributes to it a deeper purpose, a spiritual one, leading some scientists to attempt to make a bridge between science and religious beliefs. Opening, in some cases, the door to **ethical considerations,** leads to a sense of responsibility for the well-being of the planet and its inhabitants. Ideas defended by pioneering primatologists like Dr. Jane **Goodall** (646, p.227), who as mentioned in §5.3., is an ardent advocate for the ethical treatment of animals, and the foremost expert on great ape behavior. In her foreword to the **2017** book of the science philosopher Ervin Lazlo: *"The Intelligence of the Cosmos,"* she underscores the idea that:

> *"We must accept that there is an intelligence driving the process (of evolution), that the Universe and life on Earth are inspired and informed by an unknown Creator, a Supreme Being, a Great Spiritual Power."*

To clarify her viewpoint, I'll report that in an interview with Moss Stephen for the Guardian on 13/01/**2010**, she explained:

> *"I was raised as a Christian and I see no contradiction between evolution and belief in God."*

These statements encapsulate Dr. **Goodall**'s perspective on the relationship between science, evolution, and spirituality. Her work and beliefs serve as a source of inspiration for many individuals and encourage us to explore the intricate connections between science and faith.

The debate we are reporting concerns Man and Life in general, but any person adhering to the Big Bang theory (§III.2, pp.145-151) cannot but admit that the creation of the universe is evolution: the appearance of elementary particles which gave birth to the hydrogen atom H (the simplest known), which allowed the formation of helium atoms (He)...geological evolutions still operating: the disappearance of islands, the erosion of cliffs under the effect of anthropic factors, and continental drifts … It is well known that the East African Rift Valley is, since the beginning of the Miocène at least (55 million years ago), slowly stretching (few mm/year) over more than 3,000 km from the Gulf of Aden in the north towards Madagascar in the south. It will result in a few million years the partition of the African continent in two and the creation of a new ocean… The work of the

---

[705] **Swimme** Brian *(Seattle, WA-US 1950), a mathematician who defended his thesis on string theory, and a follower of P.* **Teilhard de Chardin** *teaches evolutionary cosmology at the CIIS in San Francisco. He is a convinced evolutionary,* Theist *he authored books and films on this subject. His primary area of interest lies in studying humanity's role on Earth and in the cosmos. Created in 1989 the* "Center for the Story of the Universe" *delivers Masters and Ph.D.*

prominent Professor of Geophysics Robert **White**,[706] an *Anglican,* demonstrates that the most rapid and largest effusions of volcanic rocks that we observe are due to continental rifts [Wh].

Let's note, however, that with the progress made in the field of Genetics, Man risks disrupting natural evolution by generating environmental modifications to the detriment of Nature, Animals, and Man which authorizes the *Lutheran theologian* Philip **Hefner**[707] to say that humanity is *"self-creating,"* joining the idea of P. **Teilhard de Chardin** foreseeing an ultra-human taking in his hands the destiny of Humanity. Let's hope that, if this were to happen, these super-humans would be endowed with Wisdom.

In the realm of clashes between evolutionary theory and Genesis, our exploration has journeyed through the conflicting ideologies of YEC, ID movements,.. and the answers of prominent Theistic scientists mainly biologists or geologists. Yet, this debate extends beyond life sciences; it draws insights from eminent specialists in diverse scientific disciplines. Allow me to introduce two distinguished voices from the cosmos:

- Vera **Rubin** (ref.388, p.115) *Jewish* Astrophysicist. In a 1989 interview,[708] Rubin mused on the marvel of existence, stating:

"The laws of physics being what they are, the galaxies, the stars, and the planets came to exist, and the supernovas came to exist, and humanity came to exist, and the evolution is remarkable. I think it is remarkable."

- Richard **Smalley** (ref.424, p.126), *Christian* Nobel Laureate in Physics in 1996, a staunch supporter of theistic evolution. Smalley expressed in **2004**, at the twilight of his life:

"Although I suspect I will never fully understand, I now think ..... God did create the universe about 13.7 billion years ago, and of necessity has involved Himself with His creation ever since."

---

[706] **White** Robert *(ENG: 1952). Professor of geophysics, and vulcanologist in the Earth Sciences Department at Cambridge University; Project leader of a research group studying the internal structure of volcanoes all around the earth, and Earth's crust dynamic .... Student, then Research Fellow at Emmanuel College, Cambridge, Fellow of St Edmund's College, Cambridge, Fellow of the Geological Society. .. Was elected a Fellow of the Royal Society in 1994. Author and co-author of many papers, he has a lot of distinctions among which the Gold Medal of the Royal Astronomical Society for Geophysics in 2018.* Anglican, *he has published various papers and books on Science and Religion. He is the Emeritus Director of the Faraday Institute for Science and Religion.*

[707] **Hefner** Philip *(USA: Denver, CO 1932 – 2024 Chicago, Il). Lutheran theologian, philosopher of theology, and lecturer. Ph.D. at the University of Chicago then attended the Lutheran School of Theology in Chicago where he was professor emeritus.*

[708] Alan Lightman *for AIP (American Institute of Physics.). Washington D.C. on 03/04/1989.*

Adding to this chorus of perspectives are two eminent scientists embracing evolution as an asset for religion:

- Theodosius **Dobzhansky** (ref.539, p.165) *Russian Orthodox* geneticist. Dobzhansky argued in 1973 that evolution is a scientific fact, reconciling it with religion by stating:

"*... nothing in theology makes sense except in the light of evolution*"[Do], Assertion confirmed by Charles **Combulazier**,[709] a *Catholic* priest and Doctor of Sciences:

"*Evolution is the only argument to prove the existence of God.*"

- A. **Penzias, 1978** Nobel Prize in Physics, an Orthodox Jewish (ref.364, p.108). Penzias confirmed his support for evolution, connecting it with biblical concepts. On March 12, **1978**, he stated:

"*The best data we have are exactly what I would have predicted, had I had nothing to go on but the five books of Moses, the Psalms, the Bible as a whole.*"

His nuanced views echo in 1995 and 2008, highlighting the delicate balance of the universe as consistent with a purpose underlying creation. In 1995, to the question: "*Do you think God revealed Himself at Sinai?*" he smartly answered:

"*Or, perhaps God is always revealing Himself. ... Again I think of Psalm 19: "The heavens declare the glory of God, and the firmament shows his handiwork." "All reality, to a greater or lesser extent, reveals God's purpose. There is a certain relationship between the purpose and order of the world in all aspects of human experience ...... This world is most consistent with the purpose of creation.*"

In **2008**, when interrogated on the relationship between science and genesis, **Penzias** answered:

"*Astronomy brings us to a singular event. A universe, that was created out of nothing, a universe endowed with the very delicate balance to produce exactly the conditions required for life, and a universe that has a purpose underlying it.*"

These diverse perspectives weave a rich tapestry, transcending the confines of traditional debate. Further expanding our panorama, let's embrace the words of

---

[709] **Combulazier** Charles *(Marseille 1903-1991)*. Priest, catholic chaplain, *and disciple of Pierre* **Teilhard de Chardin** *received in 1973 the Montyon Prize (awarded by the French Académie des Sciences and l'Académie Française) for his essay:* "God Tomorrow."

Antoni **Gaudí**,[710] the architectural maestro of the 19th century, who declared *"God's Servant"* by the Catholic Church saw creation not as a human invention but as a continuous discovery:

> *"The creation continues incessantly through the media of man, but man does not create....... he discovers."*

In this symphony of voices, science and faith harmonize, each note contributing to the intricate melody of our understanding.

5.5. Does scientific rationality exclude faith?

It is the Heart that feels God and not the reason: this is what faith is.
B. Pascal Pensées

"Understand, to believe, my word; believe, to understand, the word of God."
St. Augustine Sermon 43

"Science without religion is lame, religion without science is blind" [Ei3].
A. Einstein

The answer to this question is no; it is not because we cannot demonstrate something, that cannot exist. We have seen this with the presocratics, the scholastics, the physicists, the mathematicians, the cosmologists, and the biologists we have mentioned, many of whom have left their mark on the scientific field. Similarly, Vatican II (1962-65) recalled the importance of scientific research and invited theologians to take note of its results, without ignoring philosophy. In **1999**, Cardinal **Ratzinger**, better known as Pope

---

[710] **Gaudi** *Antoni (Catalonia, ESP: Reus or Riudoms 1852-1926 Barcelona), Born* **Gaudi i Cornet**. *He enrolled in the Piarists (a religious order of clerics regular of the Catholic Church) where he displayed his artistic talents. His family originated in the Auvergne region, southern France. In 1878 he graduated from the Barcelona Higher School of Architecture. He also studied French, philosophy, and economics. To finance his studies, he worked as an engineering engineer, for various architects and constructors. Initially, influenced by neo-Gothic art, he was profoundly impacted by the Nature that he could observe during his long periods of illness (he suffered since he was 6 years old of chronic arthritis). Speaking of The Sagrada Familia, one of his seven realizations registered as a cultural site on UNESCO's World Heritage List, he said:* "My model is a tree. Its branches with leaves on them. .... A tree does not need help from the outside." *The geometry of his arches, which are well-fitted by conical curves, has been extensively studied. He was a man of high spirituality. Nicknamed* "God's Architect," *his nomination as* "God Servant," *is a first step in the beatification process. According to* **Josan** I*, *who asked him:* why are you taking so long to finish La Sagrada Familia? *Gaudi replied,*" My client - meaning God - is not in a hurry."

***Josan** I. wrote "Pilgrimage - a rudimentary form of tourism" GeoJournal of Tourism Geocities, 2(4):160-168, Moore 2011).*

Benedict XVI, recalled that *"Logos"* refers not only to the *"Verb*[711] *of God"* but also to *"reason,"* insisting on the rationality of the Faith. In 2008, he wrote that the Church's mission is to *"keep alive the sensitivity to the truth"* and *"always invite reason to the search for the truth, for good, for God."*[712] God is also named the *"Word."*

Although the perception of the universe that we have today is very different from the more limited perception of the Greek philosophers[713] - who, although pagans, believed in a god superior to the others, which some contemporary philosophers call *"pre-Socratic theology"* - nothing in the current discoveries can deny the existence of God. I'll remind, chronologically, that:

- **Thales** (ref.35, p.12), the initiator of the Ionian rationalist philosophy, would have said according to what **Aetius**[714] has postulated:

"...that God is the Intellect of the world.... That through the elemental humidity, a divine force moves it."

The term humidity refers to his conception of the universe (cf. p.19). His god is not one of theists but the one who is above the numerous gods which are, according to him, spread in nature.

- If **Cicero** recalled that:

"Anaximenes establishes that the air is god, that it is generated, that it is immense, infinite and always in movement...," he did not adhere at all to this idea because for him:

"... water is the Principle of things and god is the Intellect that shapes all things from water."

- For **Al-Kindi** Sunni, it was not sufficient to believe in the Ancients' thoughts or books, it was necessary to understand and thus *"follow the path of science."*

---

[711] *The* **Word** *in Genesis is the name given to* **God.** *Word is a translation of the Greek Logos when referring to Jesus Christ who, according to the dogma of the Trinity, is God. The evangelist John says (in the prologue):* "The Word became flesh" *[Jn: 1-14].*

[712] *op. "Vatican News.*

[713] *I advise the reader interested in this subject to read* "The Atom in the History of Human Thought" *by B. Pullman [Pu] which is very enriching.*

[714] **Aetius** *Flavius (Durostorum=Silistra, BGR 395-454 Ravenna, ITA). Senator, Consul, and Roman generalissimo, it is said that he was the* "last of the Romans" *because he fought against the barbarians and especially Attila. After his assassination, the Western Empire was ruled by Roman generals of barbarian origin.*

- **-- Avicenna**, a Persian Muslim, the third Master after **Aristotle** and **Al-Fârâbî,** according to his disciples who called him: *"prince of scholars,"* praised God:

"Blessed be God the best of creators and praise to God Lord of the universe."

- **St Anselm of Canterbury,** [715] qualified with the nickname of *"Magnificent Doctor"* by Pope **Clement XI** in 1720 who declared him *Doctor of the Church*. Philosopher, Theologian, and *Benedictine monk,* he was one of the major writers of the Medieval West, trying to reconcile faith in God and reason inherited from the Greek thinkers. He was credited as the founder of scholastic theology. For him:

"...the priority of faith is not opposed to the research proper to reason... [that]...is not designated to express a judgment on the content of the faith; it would be unable to do so because it is not suitable for that. but must allow the whole understanding of the content of the faith. Its task is rather to find meaning, to discover reasons that allow everyone to reach a certain understanding of the content of the faith" (cf. [JP] - p. 59).

A concept that he summarizes as follows:

"I do not seek to understand to believe, but I believe to understand. For this also I believe – that unless I believe, I shall not understand." [716]

This statement is to be compared with that of **St Augustine** (epigraph §V.5, p.254).

---

[715] **St Anselm of Canterbury** *called Anselm of Aosta or Anselm of the Bec (Aosta, ITA 1033 - 1109 Canterbury, ENG), was canonized in 1494 by* **Pope Alexander VI** *(Rodrigo Borgia). Trained by the Benedictines, he left Aosta in 1056 to take advantage of the teaching of* **Lanfranc of Pavia** *a monk, and arrived in Avranches, FR in 1059 to follow the courses of the Abbey of Mont St Michel. In 1060, he entered the Abbey of Le Bec (Eure, FR), an important intellectual place of the time, as a monk. He taught grammar and took courses in rhetoric and dialectics. He was very influenced by the reading of* **Aristotle's** *philosophy. From 1092, he participated in the reorganization of the Welsh church, which he eventually integrated into that of England. He refused, in 1093, the king's proposal to appoint him archbishop because he thought it was the pope's attribute, but finally accepted.*

[716] *"Proslogion": by* **St Anselm** *Livre de Poche 1993. It is an Allocution on the existence of God. His ontological* argument of the existence of God is famous but seems to me difficult to defend by a scientist because it is not necessarily logical, even if R.* **Descartes** *took it up in another form.*

* According to the Grand Robert [Ro]*, an ontological argument is related to ontology (Onto in Greek = Being = God). "It aims at proving the existence of God by the sole analysis of his definition, of his essence (God is perfect, therefore he exists)."*

- **Guillaume de Conches** ([Pu] p.129), a *Catholic theologian*. According to him, the Bible has no competence in the scientific domain. Nevertheless he:

"... recognizes that it is God's action that is at the origin of the world... affirms ... that this action is exercised through the laws of nature."

- Moses **Maimonides**, a *Jewish rabbi, "the second Moses,"* wrote:

"As for the things of Astronomy and Physics, there will be, I think, no doubt that they are things necessary to understand the relation of the universe to the government of God."

- **St. Thomas Aquinas**, *Dominican friar and priest, Father, and Doctor of the Church, "Apostle of the Truth"* according to **Pope Paul VI**, placed in the foreground the harmony existing between faith and reason:

"Faith and reason can never contradict each other because they have the same origin. God who is the only truth" (cf. *[Th] pp. 331-336*).

Master of thought and in the art of practicing theology, he was not afraid to confront faith with reason.

According to him, faith can rely on reason and trust. This, among other things, led Pope **Paul VI** to write on 29/09/1974 (the 700th anniversary of St Thomas Aquinas' death) in the Apostolic Letter *"Lumen Ecclesiae"*:

"...Thomas had in the highest degree the courage of truth, the freedom of spirit, allowing to face new problems" ... to reconcile the secular character of the world and the radical character of the Gospel, thus escaping that unnatural tendency which denies the world and its values, without failing to meet the supreme and inflexible demands of the supernatural order."

Now, we will discuss different perspectives that - since the empiricism advent, during the 16th century - scientists had regarding the relationship between their beliefs in God and their scientific work. These viewpoints can range from those who see no connection between science and religion, to those who believe in an interconnected relationship, and others who find that their faith positively influences their scientific pursuits. Also, it is noteworthy that this classification is not easy, because some may assert a clear distinction between the domains of science and religion, emphasizing their separate realms and methodologies; others hold nuanced positions.

a) <u>Some scientists don't claim any connection between Science and Religion</u>. In their view, scientific understanding is based on evidence, observation, and empirical methods, while religious beliefs are based on faith, tradition, and personal experiences. It is the case for example for Galileo **Galilei** *Catholic* scientist (rfs.144&146, p.41) writing to **Christine de Lorraine,**[717] his patron:

> "*The Holy Spirit intends to teach us how one goes to heaven, not how the heaven goes.*"

He has a nuanced position, but I include him here, due to the perceived separation implied by his work and beliefs. Don't forget that he faced challenges from the Church, which could imply a perceived separation between his scientific pursuits and religious beliefs.

More incisive is R. **Feynman** (ref.361, p.108), a *freethinker* of *Jewish education, laureate of the* 1965 Nobel Prize in Physics:

> "*Religion is a culture of faith, science is a culture of doubt.*"

The renowned *Jewish* astronomer, and astrophysicist Vera **Rubin** (ref.388, p.115) known for her groundbreaking work on dark matter, who often speaks of the relationship between science and religion, declared during an interview:

> "*In my life, science and religion are separate. I am Jewish and religion is for me a moral code and historical manna. I try to do my scientific work morally, and, ideally, I believe science should be seen as something that helps us understand our role in the universe.*"

While *Catholic* biologists Jean-Marie **Pelt** (ref.583, p.191), and Francisco **Alaya** (ref.682, p.243), as well as many other scientists, refuse to mix their faith and their work as researchers, **Pelt** demonstrates in "*Dieu de l' Univers*" why:

> "*Science and Faith are two different domains, the former allows him* (the biologist or the botanist) *to understand nature and his faith answers the ultimate questions*" [Pe].

The same J-M **Pelt** also stated ([Pe2] p. 164):

> "*Science can allow us to approach the divine through contemplation and wonder at the way the world is wired.*"

The renowned *Jewish* psychiatrist Kenneth **Kendler**[718] is opposed to any association between science and religion because he believes that they have

---

[717] **Christine de Lorraine** *(Nancy, FRA 1565-1637 Florence, ITA), a* Catholic. *Daughter of Charles III, Duke of Lorraine and Bar, and Claude de France. She was the granddaughter of Catherine de Medici who raised her after her mother died in her 10th year. She married* **Ferdinand 1st Medici** *(cardinal in 1562 when he was only 14 years old, becoming grand Duke of Tuscany at 38 years old, following the death of his brother; death, which one suspects him to be the sponsor).*

[718] **Kendler** Kenneth S. *(N-Y, USA 1950). Professor of Genetics & Director of the Virginia Institute for Psychiatric and Behavioral Genetics. Co-editor of* "Psychological Medicine." *He studied medicine at Stanford University School of Medicine and completed his training in*

different foundations, just like are, knowledge and wisdom, he nevertheless recognizes their complementarity.

For the *orthodox Jewish* scientist Yeshayahu **Leibowitz**[719] belief, honesty, and immorality have nothing to do with genetics or education, they are personal decisions: *"The spiritual force is your will."* In the same spirit, he answered in June **1992** to Joshua Haberman who interviewed him:

"...there is no connection between science and religion."

In **2009**, on the occasion of her 100th[720] birthday Rita **Levi Montalcini**,[721] *Agnostic*, raised Sephardic Jewish, co-winner with Stanley **Cohen** of the 1986 Nobel Prize in Physiology or Medicine *"for their discoveries of growth factors,"* said to the writer Paolo Giordano who interviewed her:

"The human mind indeed possesses I know not what is divine, where the first seeds of useful thoughts were sown...,"

---

genetic psychiatry at Yale University in New Haven, US-CT. Confessed in an interview [Fa] that if he had not studied business he would have studied religion. Author of 1,200 scientific papers. Jewish.

[719] **Leibowitz** Yeshayahu (Riga, LVA 1903-1994 Jerusalén, ISR). He began his studies in Chemistry and Biology in Berlin in 1919 and obtained a Ph.D. in 1924. He studied Biochemistry and Medicine in Basel, defended his thesis in 1934, and emigrated to Palestine. Appointed professor of organic chemistry at the Hebrew University of Jerusalem, then professor of biochemistry in 1941, *obtained the chair of neurology in 1954. At the same time, he taught a course in Jewish studies at the University of Haifa. Retired in 1973, he taught philosophy at the university. He was an orthodox Jew, anti-conformist, and had many enemies. He recognized only voluntary submission to God and recognition of the yoke of the commandments, cf. [Hab] p.139. His supporters, on the other hand, considered him a prophet. Impassioned and unyielding debater, he refused the Israel Prize, the most prestigious in the country.*

[720] *Science and Techno Wired - Italia.*

[721] **Levi-Montalcini** Rita *(ITA: Turin 1909-2012 Rome). Although she was born into a Jewish family deeply attached to its values, at the age of 20 she obtained from her father that he let her study. She graduated in 1936 from the Medical School of Turin where she was a student of the professor of histology* Giuseppe **Levi** *who inspired her. Rita continued her academic career in neurophysiology until 1938 when the anti-Semitic laws, prohibiting Jews from Academic and professional careers, were promulgated by Mussolini. Until 1943 she continued to work in a laboratory she had set up in her home where G. Levi joined her and became her first and only assistant. In 1946 she accepted a position at St Louis-MO which she held for 30 years before returning to Rome. Director of the Institute of Cell Biology at the Italian National Research Center. She published her autobiography* "In Praise of Imperfection" *in 1988. She founded the EBRI (European Brain Research Institute) in 2002, in Roma, IT.*

Remind that *the same Mussolini had banned the Pentecostal Christian movement in 1935. She humorously wrote:* "I should thank Mussolini for having declared me to be of an inferior race. This led me to the joy of working, not anymore unfortunately in university institutes, but in a bedroom." *She also humbly said:* "If I had not been discriminated against or had not suffered persecution, I would never have received the Nobel Prize." *Cf. Internet* "Women who changed science" *@Becker Medical Library Washington University School of Medicine.* Agnostic.

"I am agnostic, but I am fascinated by the mystery of existence. I think that science and spirituality are complementary."

For the astrophysicist Hubert **Reeves** (ref.4, p.1), it was necessary to:

"Distinguish between the reasonable and the rational. The first one includes intuition and effectiveness, and the second one implies only a logical process" [Re1].

Interviewed on 13/08/**2013**, by Patrick Poivre d'Arvor:
PPDA. "Is religion compatible with science?"
H.R. *"Absolutely. I have friends who are excellent scientists while being believers. There is no conflict. Especially since I don't believe that science will ever be able to prove the existence or not of God."*

It's noteworthy that he declared to be neither Atheist nor Agnostic.

To highlight the complexity of this classification and emphasize that it is not rigid, Pope **John Paul II**'s statement, grounded on the book of Wisdom and Proverbs [JP], offers valuable insight. He stated:

"...that there is a profound and indissoluble unity between the knowledge of reason and that of faith (p.28) ......There can therefore be no competitiveness between reason and faith: one integrates with the other, and each has its field of action (p. 29)."

This quote underscores the idea that while there may be distinct domains of science and religion, they are not mutually exclusive or in competition. Instead, they can coexist and complement each other, demonstrating the interconnected nature of the science-religion relationship.

b) Scientists who make interconnected relationships. Some scientists claim clearly that they see science and religion as interconnected and complementary. They might believe that both science and religion offer different ways of understanding the world, and these perspectives can coexist harmoniously. For them, science can explain the *"how"* of the phenomena, while religion addresses the *"why"* or deeper purpose.

Lord Francis **Bacon** (ref.151, p.43) scientist, statesman, and philosopher said (Essay16, titled *"Of Atheism"*):

"A little philosophy inclines the man's mind to atheism, but depth in philosophy brings the men's minds about to religion."

In this essay, Bacon argues that atheism is irrational and contrary to both nature and reason. According to him, the contemplative atheist is rare, and most people who deny God do so out of self-interest or ignorance. I think that this last argument cannot be applied to scientists.

Baron Augustin **Cauchy,**[722] one of the most prolific mathematicians of his time defending the *Jesuit* schools said in 1838:

"Catholic, I cannot remain indifferent to the interests of religion; geometer, I cannot remain indifferent to the interests of Science."

It is another proof that one can be a brilliant scientist and a *practicing Catholic* without mixing the two. He also stated:

"I am a Christian, I believe in the divinity of Christ, like all the great astronomers, all the great mathematicians of the past" [AI].

In the annals of science, the illustrious Michael **Faraday** (ref.235-p.71) stands as a remarkable figure, known for his pioneering work in the world of experimental science. Faraday, a *deeply religious* man, maintained silence about God in his lectures throughout his career, with one notable exception in the year **1848**. This momentous occasion took place during a series of captivating conferences on *"The History of Fire,"* specifically in the sixth conference titled *"The Chemical History of a Candle."* It was none other than the renowned *Catholic* scientist Louis **Pasteur** (ref.240-p.73) a trailblazer in Microbiology, who bore witness to Faradays's unexpected revelation.[723] In a hushed auditorium, Faraday, to the astonishment and approval of his audience, uttered these profound words:

"... I have just surprised you by uttering the name of God here. If this has not happened to me before, it is because I am a representative of experimental science in these classes. But the notion and respect of God come to my mind by ways as sure as those which lead us to truths of the physical order."

On another occasion [Gs] he expressed his deep-seated belief:

---

[722] **Cauchy** Augustin, L. baron (FR: Paris 1789-1857 Sceaux). He entered the Ecole Polytechnique at the age of 16 and the Ecole Nationale des Ponts et Chaussées at the age of 18. Participated in the construction of various bridges before devoting himself to mathematics. A field in which he is recognized by his peers. His work on electromagnetic waves earned him a prize in mathematics in 1816 and admission to the Académie des Sciences. He has to his credit about 800 publications and 7 books! He was a professor at the Polytechnique until his exile in 1830, during the enthronement of Louis-Philippe I, because of his religious and royalist convictions (he believed that royalty was the support of our Christian civilization). He taught in Switzerland, Piedmont, Rome, and Prague until his return to France in 1838. He was elected in 1851 to the Collège de France, and in 1849 he became the holder of the chair of mathematical astronomy at the Faculty of Sciences in Paris.
An ardent Catholic, he was very close to the Jesuits and participated in the foundation of numerous charitable works such as the Oeuvre des Écoles d'Orient, the Oeuvre d'Irlande, the Institut Catholique de Paris, and supported the St. Vincent de Paul association...
[723] Extract from the speech of reception of L. Pasteur at the French Academy of Sciences.

"Since one God created the world, all of nature must be interconnected into one whole."

Faraday's eloquent declaration echoed his firm conviction that science and faith were not adversaries, but rather two pathways leading to the revelation of God's wisdom and power.

In **1870**, during one of his readings at the request of the Victorian Institute, the Lucasian Professor of Mathematics, Sir George **Stokes** (ref.226-p.68) a passionate Evangelist, engaged *"the religious scientists to follow scientific discovery fearlessly."* In **1883**, the *"Father of Microbiology,"* L. **Pasteur**, a devout *Catholic,* became a full member of the French Academy of Sciences. He had to pronounce, as is customary, the eulogy of his predecessor, the late Emile **Littré**,[724] whose human qualities and zeal for work he admired, calling him a *"lay saint."* Nevertheless, he could not refrain from expressing *"his disagreement with his philosophical opinions."* Indeed, the *Christian* E. Littré had become *agnostic* after reading Auguste **Comte**'s *"Système de Philosophie Positive,"* whose ideas he had embraced and adapted to his work, proclaiming that this system could be applied to experimental sciences. However, according to L. **Pasteur**, the research works of misters Comte and Littré, having to do with the study of facts of the past (History, Languages, ...), did not require any inventiveness, contrary to those of a scientific experimenter who must confront his hypotheses with the results of his experiments. For L. **Pasteur**, *"this positivist system fails ... above all, by the fact that it ... does not take into account the most important positive notion, that of infinity."* Then, followed a poetic dissertation:

"Beyond this starry vault, what is there? New starry skies. So be it! And beyond that? The human mind, driven by an invincible force, will never cease to ask: What is beyond? Does he want to stop either in time or in space? As the point where he stops is only a finite quantity, greater only than all those which preceded it, hardly does he begin to envisage it, that the implacable question returns and always, without him being able to silence the cry of his curiosity. It is useless to answer beyond are spaces, times, or sizes without limits. No one understands these words. He who proclaims the existence of the infinite, and no one can escape it, accumulates in this affirmation more supernatural than there is in all the miracles of all the religions; because the notion of the infinite has this double character of being self-imposed and of being incomprehensible. When this notion takes hold of our understanding, we have only to bow down. Again, at this moment of poignant anguish, one must ask for grace from one's reason: all the springs of intellectual life threaten to slacken;

---

[724] **Littré** Emile *(FR: Paris 1801-1881),* agnostic. *Physician, philosopher, journalist, politician, and author of the* "Dictionary of the French Language."

one feels ready to be seized by Pascal's sublime madness[725]*... The idea of God is a form of the idea of the infinite."*

The *Anglican* Lord **Rayleigh** (ref.267, p.80), 1904 Nobel Prize in Physics, expressed, during his acceptance speech his views on the relationship between science and religion, and his admiration for Christ:

> *"I have never thought the materialist view possible, and I look to a power beyond what we see, and to a life in which we may at least hope to take part. What is more, I think that Christ and indeed other spiritually gifted men see further than I do, and I wish to follow them as far as I can"* [BP].

An intriguing perspective comes from one of the most influential philosophers of the 20th century: the mathematician, biologist, and philosopher Alfred North **Whitehead** (ref.542, p.166). He posited that science and religion share a common origin, rooted in the essential principles of harmony and order required for the world's existence.

The concept of harmony takes on different meanings for polymaths like Gottfried W. **Leibniz** (ref.182, p.53) and A. N. **Whitehead**. For **Leibniz**, as elucidated in reference 182, the "pre-established harmony" idea implies that every substance or monad, whether of matter or mind, acts solely upon itself and interacts with others according to a divine preordained plan. This involves a perfect harmonization, rejecting the conventional concept of causality.

In contrast, **Whitehead**'s philosophy, complex and holistic, resists easy comprehension in its entirety, allowing for various interpretations. At its core, his *"process thought,"* proposes that both living entities (such as humans, animals, and nature) and non-living elements (found across the universe) are not static substances but rather *"events"* in constant flux. These *events* are interconnected, mutually influential, and in a perpetual state of evolution through a process of creativity and actualization, adapting to their environment. In this view, departing from the conventional Darwinian evolution theory, harmony emerges as the outcome of the freedom and responsibility of beings.

Whitehead departed from conventional monotheistic notions of God, steering clear of depicting a commanding deity akin to Caesar imposing His will. He said ambiguously:

> *"God is not before all creation, but with all creation."*[726]

---

[725] *Concerning* "Pascal's madness," *many of his enemies tried to correlate his creative genius, which no one can dispute, and his faith in God, with a sickly temperament. His visions, in particular, were attributed by some to madness. But what man of genius does not have a behavior that differentiates him from ordinary mortals? Painters like* **Van Gogh,** *who cut off his ear because he had psychic problems,* Salvador **Dali**, *who had a strong propensity for exaggeration, ... and many other artists have been talked about for their* "bouts of madness." *However, this does not detract from the quality of their work.*

[726] Quote from C. **Birch**'s 1990 Templeton Prize Address.

In essence, Whitehead posited that God operates not through coercion but through persuasion, comprehending all existence with perfect knowledge and boundless love. He envisioned God evolving alongside humanity and the universe, orchestrating a harmonious progression toward higher expressions of creativity and awareness, a concept reminiscent of **Teilhard de Chardin's** Omega Point. Although Whitehead grappled with the allure of *Catholicism*, he intentionally distanced himself from any specific religious framework. Nevertheless, he recognized an undeniable societal role for the divine. Among the myriad ways in which Whitehead conveyed his understanding of God, a particularly noteworthy perspective emerged:

> *"It is as true to say that the World is immanent in God, as that God is immanent in the World."*

This emphasizes the profound interconnectedness of God with His creation, positioning Whitehead's beliefs between theism and creationism.

This vision of a God characterized by harmony, peacemaking, persuasion, mercy, and a presence amid humanity for their alleviation contrasted sharply with initially Charles L. **Birch**'s more Puritanical perspective (ref.541, p.165). This contrast prompted Birch to rediscover a faith that had *"evaporated,"* leading him to dedicate his life to the service of humanity, encompassing nature, and animals (cf. §V.3.). In his **1990** acceptance address to the Templeton Prize jury Birch emphasized the compatibility of science and religion with the assertion:

> *"Those who say that science and religion do not mix understand neither."*

This statement reflects his conviction that these seemingly disparate realms are not mutually exclusive but, rather, coexist and inform one another.

The prominent scientist Erwin **Conklin** (ref.633-p.220) was also a religious person, believing that science and religion were compatible. *Methodist*, he respected other faiths and traditions. He claimed:

> *"I have no quarrel with any religion or with any philosophy which does not deny or belittle the facts of science or of human experience"*[Cn].

Robert **Millikan**[727] a *Christian apologist* and recipient of the 1923 Nobel Prize in Physics *"for his work on the elementary charge of electricity and on the photoelectric effect,"* stated unequivocally:

---

[727] **Millikan** Robert *(USA: Morrison, IL 1868 - 1953 San Marino, CA). 1923 Nobel Prize in Physics,* Apologist for christianity. *Master in 1891 at Oberlin College (U-OHS), Ph.D. in 1895 at Columbia Univ. N-Y, assistant in Physics in 1896, Univ. of Chicago, IL. Realized in 1909 the first measurement of the electric charge of the electron (one of the fundamental constants). Professor in 1910, he worked during the first war for the army on anti-submarine and meteorological devices. Back to civilian life, he experimentally verified the photoelectric effect theory. Measures h, another fundamental constant. 1921 appointed director of Caltech, CA. Measures the radiation that V. **Hess** has just detected and gives it the name* "cosmic radiation." *Received numerous awards such as the Franklin Medal, the Faraday, and Lecture Ship of the*

"I can categorically state that unbelief has no scientific basis. I believe there is no contradiction between faith and science."

Sir A. **Eddington** a *Christian* (ref.290, p.87), one of the 20th century's foremost astronomers had a more aggressive opinion:

"No inventor of atheism was a man of science. All of them were only very mediocre philosophers."

The *Baptist* Arthur H. **Compton** (ref.326, p.98), 1927 Nobel Prize in Physics, had a more nuanced opinion:

"Science cannot quarrel with a religion which postulates a God to whom men are as His children." [728]

In 1950, Sir Ernest **Chain** (ref.336-p.100) devout *Jewish,* and a **Nobel** Prize Laureate wrote a letter to A. **Einstein,** a pantheist and a **Nobel** Prize too. In his letter, he not only criticized Einstein's endorsement of establishing a Jewish State in Palestine but also expressed his perspectives on the interplay between Science, searching for the truth, and religion's quest for the meaning of life. He expressed his feelings:

".... Science and religion are not only compatible, but complementary, and both are essential to the full development of the human personality...."

During a speech he gave at the Royal Society of London in **1961**, the *Anglican* Sir W. **Laurence Bragg**, 1915 Nobel Prize in Chemistry (ref.296-p.89) said:

"From religion comes man's purpose; from science, his power to achieve it. Sometimes people ask if religion and science are not opposed to one another. They are: in the sense that the thumb and fingers of my hand are opposed to one another. It is an opposition through which anything can be grasped" [Sm].

The theoretical physicist Olivier **Costa de Beauregard,** [729] director of Research at the CNRS, a *Catholic*, defended in **1979,** during the Cordoba

---

Royal Society of Chemistry awarded 1 year out of 3. He made numerous contributions to scientific journals.

Son of a priest he was a committed Christian *all his life. He made lectures and authored many books on the reconciliation of science and religion. He participated in various organizations of mutual aid and defense of Peace. The quote is from Aleteia.*

[728] *According to Wikipedia: "Science cosmic Clearance" Time January 13, 1936.*

[729] **Costa de Beauregard** Olivier *(FRA: Paris 1911-2007 Poitiers). Research engineer in aeronautics,* Catholic, *he joined the CNRS in 1940 after being demobilized. He became Director of Research in 1971. A student of* **Louis de Broglie**, *he defended his Ph.D. in 1943 and contributed to the development of the Theory of General Relativity and Quantum Mechanics. He was a specialist of time (he interpreted the EPR effect) and of the* negentropic theory *(unlike*

Colloquium "*Science and Consciousness - The 2 readings of the Universe*," the thesis according to which parapsychological phenomena exist and do not contradict the laws of physics.

For the **1977** Nobel Prize in physics, Sir Nevill F. **Mott**, (ref.589, p.194) an *Anglican* who discovered the faith at 50, there is no problem [Pd]: Faith and Science can coexist, wrote in 1986 [Mot2]:

> "*I believe in God, who can respond to prayers, to whom we can give trust, and without whom life on this earth would be without meaning (a tale told by an idiot). I believe that God has revealed Himself to us in many ways and through many men and women and that for us here in the West the clearest revelation is through Jesus and those that have followed him.*"

Although one cannot classify the Soviet dissident Nathan **Sharansky**[730] - a secular *Jewish* by birth, initially a *Marxist atheist* by conviction - in the category of scientists, no one can deny his rationality, his logic, which allowed him to become, at 15 years old, the Donetz, UA chess champion, the winner in 1996 of Garry **Kasparov**, then world chess champion (from 1985 to 2000) visiting Israel. Also, I extracted from his interview in November **1991** with Joshua **Haberman** [Hab], a few messages for atheists:

> "*I developed my relationship with religion, my Judaism, in prison.*"

While he was incarcerated for the "*crime*" of defending the Refuzniks when he was exhausted, he received the news that his father had died. To "*accompany him in spirit*" he decided to start reading the book of Psalms that his wife had sent him in prison and later said that from that moment:

> "*...God came to my aid. .... This spiritual world gave me the strength and the desire to survive.*"

And in the conclusion of his confession, he delivered this message of hope:

Entropy which applies to isolated systems and is always positive, negentropy is a measure of the order prevailing in an open thermodynamic system. It is negative and allows us to interpret the self-structuring of living beings in opposition to disorder).

[730] **Sharansky** Anatoli *(Donetz, UKR 1948). Laureate in Applied Mathematics from the Institute of Physics and Technology of Moscow. He adopted* Natan *as his first name when he embraced the Jewish religion. Defender of the Refuzniks, he fought for 12 years to defend their rights and was sentenced in 1978 to 12 years in prison. Member of the Knesset from 1996 to 2006. He was successively Minister of Economy (1996-1999), Minister of the Interior (1999-2000), and Deputy Prime Minister (2001-2003). Refuzniks: Jewish Soviet dissidents (believers or not) living in the USSR, to whom the Soviets refused during the* "cold war," *for various reasons, the right to emigrate to Israel. This movement took on an international dimension and also assimilated conscientious objectors, ...*

"I had the chance to compare two ways of life, the life without God, a materialistic life, and the life that you feel will allow you to be what you should be. This is how I came to God."

Stanley **Jaki,**[731] a remarkable figure, who donned the dual roles of a physicist and a *Benedictine priest* while serving as a distinguished professor of Astrophysics, earned the prestigious Templeton Prize in **1987**. He was not merely a scholar (he was one of the first to say that **Gödel's** theorem (ref.613, p.212) applies to the TOE in Theoretical Physics); but, a profound philosopher and theologian, recognized for his groundbreaking work at the intersection of science and theology.

The Templeton Prize judges, with great reverence, acknowledged that *"he offered the world a reinterpretation of the history of science, which throws a flood of light on the relation of science and culture, and not least the relation of science and faith."* His written works meticulously explore the distinctions and harmonies between science and religion. His magnum opus *"The Relevance of Physics,"* earned praise from luminaries like Dr. Peter **Hodgson,**[732] the head of the Nuclear Physics Theoretical Group at Oxford, a *Roman Catholic*, who described it as *"unsurpassed by any other work."* For **Jaki**, there is no inherent contradiction between science and religion, nor is there a need for complete cooperation, as he believed they served distinct, yet complementary purposes. In his address during the prize reception, he eloquently stated:

"If intellectual honesty, usually taken for a fruit of scientific method, is to be had only through Christian love, science, and religion should not seem far removed from one another. Science is as closely related to religion, and especially to Christian religion, as a child is to the womb out of which it came forth and with full vitality."

Let's note that the **1999** Turing Prize (the highest honor in computer science) was awarded to the American computer scientist and engineer Frederick P.

---

[731] **Jaki** Stanley L. *(Györ, HUN 1924-2009 Madrid, ESP). After an undergraduate degree in Philosophy, Theology, and Mathematics, and Physics he defended a thesis in Physics at Fordham University, NY, under the direction of* Victor **Hess***, 1933 Nobel Prize in Physics. He was a Fellow, then a Visiting Fellow of Princeton. A few years later he obtained two Master's degrees in Theology and a thesis in Theology in Rome and was ordained Priest.*
A Benedictine priest, Theologian, *historian of science, and professor of physics at Seton Hall University in New Jersey, he taught at Oxford, Yale, and Edinburgh. The author of more than twenty books on the relationship between science and Christianity is a member of the Pontifical Academy of Sciences. 1987 Templeton Prize.*
[732] **Hodgson** Peter E. *(London, ENG 1928-2008) head of the Nuclear Physics Theoretical Group of Oxford, Lecturer in Nuclear Physics, contributed to the discovery of the K meson and authored many scientific papers and books. He was awarded his Ph.D. in 1951. Fellow of Corpus Christi College, he was an active member of the Atomic Scientists Association.* Roman Catholic, *he got involved in organizations promoting the relationship between science and religion.*

Brooks[733] *"for his landmark contributions to computer architecture, operating systems, and software engineering."* Although *"he grew up a churchman"* [CD] he had doubt about the trustworthiness of the Scripture and became a true *Christian - he is an apological evangelist -* when he met at work, Gerry **Blaauw**, IBM engineer, *"a strong Christian"* who invited Fred, his wife, and other couples for weekly home Bible studies. But C.S. **Lewis**[734] was for him *"the most helpful theological writer"* followed by the *theologians and preachers* Martin **Luther** (ref.136, p.38), and John **Westley**,[735] the founder of the Methodist Movement. When **Lewis** gives lectures he always starts with Mere Christianity alternating with the Bible. He defines him as a *"servant of Jesus Christ."* At the occasion of

---

[733] **Brooks** Jr. Frederick P. *(Durham, US-NC 1931)* evangelical apologist, *1999 Turing Prize. At the age of thirteen, seeing on the front page of Time magazine the first programmable computer, an electromechanical device, making 3 operations/s, he discovers his way. He obtained a physics and mathematics course at Duke University (Durham, NC-US), and from 1953 to 56 a Ph.D. at Harvard, Cambridge, US-MA. His supervisor was* Howard **Aiken** *the architect of the first machine! Appointed by IBM he became the development manager for the IBM 360 computers. He left in 1964 to create the computer science department at Chapel Hill, University, US-NC. Elected at the National Academy of Engineering (1976), American Academy of Arts and Sciences (1976), Royal Academy of Engineering of UK (1994), US National Academy of Sciences (2001)....... He has authored and co-authored many books. The most famous is* "The Mythical Man-Month" *in which he establishes Brooks' law:* "Adding manpower to a late software project makes it later"

[734] **Lewis** Clive Staples *(Belfast, IRL 1898 - 1963 Oxford, ENG). Academic, writer, literary critic, and apologist for* Christianity. *Fellow and tutor in English Literature at Oxford until 1954, when he was unanimously elected full professor of Medieval and Renaissance Literature at Cambridge University. An atheist based on arguments that seemed logical to him, it was the reading of* Gilbert K. Chesterton *(ENG: Kensington 1874 - 1936 Beaconsfield), writer, journalist, and* Christian apologist, *that led him to convert to* Catholicism. *Author of more than thirty books including* "Mere Christianity." *In this book,* Lewis *argues that the concept of* "Right and Wrong" *and its resulting* "Moral Law" *are not cultural, that they have always existed, even in the earliest times, because they are imbued in us by the Spirit of God. Despite this, they are not always followed by humans, hence all the difficulties we have in overcoming evil in our earthly existence. For him, this concept of Good and Evil is as universal as the law of gravitation.* George Bernard Shaw *would have said of him:* "he was a man of colossal genius."

[735] **Wesley** John *(ENG: Epworth 1703-1791 London). Born into a family of* Anglican *pastors, he entered the clergy and was ordained as an* Anglican priest *in 1735. His initial mission to evangelize the Native Americans in Georgia with his father proved disappointing, leading to his return to England. However, a transformative encounter with* Moravian evangelists *(ref.747, p.277) deeply impacted Westley, prompting a reevaluation of his faith. Questioning the prevailing doctrines of predestination and distancing himself from Lutheran and Calvinist teachings, Westley found inspiration in* **Paul**'s *Epistle to the Ephesians. Motivated to return to the roots of the* Christian faith, *he initiated the* Methodist movement *within the Church of England. Early collaborators in this venture included his younger brother* **Charles** *and George* **Whitefield** *a friend. Wesley's commitment to spreading the movement was unparalleled; his tireless travels across England, covering an astonishing 400,000km on horseback, became instrumental in the enduring success and reach of Methodism. John Wesley's legacy extends beyond his theological contributions, his dynamic approach to faith and evangelism left an indelible mark on the landscape of Christianity, fostering a movement that transcended traditional boundaries and continues to influence believers worldwide.*

the reception of his prize on February 27, **2000**, in Austin, US-TX, he delivered a lecture entitled *"The Design of Design."* I retained:

"Design is a way of expressing our values and our vision. It is a way of honoring God, who made us in His image and, gave us the possibility to create. ... Science is not enough to understand design, because design is not a science. It is a creative process, involving both intuition and rationality."

During a conference entitled *"Science and Religion: Conflict or Convergence?"* held October 12, **2000** Baltimore, the professor Emeritus of Human Genetics and Psychiatry at Virginia Commonwealth University Lindon J. **Eaves**,[736] an *Anglican* priest, said:

"There is a great correspondence between the love of God and the love of truth. Thus, the rules and passions, that we use to know the love of God, the scientist embodies, in his passion for truth .... What keeps the scientist working long hours through the night? It is the hope that there is a truth that will be very beautiful when he discovers it."

Kenneth **Miller**[737] professor of biology at Brown University, Providence, RI. *Catholic,* author of *"Finding Darwin's God,"* claims in the introduction:

"Science is not the enemy of religion. It is not a threat to faith. It is not a challenge to believe. Science is, instead, a way of exploring the natural world that can enrich our understanding of God's creation and deepen our appreciation of the divine."

In this book, he explains his main arguments, and challenges creationists and *"new atheists."* He also shares his journey as a scientist and a *Christian,* and how he reconciles his scientific knowledge with his religious faith. Sharing Darwin's ideas on the theory of evolution, he intervenes with American high school and university students to defend his point of view against, in particular, the supporters of *"Intelligent Design."*

Reverend Canon Dr. Arthur **Peacocke** (ref.651, p.233) was a remarkable individual who seamlessly blended his activities as a physical biochemist and *Anglican priest*. In addition to his groundbreaking scientific work, he dedicated his lifetime to exploring the intricate relationship between science and theology.

---

[736] **Eaves** Lindon J. *(Walsall, ENG. 1944), studied genetics at the University of Birmingham and theology at Rippon College in Cuddesden (5km from Oxford). Professor of Human Genetics and Psychiatry at Oxford from 1981; Professor Emeritus at Virginia University in Richmond, VA. Author of numerous scientific articles.* Anglican priest.

[737] **Miller** Kenneth R. *(Rahway US-NJ 1948). After a thesis in Biology in 1974 at the University of Boulder in Colorado, he taught for 6 years at Harvard before being appointed professor in 1980 at Brown University, Rhode Island (US-RI). Recipient of various awards.* Catholic.

His journey into this intersection began in 1973 when his book: "*Science and the Christian Experiment*," was honored with the prestigious Leconte du Noüy prize. This pivotal moment marked the start of his transformative impact as a leading figure in the creative dialogue between Science and theology. His significant contributions were recognized in **2001** when he was awarded the Templeton Prize (ref.434, p.129).

The Templeton Prize jury, in their nomination, aptly acknowledged Peacocke's unique achievement, stating that he: "*had created a new understanding of the relationship between theology and science.....generating a new theology for a scientific age.*" During his acceptance address lecture, **Peacocke**, drawing upon his background as a biologist with a deep understanding of the physical biochemistry of DNA, made a thoughtprovoking analogy. He eloquently described:

> "*The complementary double-strands of the DNA on which our lives and inheritance depend proves to be an appropriate symbol for that double quest which constitutes human life: the searches for intelligibility and meaning.*"

He concluded his lecture with an inspiring message:

> "*The sciences have drawn back the curtain for us now to be able to perceive the natural, evolved commonalities of humanity in a new light, in the light of a perception of that Ultimate Reality, God who, giving everything its existence, is both the common agent through nature of our created creativity and will be the destiny of our fulfilled humanity.*"

The **2002** Templeton Prize honored John **Polkinghorne** (ref.571, p.184), a remarkable individual whose career trajectory and contributions have uniquely bridged science and technology. Initially serving as a Professor of Mathematical Physics at the University of Cambridge in London, he made significant contributions, during the 1970s, in the field of particle physics. He was notably part of the team that discovered the quark, a fundamental particle entering the composition of protons and neutrons (cf. Chapter II).

Polkinghorne made a pivotal decision in 1979, resigning to embark on a profound journey of theological exploration. By 1982, he had embraced the vocation of an *Anglican priest*, effectively uniting his scientific and spiritual passions. This significant shift in his career path demonstrated a remarkable commitment to integrating his rational scientific sensibilities with the profound mysteries of theology. Polkinghorne championed a modern and harmonious approach to understanding the Bible; effectively advancing the scholarship discourse on the intricate relationship between science and religion. His tireless effort through his written works, lectures, and public engagements; propelled him to international acclaim, earning him notoriety in scientific and theological circles. His legacy not only endures but inspires countless individuals seeking a deeper understanding of the interplay between science and faith. During the reception of his Prize, he stated:

> "*I want to take science and religion with great and equal seriousness, I see them as complementary and not as rivals. The most important thing they share is that both, believe there is a truth to be sought and found, a truth whose attainment comes through the pursuit of well-motivated belief. Of course, the two forms of inquiry view reality from different perspectives, science studies the processes of the world, while religion is concerned with the deeper issue of whether there is a divine meaning and purpose behind what is going on. I believe I need the binocular approach of science and religion if I am to do any justice to the deep and rich reality of the world in which we live. I think of myself and some of my colleagues in this task as* "two-eyed" *scientist-theologians.*"

In **2009**, *Jesuit* astronomer George **Coyne,** [738] a defender of the Weak Anthropic Principle (ref.648, p.229), received the George Van Biesbroeck Award for being *an outstanding living scientist in astronomy*; among other honors. This is proof, that it is possible to manage a career as both, a *Catholic theologian* and a recognized scientist.

In his book "*Darwin's God*" [Mg2] published in **2015**, Alister **McGrath,** [739] Biophysicist, *Anglican priest*, and *apologist* of the *Christian* religion, challenges **Darwin**'s views on the nature of faith, the origins of religion, and the role of science in human life. In his introduction, he recounts his journey from atheism to Christianity and explains his motivations for writing the book.

> "*Atheism, I began to realize, rested on less-than-satisfactory evidential bases. The arguments that had once seemed bold, decisive, and conclusive increasingly turned out to be circular, tentative, and uncertain.*"

---

[738] **Coyne** *George V. (Baltimore, MD-US 1933), Astronomer. After a degree in philosophy obtained at Fordham University in New York, he obtained a Ph.D. in Astronomy in 1962; did research at Harvard, then was appointed lecturer in 1963. He became a teacher-researcher in 1965 at the Jesuit University of Scranton-PA, then was seconded as a professor at the University of Arizona. From 1978 to 2006, he was director of the Vatican Observatory, where he led several scientific and educational projects. At the same time, from 1979 to 1980, he was acting director of the Steward Observatory in Arizona and head of the astronomy department. Member of the Pontifical Academy of Sciences.*
Theologian. *He joined the* Jesuits *in 1961, obtained a license in theology in 1965, and was ordained a* Catholic priest.
[739] **McGrath** Alister *(Belfast, IRL 1953) Studied at Oxford, Ph.D. in Chemistry and Molecular Biophysics. Atheist, he converted to* anglicanism. *Ph.D. in Divinity and Letters from Wadham College, Oxford. Professor of Theology, Chair at King's College, London; also taught at Cambridge,* ordained Priest, *he is a* Christian apologist,

He had previously:

> *"rejected Christianity because it was a position of faith that could not be proved to be right. But I couldn't prove my atheism. I went to Oxford, to study chemistry and discovered that Christianity was a much more interesting and attractive intellectual option than atheism."*

According to him:

> *"The English experience suggested that nobody doubted the existence of God until theologians tried to prove it."*

The same **2009** year, **Bernard d'Espagnat**[740] a non-dogmatic *Catholic* and distinguished Theoretical Physicist and Philosopher, was awarded the **Templeton Prize** for *"his exploration of the philosophical implications of quantum mechanics."* His work has reshaped our understanding of reality and pushed the boundaries of knowable science. For d'Espagnat, quantum physics represented a revolutionary departure from the classical physics worldview, urging us to abandon the preconceived notions that governed our perception of the world and our relationship with it. As a pivotal figure in theoretical physics from the mid-1960s to the early 1980s, d'Espagnat played a crucial role in debunking the Einstein-Podolsky-Rosen (EPR) Paradox, ultimately confirming the existence of the phenomenon of *"non-local-entanglement."* At the 2009 Templeton Prize reception, Alain **Aspect**, destined to receive the **2022 Nobel** Prize in Physics (ref.324, p.96) acknowledged the profound influence of Bernard d'Espagnat's work on his own. Indeed d'Espagnat's contributions laid the groundwork for subsequent advancements in *quantum information science.*

---

[740] **Bernard d'Espagnat** *(FRA: Fourmagnac 1921-2015 Paris), distinguished theoretical physicist.* Professor Emeritus of Theoretical Physics at the University of Paris-Sud, France. *As a teenager, he navigated between his interests in Arts, Philosophy, and Science, ultimately choosing science as an essential tool for philosophical exploration. In 1939 he enrolled at the Ecole Polytechnique in Paris; but, World War II disrupted his academic pursuits. After the war, d'Espagnat embarked on a research career at the CNRS from 1947 to 1957. He completed his Ph.D. in 1950 under the guidance of **Louis de Broglie**. Subsequently, he served as an Assistant to E. **Fermi** in Chicago and collaborated with N. **Bohr**. Transitioning to academia, he became a Professor of Theoretical Physics at the Sorbonne in 1959, maintaining this position until he was appointed Emeritus Professor in 1987.*
Throughout his career, he collaborated with the luminaries of quantum physics. D'Espagnat's distinctive contribution to both scientific and philosophical realms focused on what he termed "veiled reality," an obscured yet unifying underlying conventional notions of time, space, matter, and energy, concepts challenged by quantum physics as potentially illusory. Beyond his scientific papers, he authored several books and articles exploring the intersection of science, religion, and spirituality. In recognition of his impactful work, Bernard d'Espagnat was awarded the Templeton Prize in 2009. His contributions were acknowledged by various Academies, including the French Academy of Moral and Philosophical Science.

As a visionary in the quantum physics research community, d'Espagnat introduced the concept of the *"Veiled reality,"* proposing the existence of a deeper level of reality beyond our direct observation and measurement:

"The concepts we use, such as space, time, causality,... are not applicable to ultimate reality,"

"Science aims, to make an account, of reality as it appears to us, accounting for the limitations of our mind, and our sensibilities."

From a young age, he marveled at the world's beauty, associating it with a "Being" he named *"God."* Throughout his life, d'Espagnat endeavored to bridge the gap between his scientific pursuits and his spirituality. Upon receiving the Templeton Prize he boldly asserted:

"Some will be surprised but I claim that there are forms of higher spirituality that are compatible with what emerges from contemporary physics."

Robert **Pollack,**[741] *Jew*, the awarded Biologist who discovered a new form of cancer chemotherapy. Explaining his motivation for exploring the relationship between faith and reason, and how they can both inform our understanding of ourselves and the world, he stated [Pk]:

"... the connection between Faith and Reason runs deeper, as deep as any part of the human mind."

In an article [Pk1] published in **2016**, he posited:

"... scientific hypotheses about the unknowable are, by definition not disprovable, and therefore, are no meaningful" [Pk1],

"It is not worth a moment of anyone's time to seek the proof through the science of any religious belief."

Emphasizing his position, that science and religion are different domains of inquiry, and cannot be reduced to each other.

---

[741] **Pollack** Robert E, *(Brooklyn 1940)* Jewish. *Professor of Biological Sciences at Columbia University, Dean in the 1980s. 1966, Ph.D. in Biological Sciences from Brandeis University, Waltham MA-US. Postdoctoral fellow at NYU Medical Center and at the Weizmann Institute in Israel. Recruited by* James **Watson** *then tenured Associate Professor of Microbiology at Stony Brook University before joining 1978 Columbia University. where he is a professor of Biological Sciences and Director of Seminars.*
Creating 2014 "The Research Cluster on Science and Subjectivity," *he authored various research projects, reviews, articles, and books, and won the Lionel Trilling Award and others for his scientific work.*

José **Funes**,[742] a *Catholic priest*, a doctor in astronomy, and director of the Vatican observatory from 2006 to 2015, co-authored many scientific papers. Welcoming Pope Francis on July 16, **2013,** he said, according to Vatican Radio:

> *"We go back, in the sense that we also explore the beginning of the universe from the point of view of science, but we also go far away, because we also study the farthest, the most distant galaxies... And this brings up questions that we should ask about the relationship between science and faith. I think this is the mission of the Observatory: go out to the truly most distant boundaries, the boundaries of the universe, that is always a gift of God."*

October 9, **2015**, the *Protestant* William C. **Campbell**,[743] co-laureate with Professor Satoshi **Omura** of one half of the **Nobel** Prize in Physiology or Medicin *"for their discoveries concerning a novel therapy against infections caused by roundworm parasites"* answering an interview (by **Darragh** Murphy: *"Meet Ireland's new Nobel Laureate, William C. Campbell"*) for The Irish Times, said:

> *"I believe in God, I pray every single night of my life, but I have a very complicated sense of religion, and I am pretty fuzzy in that segment of my life.... My faith, as that of millions of others, has evolved, if that is the right word, as civilization has evolved. Evolved but not abandoned. Religion and science can coexist. At least, that had better be true. There are certain intangibles."*

Their discovery of Avermectin and Artemisinin, drugs and pesticides, had a huge impact against the parasitic roundworm (spread by black flies), which was very active in Africa, and Latin America provoking itching, skin lesions, visual impairment, and blindness. The fact that W. Campbell floated the idea that it would be distributed freely wherever it was needed (Africa, Latin America) was

---

[742] **Funès** José G. *(Cordoba, ARG 1963). He obtained a master's degree in Astronomy in 1985 (University of Cordoba), joined the* Jesuits, *studied theology at the Gregorian University of Rome, and was* ordained a priest *in 1995. In 1995 he specialized in astronomy in Padua and obtained a doctorate in 2000. He joined the Vatican Observatory and became its director in 2006.*

[743] **Campbell** William C. *(Ramelton, North Ireland, UK 1930). Physician, Biologist, and Parasitologist. After a BA from Trinity College, Dublin, IRL he graduated with a doctorate in 1957 from the University of Wisconsin, Madison, US-Wi, then he was appointed until 1990 to the Merk Institute for Therapeutic Research, becoming director in 1984 for Assay Research and Development.* He moved to Drew University where he was appointed Professor in Parasitology and in History of Biomedical Science; he could do real research with his students. 2015 **Nobel** Prize for Physiology or Medicine.
Protestant, member of the United Church of Christ *(formed in 1957 by the merger of the Congregational Churches, the Christian Churches, the Evangelical Synod, and the Reformed Church, it joins Lutheran, Calvinist, Anabaptist, and Congregational traditions),*

worthy of the humanist that he is. The idea validated by Merck Research Laboratories allowed fighting, effectively, this illness.

Ian H. **Hutchinson,**[744] a *Christian*, is a recognized nuclear physicist and a member of the Biologos Foundation, and as such defends religion against what he calls "*Militant Atheists.*" He replies to the scientists who "*slyly*" equate scientism with science [Hu1] that just as one cannot scientifically demonstrate the love of a father for his children, one cannot scientifically demonstrate the love of God for his creatures, and that one should not try to demonstrate the existence of God experimentally. It is enough for him to believe in Jesus Christ to be convinced of this. For him [Hu2]:

> "*The resurrection of Jesus is the central miracle of Christianity. It is not a scientific hypothesis, but a historical claim based on eyewitness testimony and circumstantial evidence.*"

He demonstrates that belief in the Bible - not forgetting that it is full of metaphors and earliest to Science - and especially in *Christianity*, are perfectly compatible with Science. Science and Theology each have their role in our understanding of God and the World. According to him:

> "*Miracles are by definition, abnormal and non-reproducible, also they cannot be proved by scientific methods......... They are not violations of natural law; but they are the actions of a personal agent, who is not bound by the laws he established for the natural world............Miracles are possible but rare. They are signs of God's special intervention in history, no arbitrary disruptions of nature.*" [Hu2]

Member of the All Saint's Church of Belmont (US-MA), an *Episcopal Protestant* Church, he defends *Christianity*, as a whole.

Closer to us we find some rationalist believers such as the *Christian priest* Father Philippe **Deterre,**[745] member of "*la Maison de France*," director of Research at the CNRS, an expert in Biology, managing without restriction the two aspects (extract from his biography):

---

[744] **Hutchinson** Ian Horner (*ENG 1951*). *He obtained his B.A. at Cambridge in 1972 and defended his thesis in 1976 at the Australian National University. Hutchinson joined MIT in 1983 as a Professor of Nuclear Science and Engineering and worked on the first Tokamak (a device with a very large (E, H) field, to produce energy by nuclear fusion) operating outside Russia. He headed the Department of Nuclear Physics and Engineering from 2003 to 2009.* Christian. *He argues against New Atheism, a sect advocating that Religion, superstition, and irrationalism should not be tolerated.*

[745] **Deterre** Philippe *(Troyes, FR 1952). Biologist, Research Director at CNRS, Faculty of Medicine of the Pitié Salpêtrière, Paris. 1983 Doctorate in Biology, Ordained priest (Catholic), 1995 Master in Theology. Member of "La Mission de France" (community of working priests engaged in dialogue with contemporary culture, as well as deacons, lay couples, and celibates,....).*

"I find a great analogy between the relentless search of the scientist and the search of the believer in God, or even the spiritual search of the person who has no faith."

c) Faith as an Inspiration in Scientific Work. Some scientists find that their faith and religious beliefs, inspire and influence their scientific work. Their faith may provide them, with a sense, a motivation, or a framework through which they interpret scientific discoveries. They may see their scientific pursuits as a means to understand and appreciate the intricacies of the natural world, which they believe was created by a High Power.

René **Descartes** (ref.152, p.43) scientist, and *Christian* philosopher, introduced in *"Meditation Third"*[746] the concept of *"objective reality,"* the reality or existence that an idea represents, stated:

"...the (idea = image at that time) by which I conceive a God – sovereign, eternal, infinite, immutable, all-knowing, all-powerful, and the Creator of all things that are out of Himself – this, I say, has certainly in its more objective reality, than those ideas by which finite substances are represented."

He emphasized that his idea of God, with his divine attributes, possesses a higher level of objective reality than ideas representing limited and finite things, reflecting his prioritization of the divine and the infinite in the hierarchy of reality within his philosophical framework.

Blaise **Pascal**'s brilliant and inquisitive mind who, using reason, made remarkable discoveries in the field of science, at an early age (ref.170, p.48), was according to Pope Francis *"a tireless seeker of truth, and a restless spirit"*[]. Living at a time when skepticism about religions was matched only by the scientific discoveries, to which he made a major contribution, he decided in the second part of his short but very fruitful life, to devote himself to theology and search for a Christian answer to the place of man in this universe.

"Man must not give up seeking God of whom he has an intuition: I see well that there is in nature a necessary, eternal and infinite Being,"

"The faithful ... see immediately that all that is nothing other than the work of the God whom they adore" [Pa 242-781].

Pascal clearly distinguished between physics and metaphysical speculations, although he judged them both necessary. With him, the man of faith, and the man of reason, walk the only path of truth.

The "**Great Mr. Newton**" (refs.185&204, Pp.59&61), an *Anglican* one of the Masters of Physics, wrote in Philosophia Mathematica [Ne2]:

---

[746] *Third meditation in Descartes Œuvres et Lettres [Br] Gallimard - Paris, 1953 p. 289.*

"God governs all things, not as the soul of the world, but as the Lord over all."

The distinguished *Lutheran scholar* and *theologian* Gottfried **Leibniz** (ref.182, p.53), an exceptional polymath, and one of the founders of calculus has, according to Father Thonnard, as his guiding idea, that:

"... there is a universal science of a mathematical type, capable of giving reason a priori to all that exists."

Likewise, the devoted member of *the Moravian Church*[747] Emil Theodor **Kocher,**[748] 1909 **Nobel** Prize in Physiology *"for his work on physiology, pathology, and surgery of the thyroid gland"* is a good example of a great scientist devoted to science and his faith. He attributed all his successes and failures to God. For him, the rise of materialism, particularly in science was an evil.

The prolific scientific inventor Thomas **Edison,**[749] *a Deist,* who mastered the bulb light, in awe of God's work said:

"My great respect and admiration for all engineers and especially for the greatest of them all, God."

Sir Joseph **J. Thomson** (ref.308, p.91), *Anglican,* 1906 **Nobel** Prize in Physics, addressing the British Association:[750]

"Great are the Works of the Lord."

The *Lutheran* Georg **Cantor** (ref.174, p.51), the father of the *"set theory"* (which is a fundamental theory in mathematics), also famous for his work on real

---

[747] Moravian Church, *formally the* Unitas Fratum *or* Unity of Brethren, is *a Christian movement dating back to the Bohemian Reformation of 1457 refuting some practices and doctrines of the Catholic Church. It is admitted to be the first Protestant movement supporting, among other things, liturgy in the Czech language, the marriage of priests, the withdrawal of indulgences, and refuting the existence of Purgatory. Author of about 250 articles and supervisor of a hundred doctoral candidates.*

[748] **Kocher** Emil Theodor *(Bern, CH 1841-1917)* Christian *Physician. 1990* **Nobel** *Prize in Physiology or Medicine. After defending his thesis in Bern (1865), he started a journey through Europe, meeting the most famous European surgeons, of the time (he met L.* **Pasteur** *in Paris). Returning to Bern, he was awarded the position of Professor of Surgery and Director of the University Surgical Clinic. It was under public pressure, that he could succeed G.* **Lücke** *as a Professor of Surgery and Director of the Clinic in 1872 (only German could be appointed professorship). He radically transformed the field of surgery, by implementing antiseptics, monitoring anesthesia, and minimizing blood loss. His contributions to neuro and thyroid surgery, and hemostasis... were important. His publications (about 250) too. He was an honorary member of numerous academies and medical societies. Was an active member of the* Moravian Church.

[749] **Edison** Thomas *(USA: Milas, OH 1847–1931 West Orange, NJ)* Deist. *Self-taught, took more than 1,000 patents.*

[750] Nature, *1090, Vol. 81 p. 257.*

numbers and the theory of infinity said:[751]

>*"I entertain no doubts as to the truths of the transfinite, which I recognized with God's help and which, in their diversity, I have studied for more than twenty years; every year and almost every day brings me further in science."*

The concept of transfinite numbers (Aleph and Beth) introduced by Cantor, in the late 19th century, to describe the sizes of infinite sets necessitates a good background in mathematics; also we just remember that Cantor's work on transfinite numbers revolutionized our understanding of the infinite, and laid the foundation for set theory and modern mathematics.

During a speech titled *"The Essence of Matter,"* in **1944**, on the occasion of the 25th anniversary of the Kaiser Wilhelm Society in Florence, Italy, Max **Planck** (ref.255, p.78) shared a profound insight that resonates until this day. With an unwavering dedication to the relentless pursuit of the most rigorous scientific endeavors, the meticulous study of matter itself, Planck unveiled a revolutionary truth: matter, as we perceive it, is but, a fleeting illusion. Intriguingly, he posited that the very fabric of our physical reality is woven by an invisible force, orchestrating the delicate dance of atomic particles. He suggested that this force is the handiwork of a conscious and intelligent Spirit, an elusive, yet omnipresent matrix, that breathes life into every atom, creating the exquisite, albeit minuscule, solar systems within. Max Planck's words unveiled, not just the core of his research, but also a profound philosophical revelation.

This was among the last public statements of a physicist who had dedicated his lifetime to unraveling the mysteries of the universe. His groundbreaking work in quantum theory (cf. chap.2) – encapsulating the revelation that matter is an expression of energy and vibration- had forever altered our understanding of the cosmos. Yet, **Planck** ventured further, suggesting that behind this cosmic symphony of energy and matter, an all-encompassing intelligence orchestrates the grand cosmic opera.

These profound thoughts echo in his autobiographical work, *"A Scientific Autobiography"* [P13], where he contemplated the existence of universal constants. For him:

>*"The existence of these constants is a palpable proof of the existence in nature of something real and independent of every human measurement."*

Max **Planck** was more than a scientist; he was a seeker of truth and a harmonizer of faith and reason. His journey was a relentless quest to bridge the chasm between science and spirituality, ultimately to fathom the profound depths of reality. He encapsulated this vision with eloquence, proclaiming:

---

[751] *www.azquotes.com/author/2424-Geeorg_Cantor.*

"Religion and natural science are fighting a joint battle in an incessant never relaxing crusade against skepticism and dogmatism, again disbelief and again superstition, and the rallying cry in this crusade has always been, and always be: "On to God."

In those words, he illuminated a path toward a richer, more harmonious understanding of the universe and our place within it.

The physical chemist Pierre **Duhem** (ref.474, p.144), a *Catholic*, affirmed that between:

"2 judgments which do not have the same terms, which do not concern the same objects, there can be neither agreement nor disagreement."

For him, the *Catholic* religion favored scientific progress [Du]. This statement can surprise some, but it is important to understand that **Duhem**'s perspective was influenced by the historical context in which he lived, his *Catholic* faith, and his belief that the Church's acceptance of reason, and the pursuit of knowledge contributed to an environment conducive to scientific exploration. He sincerely thought that the *Catholic* worldview provided a stable foundation for scientific inquiry by encouraging a rational understanding of the universe.

With the faithful *Christian* George Washington **Carver,**[752] nicknamed *"The black Leonardo"* by the Times (**1946**), we discover a man who - although living at a time when the rights of blacks were not always respected - was able, thanks to his faith in Jesus, and his pugnacity to overcome the obstacles and gain recognition from the scientific community. He liked to say, that *"Jesus was supporting him in his research work."* He was a distinguished *Christian* and an Afro-American agricultural scientist. An environmentalist before his time, he co-invented techniques to improve the quality of soils depleted by intensive cotton cultivation and, advised the farmers to grow peanuts (giving them 46 recipes), and sweet potatoes... to diversify their food and improve their life quality. *Christian*, he integrated faith and science into his life, humanitarian he promoted racial harmony. Among the various quotes of **Carver** in the George Washington Carver National Monument erected in Newton County (SW-MO) one can read:

"...... Never since, have I been without this consciousness of the Creator speaking to me through flowers, rocks, animals, plants, and other aspects of His creation."

---

[752] **Carver** George Washington *(USA: Diamond MO 1864-1943 Tuskegee, AL)* Christian. *Agricultural chemist, teacher, and environmentalist, Born a slave, he was raised like his own child by Moses* **Carter***, a German immigrant who had bought his mother as a slave. In 1891 he became the 1ˢᵗ black student to attend the University of Iowa in the Agricultural College. He could be awarded a Master's degree in science in 1896; but, because of segregation, he could not prepare for a Ph.D.. Nevertheless, the same year, he headed the Agricultural Department of Tuskegee University, where he taught until his death. Theodore Roosevelt, President of the USA praised his work. He was a member of the Royal Society of Arts in London in 1916.* Catholic.

Let me report an anecdote [Mi] about Sir James ***Jeans,*** [753] the famous devout *Anglican,* Cambridge Astronomer, who, on a rainy Sunday in 1909 was walking, with a bible clutched under his arm and an umbrella closed in his hand when Inayatullah **Mashriqi,** [754] a student in Cambridge, recognizing him asked to him why he had not opened his umbrella and, 2nd question: "*I would like to know what a man of universal fame such as yourself is doing-going to pray in Church ?.*" J. **Jeans** answered: "*Come and have tea with me this evening.*" At 4 o'clock Lady James opened the door, and the tea was ready. After having his questions repeated to him:

> "*he [Jeans] went off into an inspiring description of the celestial bodies and the astonishing order to which they adhere, ...When I behold God's marvelous creation feats, my whole being trembles, in awe at His majesty. When I go to church I bow my head and say "Lord how great you are".... I obtain incredible peace and joy from my prayer.*"

Khan **Mashriqi** nicknamed Allama (learned) Mashriqi, a *Muslim evolutionist,* accepted some of **Darwin**'s ideas, and was the 1st student of any nationality to achieve honors in four different Tripoes (high-level competitions). Deeply impressed, he replied with a verse (35:27-28) of the Koran:

> "*In the mountains, there are red and white, of diverse hues, and pitchy black; and human beings, and beasts, and cattle, diverse are their hues. From among His servants, only those who know fear Allah.*"

Sir J. **Jeans** exclaimed:

> "*It is those alone who have knowledge who fear God.*"

---

[753] Sir **Jeans** James H. *(ENG: London 1877-1946 Dorking, Surrey)* Anglican. *British physicist, astronomer, and mathematician. A brilliant student at Trinity College, Cambridge, he is ranked second in the competition, for the famous Cambridge Mathematical Tripos. Elected a Fellow of Trinity College in 1901 he taught at Cambridge, then went to Princeton University in 1904 as a professor of applied mathematics. He returned to Cambridge in 1910. He made important contributions in many areas of physics, including quantum theory, radiation theory, and stellar evolution. He is, with Arthur **Eddington**, one of the founders of British cosmology. He received a lot of honors.*

[754] Khan **Mashriqi** Inayat Ullah *(Punjab, PAK: Amritsa 1888–1963 Lahore)* mathematician, logician, and Islamic scholar (theistic evolutionist). *Born in a Rajput (son of a King) family, he was first educated at home but completed his mathematical master's degree from the University of Punjab. He moved to England in 1907, registered at Christ's College, Cambridge, and was ranked 1st class honors in Mathematics in 1909. Then he read for two years oriental languages. In 1913 he was conferred with a D.Phil. in mathematics. On his return to India in 1913, he turned down the offer of Maharadja to run the city of Awar and was appointed vice principal of Islamia College, becoming principal in 1915. He was successively under secretary to the Government of India in the Education department, headmaster of Peshawar High School, ... In 1930 he laid the foundation of the Khaksar Movement, which opposed the partition of India, and was a political activist. Evolutionist Muslim.*

Adding: *"You can record my testimony that the Quran's an inspired book. Muhammad was illiterate. He could not have learnt this immensely important fact on his own. God must have taught it to him. Incredible! How extraordinary!"*

Former Academy of Sciences President, Alber McCombs **Winchester,**[755] a *Baptist* who specialized in human heredity, stated:

"Today, after many years of research in the scientific fields, I am happy to say that my faith in God, rather than having been shaken, has been strengthened and has acquired a stronger foundation than ever. Science brings a glimpse of the majesty and omnipotence of the Supreme Being that grows with each discovery" (cf. Wikipedia).

In **1946**, in the introduction of his article *"A Physicist Faith"* [Hs] the *Roman Catholic* Victor F. **Hess** (ref.332, p.99) asks the question that is the subject of this paragraph: *"Can a good scientist believe in God?,"* and answers: *"I think the answer is: Yes."* I have quoted some of his arguments:

- *"It is not hard for a scientist to admire the greatness of the creator of nature. From this, it is only a step to adore God,"*

- *"In all my years of research in physics and geophysics, I have never found one instance in which scientific discovery conflicted with religious Faith."*

The Methodist Ernest **Walton** (ref.344, p.103), co-winner of the 1951 Nobel Prize in Physics posited that:

"Scientists seek the truth, Christians seek the truth and in the end, the truth cannot conflict with the truth" [MB2].

After winning the prize, he gave many lectures in various countries to explain the relationship between Science and Religion and to encourage the development of Science as a:

"path to knowing more about God."

---

[755] **Winchester** Albert *(Waco, TX-US 1908 - 2007). Ph.D. at the University of Texas in 1934, and later worked for the Universities of Chicago, Harvard, Michigan, and Munich. Head of the Biology Department at Oklahoma Baptist University in Shawnee (193542), Professor of Biology at Baylor University (1942-46), Head of the Biology Department at Stetson, Deland University, FA from 1946-61, Professor of Biology at the University of Colorado at Greenly since 1962. Has served as President of the Florida Academy of Sciences. Specialist in Heredity and Biology. Member of various scientific associations especially in genetics and author of 350 publications.* Baptiste.

According to the *Catholic* Physicist Vincent J. **McBrierty**,[756] who wrote his biography [MB], **Walton** applied his religious principles in his professional life, because it was for him:

> *"a way to learn the mind of the Creator and to study His creation. It is our duty to study His work and this must apply in all fields of human thought. A refusal to use our intelligence honestly is an act of contempt for Him who gave us that intelligence...."*

In December **1968** Apollo 8 was the first scientific mission to put a manned space capsule in orbit around the Moon. *Christian* Frank **Borman**,[757] one of the 3 astronauts with William **Anders**,[758] and James **Lovell**,[759] scientists and engineers confirmed, read live on this Christmas Eve, for the televisions of the whole world, a passage of Genesis.

On 16/07/**1969** W. **Von Braun**, *Lutheran*, designer of the rocket that put into orbit the Apollo 11 mission, which allowed Niels **Armstrong**,[760] *Methodist*, even

---

[756] **McBrierty** *Vincent J. (Belfast, IRL 1942)*. Catholic *physicist, Knight of the Sacred Military Constantinian Order of St George. Belfast: BSc in 1962 and Ph.D. in 1965. ScD at Trinity College Dublin in 1969. He passed through all the stages: Fellow, Fellow, Lecturer, Professor, Dean, Professor Emeritus. Specialist in Soil Management and Environmental Sciences. He is the author of various scientific publications and books and has been a member of the Royal Irish Academy since 1983.*

[757] **Borman** Frank *(Gary, IN-US 1928)*, Christian *Astronaut. Professor of Fluid Mechanics and Thermodynamics at West Point.*
Fighter pilot since 1950, then a test pilot, he joined NASA. Participated with James **Lovell** *in the Gemini 7 mission in 1965. Commander of the Apollo 8 mission (21 to 27/12/1968). CEO of Eastern Airlines allowed Airbus to penetrate the American market. Colonel of the USAF (United States Air Force).*

[758] **Anders** William *(Hong Kong, CHN 1955). Graduated from the Annapolis Naval Academy in 1955, then in Nuclear Engineering in 1962. Astronaut (selected in 1964 by NASA) he piloted the lunar module during the Apollo 8 mission. USA Ambassador to Norway, Major General in the USAF.*

[759] **Lovell** James *(Cleveland, US-OH 1928). Graduate of the Annapolis Naval Academy, selected as an astronaut in 1952, participated in 1965, with* Franck **Borman**, *in the Gemini 7 mission (4th manned mission in space). Commander of the Gemini 12 mission in 1966, he participated in the Apollo 8 mission in 1968 and the Apollo 13 mission in 1970, which only went around the moon due to a technical incident. Appointed Deputy Director of the Johnson Space Center in Houston in 1971, he left the Navy in 1973 but remained in Houston as a Corporate Executive.*

[760] **Armstrong** Niels *(US-OH: Wapakoneta 1930-2012 Cincinnati.). Astronaut,* methodist *with* deist *tendencies. With a degree in aeronautics, he interrupted his studies in 1950 to participate in the Korean War as a fighter pilot. In 1955 he joined the NACA (precursor of NASA) as a test pilot. In 1962 he joined the NASA astronaut corps. First space flight in 1966 (Gemini 8), commander of Apollo 11 in 1969, he pronounced, while putting his foot on the Moon, a sentence that will remain famous:* **"One small step for man, a great leap for mankind."** *He left NASA and became a university professor.*

deist, to be the first man to set foot on the Moon, and Buzz **Aldrin,** [761] *Presbyterian, freemason*, to be the second, said before the launch:

> *"We have prepared everything, calculated everything, we now rely on the grace of God."*

The bet was tried by the Americans, following the announcement made on 12/09/1962 by John F. **Kennedy,** [762] then President of the United States of America, in front of the students of the University of Rice in Houston, US-TX, to see a man walking on the Moon was then far from being won. This could justify the fact that - despite his indisputable capacities in rocket technology, the potential of the scientists who participated in this extraordinary adventure, and their knowledge of the laws of the Universe - Von **Braun** finally put his trust in God.

For the *Baptist* James B. **Irwin,** [763] an American Astronaut, who took part in the Apollo XV flight, on July 26, **1971**, his exploration of the moon, in *"jeep,"* was more than a scientific enterprise, it was a spiritual experience that gave him back his faith to the point that he decided, to dedicate his life to making God

---

[761] **Aldrin** Buzz *(Glen Ridge, US-NJ 1930)*. Presbyterian, *Astronaut, he took communion on the Moon in 1969! Ranked 3rd at Westpoint in 1951, he obtained a Master's degree in mechanical engineering, then participated in the Korean War as a fighter pilot. He defended a thesis:* "Line-of-Sight Guidance Techniques for Manned Orbital Rendezvous," *hence his nickname of* "Doctor Rendezvous." *In 1966, he participated in the Gemini 12 mission. Retired in 1972? with the position of colonel, Aldrin has the 33rd degree of the "Montelais Lodge N 144 of NJ." It is less known that he took with him to the Moon the flag of the Ancient and Accepted Scottish Rite, called Scottish Rite, created between 1735 and 1737, and that he brought it back to Earth where he had his picture taken with it. The REAA adopted 12 Principles, the first of which is the recognition of* "the Great Architect of the Universe: the Higher Principle, creator of the world." *Every applicant must take an oath on* "The Great Lights" *which are 1. the Volume of the Sacred Law (VSL) which can be the Bible, the Vedas, the Koran, the Tao Te King, ... 2. The Compass symbolizing the open- mindedness, the fraternal and universal Love that unites them, 3. The square symbolizes discipline, rectitude, reason, and uprightness.*

[762] **Kennedy** John F. *(Brookline, US-MA 1917-1963 Dallas, US-TX murdered) known as JFK, 35th President of the USA (19611963) was a senator (1953-60) and representative (1947-53) of the USA for Massachusetts. Well known for the Cuban missile crisis, the catastrophic landing in the Bay of Pigs, his space race, his action for the limitation of nuclear tests as well as for the rights of African Americans, ...* Officially catholic *but having led such a dissolute life I will not highlight his faith. At the beginning of the 60s, scientists had to rely a lot on the intellectual capacities of scientists, including African American women (cf. the film* "The Hidden Figures" *by Th. Melfi) to calculate the trajectories of spacecraft.*

[763] **Irwin** James B. *(USA: Pittsburg, PA 1930-1991 Glenwood Springs, CO).* American Baptist astronaut, *the pilot in 1971 of the lunar module during the Apollo 15 mission. He set foot on the moon two years after* **Armstrong** *and said on his return:* **"Jesus walking on the earth is more important than man walking on the moon."** *He graduated from the Naval Academy in 1951 and joined the Military Astronaut Class 4 in 1963. He was part of the Apollo 15 mission in 1971 and spent three days on the lunar surface in the first lunar rover. Leaving the Air Force in 1972 with the rank of colonel, he founded* "High Flight," *a* Christian *organization intended to spread the faith. From 1973 he undertook several expeditions to Mount Ararat [Ir3] (Turkey) hoping to find the remains of Noah's Ark.*

known throughout the world. He tells us (*[Ir1] Pp. 18, 21, 119. P. 243...*) that during the mission they were very occupied, but that later he realized that the flight had been:

> "*...a revelation for me. I had previously become skeptical about God's action on me and had lost the feeling of his closeness. On the Moon, the representation of the power of God and His Son Jesus Christ became very clear to me. I felt the presence of God invade me. I felt His spirit, closer to me than it had ever been on earth.*"

The epigraph of §5.1.2, attributed to the notorious physicist, historian of sciences, and essayist, a *Catholic*, Louis **Leprince-Ringuet** (ref.597, p.201): "*God, in his permanence, takes on different faces for different humanities,*" means that God reveals himself to different cultures and times in different ways, but remains the same. He also argues that science does not contradict the existence of God, but rather reveals his wisdom and power. He also emphasized that:

> "*Our era, more than any other, yearns for the guiding light of the Gospel, ...*" [LR].

Biologist Humberto **Maturana** (ref.685-p.244), initially raised in the *Catholic* faith, underwent a transformative journey where, despite becoming critical of the Church and its dogmas, he maintained a keen interest in the mystical facets of Christianity. Intriguingly, he also delved into other religions exploration, notably Buddhism and Hinduism. Collaborating closely with, Francesco **Varela** (ref.684-p.245) his student, who initially shared a *Catholic* upbringing but, later embraced *Buddhism*, the duo embarked, on an exploration of the intersections between science (they worked on cognition and consciousness) and spirituality. Their collaborative efforts culminated with the development of the groundbreaking "*theory of autopoiesis,*" which proposed the concept of self-organization of living systems in 1972. Not only, does this innovative theory, reflect their scientific pursuits, but also showcases the profound connections they discovered between scientific inquiry and spiritual contemplation.

In **1982**, the *Deist* Sir John **Eccles** (ref.164, p.46), 1963 **Nobel** Prize in Physiology or Medicine, affirmed:

> "*The more I study the brain, the less I am a materialist*" [Ec].

It should be noted that he was not at all a materialist in the sense that we understand it. For him, materialism is the breeding ground of the consumer society, that is to say, of the search for power and money, degrading the humanistic values of love, and truth, ... and leading to the disintegration of the family unit, making man a slave.

While the *Catholic* Physicist **McBrierty** said to us:

> "*My reflection on my professional life reveals a progression from a strong inherited faith to a perception that this faith has blossomed over*

time and been strengthened by the influence of many scientists and non-scientists past and present" [MB2].

He believes that his many contacts, with atheists, agnostics, and believers in various religions, have contributed greatly, to strengthening his *catholic* faith.

Professor of biology Carl **Feit,** [764] *Rabbi,* and Talmud scholar, wrote in reference to **Maimonides**:

"... the best way to develop one's love for God and to know him is to study the work he has done with his hands. There are certain graces that a Jewish believer offers every day. Some of these are that the sun rises and sets regularly, that all the stars move in their orbits, and that our physiological functions work properly. With my knowledge of human physiology, I have a very different, and I think more important, appreciation when I express this gratitude every morning."

Jacob **Bekenstein** (ref.385, p.114), a *Jewish,* theoretical physicist, known among other things for the Bekenstein-Hawking entropy law, a specialist in the thermodynamics of black holes said (Wikipedia biography):

"I look at the world as a product of God. HE establishes very specific laws, that we have the pleasure to discover through our scientific work."

Arno **Penzias** (ref.364, p.108) *a conservative Jewish,* 1978 Nobel Prize in Physics interviewed in Jerusalem in **1991**, by *Rabbi* Joshua O. Haberman said [Hb pp. 176-]:

"Religion is that which we know beyond what we can prove."

If p. 180 of [Hb] he stated:

"...science and religion are complementary competencies,"

P. 184, referring to his work, he posited:

"by looking at the order in the world, we can infer purpose, and from the purpose we begin to get some knowledge of the Creator, the Planner of all this."

These citations, added to the ones of §5.1&5.4., are clues for his positioning in this section.

---

[764] **Feit** Carl *(Teaneck, US-NJ. 1945). Studied at Yeshiva College (Upper Manhattan) where he taught health sciences. Obtained his Ph.D. in Microbiology and Immunology at Rutgers University (NJ). A researcher in the Immunodiagnostic Laboratory at the SloanKettering Institute, he is a specialist in cancer. From 1985 until his retirement in 2016 chaired the Science Department at Yeshiva College. He taught Talmud for many years and is a co-founder of the International Society for Science and Religion. Member of the Editorial Board of Cancer Investigation.* Rabbi.

The renowned *Christian* astronomer Allan **Sandage**[765] confessed in **1998,** during an interview [Sa]:

> *"As a child, I was an atheist. It was my science that led me to the conclusion that the world is far more complicated than can be explained by science. It is only through the supernatural, that, I understand the mystery of existence."*

In the same interview, Professor of Theoretical Physics Mehdi **Golshani,**[766] a *Muslim philosopher*, said:

> *"Natural phenomena are the signs of God in the universe and, studying them, is practically a religious obligation. The Qur'an asks humans to walk the earth and see how He initiated the creation. Research is an act of worship since it reveals the wonders of God's creation."*

The sentiment expressed by Golshani resonates with the Quranic emphasis on reflection and observation as a means of recognizing the Creator's wisdom and power. Some verses of the Quran, such as those in Surah Al-Imran (3:190-191):

> *"Indeed, in the creation of the heavens and the earth and the alternation of the day and night, there are signs for people of reason,"*

highlight this importance.

According to the Italian mathematician Carlo **Cercignani,**[767] known for his contributions to the studies on the kinetic theory of gases and **Boltzmann**'s biography [Ce] published in **1998,** this later acknowledged the existence of God while highlighting the limitations of human conceptions of the divine; underscoring the complexity and depth of Boltzmann's philosophical stance:

> *"It is certainly true that only a madman will deny God's existence, but it is equally the case that all our ideas of God are mere inadequate*

---

[765] **Sandage** Allan *(USA: Iowa City IA 1928-2010 San Gabriel, CA), Christian. BA in Physics in 1948, PhD in Cosmology in 1953. Assisted Halley from 1950 to his death in 1953. Participated in measurements to improve the value of Halley's constant, was involved in measuring the expansion velocity of galaxies, and studied Quasars (QUAsi Stellar Astronomical Radiosource, having a very bright intensity). Discovered the Faith in 1978 at the age of 50.*

[766] **Golshani** Mehdi (Isfahan, IRN 1939). *Theoretical physicist, cosmologist, and* Muslim *philosopher open to Christianity. He obtained his B.Sc. in 1959 at the University of Tehran. In 1969 he defended his thesis in particle physics at the University of California at Berkeley. Member of the Templeton Commiee and attended its courses. Co-founder of the Institute for Studies in Physics and Mathematics in Tehran. Professor at the University of Tehran.*

[767] **Cergninani** Carlo *(ITA: Teulada 1939-2010 Milano). A mathematician and a prolific author of about 300 scientific papers and several monographs; he is a member of the French and Liceo, one of the oldest [16ᵗʰ century], and most prestigious European institutions, located in Roma. This secondary school, provides a classical education, mainly in humanities, literature, and philosophy.*

anthropomorphisms, so that what we thus imagine as God does not exist in the way we imagine it...." [Ce].

In 2005, Charles Hard **Townes** (ref.355, p.106) the 1964 **Nobel** Prize in Physics, a progressive *Protestant*, and a fervent supporter of the rapprochement between science and religion, received the Templeton Prize. Townes was interviewed by Bonnie A. Powell (B.P.), News Center (June 17, **2005**). To the question: B.P.: "How do you categorize your religious beliefs?."

Ch. Townes: "*...I am a Protestant Christian, I would say a very progressive one ... quite open-minded and willing to consider all kinds of new ideas,... At the same time..... I feel the presence of God. I feel it in my own life as a spirit that is somehow with me all the time.*"

B.P.: "You've described your inspiration for the maser as a moment of revelation, more spiritual than what we think of as inspiration. Do you believe that God takes such an active interest in humankind?"

Ch. T.: "*It was a new idea, a sudden visualization I had of what might be done to produce electromagnetic waves, so it's somewhat parallel to what we normally call revelation in religion. Whether the inspiration for the maser and the laser was God's gift to me is something one can argue about. The real question should be, where do brand-new human ideas come from anyway? To what extent does God help us? I think he's been helping me all along. I think he helps all of us - that there's a direction in our universe, and it has been determined and is being determined. How? we don't know these things. There are many questions in both science and religion and we have to make our best judgment. But I think spirituality has a continuous effect on me and on other people.*"

Ch. **Townes** wrote various articles throughout his life to defend this rapprochement. In particular: "*The convergence of Science and Religion*" [To], was the subject of the plenary conference he gave on August 02, **2002** at the "*ASA Annual Meeting at Pepperdine University*" in Malibu, CA during which he said:

"*I see religion as an attempt to understand the purpose of our universe and science as an attempt to understand its nature and characteristics.*"

Further on, he hopes for the:

"*possibility of a fusion into a unified understanding of the purpose and nature of the universe.*"

Although calling himself agnostic, Astrophysicist Robert **Jastrow** (ref.367, p.109) seems to have wavered towards Deism following the 1975 discovery by Penzias et al. of "*fossil radiation*" supporting the Big Bang theory. Given his notoriety, he was interviewed on various occasions but I have retained this one from June 26, **2012** (YouTube):

> *"Although agnostic, I must recognize that nature has a beginning and do not see how it could have created it alone. I must admit, that a supernatural force, that we can call God, outside of space and time, did this* [Ja2].*"*

The 7/2/**2013** Antony **Hewish** (ref.371, p.111), a devout *Christian*, and a 1974 **Nobel** Prize in Physics gave a lecture: "*My Life in Science and Religion–A Personal Story*" at The Faraday Institute for Science and Religion. We have noticed that according to him:

> *"Science is a great help in understanding religion. The concept of God is mysterious, especially when he chose to reveal himself through a man two thousand years ago. Even the most basic physics is equally mysterious, but that does not mean that physics is not correct."*

After supporting his point with a simple example[768] he added:

> *"Physics helps me get into the right frame of mind to realize that religious mysteries can exist and may be reasonable, without undermining common sense. If matter, even at its most basic level, is more complex than you could ever dream, then we shouldn't be afraid of mystery when it comes from God."*

Sir J. **Houghton** (ref. 409, p.121), a prominent specialist in Environmental Physics, a tenacious advocate for climate action, and a devout Evangelical *Christian*, not only firmly believes that there is no opposition between science and religion:

> *"I was baffled that science and faith should be often seen as contradictory, if not in direct opposition with one another… … Science and religion are not at odds. Science is simply too young to understand*" [Ho2],*.* but asserts that religion is intertwined with, and greatly enhances his work:

> *"And I believe that you know, very much in retrospect, it's not always in prospect, of course, but in retrospect, I can look and see the way God has helped me to do a variety of things and has also …. answered prayer*

---

[768] Hewish example: *A glass of water, does not seem mysterious, at all. But looking at it scientifically, we learn that water contains a lot of hydrogen - the simplest element; Quantum physics tells us, that an electron orbiting a hydrogen atom behaves strangely, being affected by* "virtual particles." *This is hard to believe, common sense says that this behavior cannot happen, but measurements say, that it does.*

in regards to the way my scientific work has gone, not only my prayers but other people's prayers too."[769]

In recognition of his lifelong commitment to demonstrating how religious faith can invigorate and stimulate rigorous scientific research, Francis **Collins** was awarded the **Templeton Prize** (ref.434, p.129) in **2020.** Collins, the Director of the NIH[770] since 2009, has been a prominent figure in genetics (refs.433&435, p.129). Throughout his illustrious career, he consistently advocated for the synergy of faith and reason. In **2007,** he took a pioneering step. He created the association *"Biologos,"* a bridge connecting the worlds of science and religion. This commitment to unity is not limited to his organizational efforts but, extends to his public speaking engagements and popular writings, such as his best-selling 2006 book: *"The Language of God"* [Col]. Collins passionately sought to inspire religious communities to embrace the latest advancements in genetics and biomedical sciences as insights to enrich and expand their faith:

"There is no conflict between being a rigorous scientist and a person who believes in a God who takes a personal interest in each of us. Science's domain is to explore nature. God's domain, is in the spiritual realm, a realm that it is impossible to explore with the tools and language of science. It must be explored with the heart, mind, and soul - and the mind must find a way to connect these two realms" [Col].

During the award reception, he conveyed:

"As a Christian for 43 years, I have found joyful harmony between the scientific and spiritual worldviews, and have never encountered an irreconcilable difference."

I now introduce a distinguished cosmologist, **Ellis** [771] George, a *devout Quaker.* As **Bernard d'Espagnat** (ref. 740, p.272), an accomplished *Christian*

---

[769] *As stated during an interview for the British Library's Voices of Science in **2014**, and in the obituary published by* "The International Society for Science and Religion" *in **2020**.*

[770] NIH: The *National Institute of Health, is a part of the US Department of Health and Human Services. 80% of research activities are conducted by scientists working in every state, and around the world. National American agency responsible for public health and biomedical matters, comprising 27 Institutes and Research Centers to which 1,200 researchers and more than 4,000 postdoctoral fellows are attached, making it the largest Biomedical Research Institute in the world.*

[771] **Ellis** George *(Johannesburg, S-Africa 1939) is not only one of the world's leading authorities on galaxy formation, Dark Matter, and Dark Energy; but also an active Quaker and fearless opponent of apartheid. Ellis's intellectual odyssey began at the University of Cape Town, where he earned a B.Sc. in Physics in 1960. Subsequently, he delved into the realms of mathematics and theoretical physics, at the University of Cambridge, securing a Ph.D. in 1964. He occupied prestigious positions in Cambridge, Chicago, and Hamburg... before accepting an appointment as Professor of Applied Mathematics at Cap Town in 1974, leaving an indelible mark until he became Emeritus Professor. As a testament to his global impact, Ellis served as the President of*

cosmologist, Ellis has made significant contributions to our understanding of the nature of the Universe. While Bernard d'Espagnat, a specialist in quantum mechanics, introduced the concept of *"unveiled reality"* to express the limitations of our perception due to the quantum nature of our universe, Ellis, straddles the worlds of cosmology and social activism. Ellis points attention to the recurring ethic of nonviolence, and self-sacrifice prevalent in major religions and the lives of iconic figures like Mahatma **Gandhi**, Nelson **Mandela**, Desmond **Tutu**, Martin **Luther King**, and **Jesus**. He coined the term *"kenotic ethic,"* to encapsulate the idea of self-emptying, inspired by Jesus's sacrifice for the freedom of mankind.

In **2004**, **Ellis** received the **Templeton Prize** from the Duke of Edinburg. During the ceremony, he stated:

> *"Ethics is causally effective and provides the highest level of values that set human goals and choices. Consequently, a crucial issue is the origin of ethics, on the one hand, and the nature of ethics, on the other. I am a moral realist, that is, I believe that we discover the true nature of ethics rather than inventing it. Indeed it is only if ethics is of this nature that it has a truly moral character, that is, it represents a guiding light that we ought to obey."*

In an interview with Krista Tippett for *"On Being,"* Ellis explained how he integrates his convictions about cosmology, faith, and ethics. As a Professor of Applied Mathematics and head of the eponym department, involved in the Quaker service fund, of Cap Town, he extended his mathematical knowledge beyond theoretical physics to fields like economics striving to improve, the medium-term quality of life of autochthons in impoverished conditions.

Reflecting on post-apartheid South Africa, Ellis remarked that the events after 1994 *"confounded the calculus of reality."* Despite the expectation of revenge after the end of apartheid, he witnessed a deviation from the *"rationally"* anticipated bloodshed, prompting *kenotic ethics* to be there, waiting to be discovered.

When asked whether *ethics is something to be discovered in the universe,* Ellis draws a parallel with mathematics, stating that just as mathematicians discovered previously unknown concepts (the existence of irrational numbers like

---

*the International Society on General Relativity and Gravitation. Beyond the confines of academia, Ellis's unwavering dedication to justice and humanity earned him The Order of the Star of South Africa, a prestigious honor presented by N. **Mandela** in 1999. Ellis, tireless efforts, in straddling the scientific and spiritual realms, culminated in the 2004 **Templeton Prize**.*

With a bibliography boasting about 500 scientific papers, Ellis's intellectual legacy is profound. Co-authoring with S. Hawkins, in 1973, the seminal colloquium: "The Large Scale Structure of Space-Time," he has not only contributed to the academic discourse but also explored the philosophical and spiritual dimensions of cosmology in popular books.

Ellis has shared his wisdom in numerous public lectures, engaging in meaningful dialogues with religious leaders and thinkers. His magnetic presence extends to documentaries and podcasts, where his insights captivate audiences.

the number $\pi = 3,1416...$, or the square root of two equal to $1,14...$ are an example), *kenotic ethics* is waiting to be discovered. He asserts that while planets, ethics lies outside the scope of science due to the inability of scientific experiments to distinguish between good and bad.

In conclusion, both cosmologists, **d'Espagnat** and **Ellis**, underscore the limitations of human understanding and the profound nature of the universe. Their insights invite humility in our attempts to comprehend the nature of reality, beyond our immediate perception, a perspective elaborated upon in Chapter II.

Dr. Jane **Goodall** (ref.646, p.227), the trailblazing primatologist renowned for her groundbreaking work (cf. §5.3&5.4), has been honored with numerous accolades for her remarkable contributions to, both environmental and humanitarian causes. Notably, in 2002, she received prestigious nominations as a *"Dame Commander of the Order of the British Empire,"* and was recognized as a *"United Nations Messenger of Peace."* Her most recent achievement is the **2021 Templeton** Prize, a testament to *"her remarkable career, which arose from and was sustained by a keen scientific and spiritual curiosity."* In her acceptance speech, delivered while reflecting on her experience in the lush Tanzanian rainforest, Dr. Goodall expressed a profound revelation:

> *"In the rainforest, I learned about the interconnection of all species, each with a role to play. I felt a strong spiritual connection with the natural world."*

According to Dr. Goodall, a belief resonates within her that it exists a *"spiritual power in every living thing."* Her contemplative reflections extend to aspirations for the world, as she eloquently states:

> *"I understand that the deep mysteries of fife are forever beyond scientific knowledge and now we see through a glass darkly; then face to face."*

Goodall's pioneering discoveries have indelibly modified humanity's comprehension of our place in an intricately interconnected world. Her tireless advocacy fervently points towards a greater purpose for our species: the stewardship and nurturing of life on this cherished planet.

Dr. Frank **Wilczek** (ref.381, p.113), a luminary in theoretical physics, and a recipient of the 2004 **Nobel** Prize, stands as a shining example of a leading scientist who harmoniously bridges the worlds of science and religion. His remarkable contributions to theoretical physics are only rivaled by the profound philosophical insights he offers, making him a true visionary of our time. Through his captivating lectures and enlightening writings, Wilczek unveils a universe characterized by mathematical beauty, inviting all to share in his transcendent vision. Despite his Roman Catholic upbringing, **Wilczek**'s exploration of the writings of the atheist philosopher Bertrand **Russell** had led him to shed conventional religious beliefs. He candidly shared that:

"A big part, of (his) later quest, has been trying to regain some of the sense of purpose, and meaning that was lost."

As a result, he identifies as *a Pantheist*, emphasizing his deep connection with the universe as a source of spiritual inspiration.

This journey, not only highlights the seamless integration of science and faith; but, it also illustrates, how faith can catalyze brilliant minds, driving scientific exploration and discovery. It also underscores that the quest for knowledge can be a profound adventure into both, the natural world marvels and the divine enigmas that underlie it. For a more detailed justification of the classification of notorious scientists who, like, G. Galilei, I. Newton, L. LePrince Ringuet, A. Penzias, and others have integrated their religious beliefs with their scientific endeavors, or to explore the perspectives of other prominent figures such as Sir W. Herschel, F. Dyson, A. Salam, and A. Grothendieck, for example, whose views align with the theme discussed in this section, I encourage you to delve into the references and the preceding chapters, specifically §5.1 and 5.2. These paragraphs provide comprehensive insights into the diverse spectrum of scientists and their beliefs, contributing to a more nuanced understanding of this intricate interplay between science and faith.

In conclusion of this paragraph about the relationship between science and religion I'll say that each scientist's perspective on the relationship between their faith and scientific work is unique and influenced by their personal beliefs, experiences, and interpretations. If we zoom in on scientist priests, we can observe that their lives and works represent a harmonious fusion of scientific exploration and theological insight, making them pivotal figures in bridging the gap between these two realms and shedding light on the profound relationship between science and spirituality. It's essential to respect and appreciate the diversity of views within the scientific community and recognize that individuals can integrate their faith and scientific endeavors in various ways. I'll let the word to Jacques **Arnould**, a theologian and science historian, who declared during the 2008 colloquium *"Sciences of secularism"* at the Collège de France:

"Science a chance for religion."

5.6. Is this world the best of all worlds?

"The most beautiful masterpiece is the one made by God,
according to the principles of quantum mechanics."
E. Schrödinger[772]

In contemplating whether our world is the best of all possible worlds, we encounter a range of philosophical perspectives. For **Plato**, a *pagan theist*, this

---

[772] Elisabeth **de Lavigné** *translator of* "25 scholars confess their faith in God" *Aleteia of 21/06/2014.*

world represents beauty and perfection, yet he believed that the intelligible world of Ideas surpasses it in greatness. Plato's god, a mathematician demiurge, does not have unlimited powers, unlike the omnipotent God of Abrahamic religions. The demiurge "*modeled*," rather than "*created*" the world, using eternal materials, such as Empedocles' four elements and the fifth element (cf. p.17). Due to these limitations, and the fact his world had an anthropic purpose, the demiurge could not be solely responsible for the evils that afflict us. **Timaeus** expressing **Plato**'s ideas states that our universe [Pla]:

"*...is as excellent as its nature permits it to be,*"

implying that the divine Craftsman constructed this world with utmost goodness and intelligence, filling every gap in the universe.

Gottfried W. **Leibniz,** two millennia later, shared a similar view. In his letter (1673) to Mr. Arnault [Lz21194], he stated:

"*God, having chosen the most perfect of all possible worlds, had been led by his wisdom to allow the evil that was appended to it, but that did not prevent this world, all counted and discounted, from being the best that could be chosen..,*" because:

"*the little evil that is there is required for the height of the immense good that is there.*"

M. Maimonides, in earlier times, also professed ([Ma1], chap. XX p.172) that:

"*God made all things by willing this universe,*"

a notion shared by the *Jesuit* mathematician Roger **Boscovitch** (ref.619, p.212), who refuted chance and posited that:

"*... the world cannot be the product of necessity, it can only be contingent.*"[773]

"*It is not contrary to the goodness and wisdom of God not to choose the best, for the best does not exist. If there were optima, the creation would be necessary, the divine freedom would be suppressed, the creatures would have in their number and in their merit, a true right to the existence.*"

A question arises: if God is omnipotent, could He have created a different universe? Some philosophers, theologians, and scientists have pondered this possibility. B. **Pullman** ([Pu] note 3 p. 69) suggests an answer:

---

[773] Contingent *should not be taken literally (ref.578, p.161) but should be seen as the author's thought that God was not obliged to create the world in which we find ourselves, thus eliminating the anthropomorphic idea of creation.*

"...some (including Einstein) will raise the question of whether God had the possibility of creating a universe other than the one He did, the argument being that the existing universe may be the only one to possess complete logical coherence."

It is worth noting that eminent physicists today seriously consider the existence of other universes as discussed in sections IV.2 and IV.3.

Elie **Wiesel**,[774] the recipient of the 1986 **Nobel** Peace Prize and an adherent of *Hassidic Judaism,* partially addresses this question in *"Job ou Dieu dans la tempête"* [EW]:

"According to the Talmud, this world in which we live is not the first: God created many others before, but he did not love them and destroyed them."

His interlocutor Rabbi Josy **Eisenberg** [775] replies:

"Yes, but He stopped at ours: this one pleases Him."

Furthermore, the Talmud Babli Pessa'him (54 a) teaches us that:

"seven things were created before the world was created: the Torah, the Techuvah, the Garden of Eden, Gehenna, The Throne of Glory, the Temple, and the name of the Messiah."

Arno **Penzias** (ref.364, p.108), a conservative *Jewish,* 1978 Nobel Prize winner in Physics, and co-discoverer of the Cosmic Microwave Background radiation (CMB) supporting the Big Bang theory, expressed his belief in a good world created by God. He stated:

"He did it elegantly. "Elegantly" means He doesn't have to monkey with it as time goes on."

"I assume that the world, as it was created, is a good world, otherwise the whole thing would be a cruel joke" [Hb],

In his words, he demonstrates his faith as a prominent scientist aligning with scriptural teachings. However, if this world is indeed good, why does it not appear so for everyone?

---

[774] **Wiesel** Elie *(Sighetu Marmatiei, ROU 1928-2016 New-York, USA). Hasidic Jew [EW] he was deported in 1944 to Auschwitz and then to Buchenwald before studying literature in France. He wrote about fifteen novels [including "The Night," the most famous, in which he talks about his experience in the various concentration camps where he was incarcerated during the Holocaust], and 4 plays. He emigrated to the United States and became an American citizen in 1963 before teaching in many universities, and Nobel Peace Prize in 1986.*
[775] **Eisenberg** Josy *(FR: Strasbourg 1933–2017) French Great Rabbi. Producer, director and presenter of the program "La Source de vie (spring of life)" on France 2 (French TV) on Sunday morning.*

In contemplating the intricacies of God's harmonious and evolving universe, scientists believers or not, have found marveling at its meticulous design, guided by divine specifications beyond our perception. God entrusted mankind with the management of Earth, a mere fraction of this vast cosmos, granting us the profound gift of Free Will and, in doing so, absolving himself of direct responsibility.

During the reception in 1990 of the prestigious **Templeton Prize** (ref.434, p.129) the prominent biologist and geneticist Charles **Birch** (ref.541, p.165), a leading campaigner for an environmentally safe universe, a lay *Anglican theologian* and philosopher, offered a perspective aligning faith with environmental advocacy. He envisioned:

> "*a deeper religion no longer envisions God as an omnipotent creator outside a mechanical universe. It has two emphases; namely God's presence in the world, and the presence of the world in God. God is present in the world as the divine eros, drawing everything to him. This is the persuasive God who ever lures the world to fulfillment of its possibilities and never manipulates, leaving the entities of creation with their degree of freedom to respond or not to respond.*"

Yet, despite this divine trust, what have we made of our earthly heritage? The Earth, entrusted to us to pass on in better condition to future generations, teeters on the brink of imbalance and deterioration. Some view our planet as being in great danger, despite the warnings from scientists spanning decades.

For believers, the Genesis account assigned mankind the mission to foster prosperity. Regrettably, our trajectory seems to be one of self-inflicted decline. Driven by self-interests such as money, power, pleasure, and others, we are collectively hurtling toward a point of no return. The consequences of this reckless path extend beyond our planet, as we contemplate spreading disorder to other celestial bodies while leaving chaos in our wake within the solar system.

According to the protestant scientist Théodore **Monod** (ref.579, p.185), *Christians* bear a special responsibility due to Christ's principal commandment, which emphasizes the love for God and our neighbors as ourselves. Monod underscores that this love for God also entails a profound love for nature:

> "*The contact with nature is the contact with the divine because in nature the divine element is present it is the only one productive and creative*"
> [Md4].

While I concur with Monod's perspective, it's important to note that this sense of responsibility is not exclusive to Christianity. Many other religious traditions share a similar ethos, recognizing humanity's duty toward the environment and acknowledging the divine presence in the natural world. As stewards of the Earth, individuals across various faiths are called upon to honor, protect, and preserve the interconnected web of life. Therefore, the responsibility for fostering harmony with nature extends beyond any religious framework, as we saw during this work.

Pioneering scientist environmentalist **Birch**, engaged in adventurous reflection on questions of science and faith throughout his career with another environmentalist John **Cobb**,[776] his mentor, who argued that *"humanity's most urgent task is to preserve the world on which he leaves and depends,"* emphasize the urgency of preserving the world we rely on. **Birch**, as for him, stated:

> *"In the ecological worldview, love, compassion, and rights extend beyond the human to all that participates in life. Each entity from protons to people has intrinsic value to itself and God because each is a subject. A biocentric ethic calls for a bill of rights for nature. It also has profound implications for an ecologically sustainable global society that is also just."*

Prominent quantum physicist and cosmologist John A. **Wheeler** (ref.280, p.84), a *Unitarian* having a great interest in the deep philosophical and metaphysical questions that arise from studying quantum physics realm and nature reality, challenges us through his rhetorical question: *"Here is a human being so what must the universe be."*[777] Often interpreted as a thought-provoking statement, he prompts us to consider the intricate connection between our consciousness and the nature of the universe itself.

In the face of an exponentially growing world population (almost 33% since the start of the 3rd millennium), reaching 8,1 billion by the end of 2023, and limited resources, we are compelled to reassess our lifestyles swiftly. This necessitates a collective move towards a responsible, less resource-intensive, and more fraternal society. Regardless of wealth, the current trajectory suggests that no one can envision a future on another planet offering the same pleasures enjoyed today.

The 1955 **Nobel** Prize for Peace, Sir J. **Houghton**, a devout *Christian* and a prominent atmospheric physicist underscored in **2017** - during an address to the International Society for Science, whom he was a founding member – our responsibility:

---

[776] **Cobb** John B. *(Kobe, JPN 1925), Environmentalist. Raised* Methodist *by his parents who were Methodist missionaries, in 1940 he moved to Georgia, US to finish High School. In 1943, he joined the US Army and served in the occupation army of Japan. In 1947 he entered the University of Chicago where his faith "evaporated." To reconstruct his Christian faith, which was more compatible with scientific and historical knowledge, he entered the University of Chicago Divinity. He was influenced by Alfred* **Whitehead***'s ideas, a mathematician and philosopher whose holistic philosophy interconnects everything. He defended his Ph.D. in 1952 and then shared his time between teaching and theological pastoral activities. He co-founded the "Center for Process Study" in Claremont, CA, making it the center of Whiteheadian process thought. Author of about 50 books he visited many universities worldwide and received six honorary doctorates.* Christian.

[777] *This sentence was cited during Birch's 1990 address when he received The Templeton Prize.*

"The Bible tells us that God's creation is good, but also that it is groaning because of human sin. We have a responsibility to care for God's world, and not to abuse it. Climate change is one of the most serious threats to God's creation and well-being of our fellow humans, especially the poor and vulnerable."

He ended his lecture with a statement on global warming:

"three qualities ... should guide our stewardship — honesty, holism ... and humility. The alliteration of the three Hs assists in keeping them in mind."

Let's, therefore heed the call to collective responsibility, and work towards a future where our actions align with the divine trust placed upon us and ensure a thriving legacy for generations to come.

As we transition to the next section, we'll delve deeper into evil's origin question. Is it attributable to God or have we projected human qualities onto a divine entity? Meantime, let's the word to:

- the wisdom of **Democritus** who stated ([Bt] p. 55):

"... the evils not being imputable to them (the gods of the pre-Socratics), but are only the consequence of human blindness and lack of discernment,"

- and the reflections of the Christian philosopher C.S. **Lewis** (ref.734, p.268) who claimed that:

"God is not a person who is good and then tells us what goodness is, rather, goodness is the actual nature of God, and we know it because God shares it with us. We may think God is unfair, but that is because we are applying our standards of fairness to a Being who knows infinitely more than we do. When we are in the water, we feel the wetness, but the fish does not, in the same way, our assessment of good and evil is not necessarily God's" [Ls],

Provide intriguing perspectives on this matter.

5.7. Why did God allow Evil?

*"What amazes me is not the existence of evil: it is natural,
and we understand that it exists and why.
What amazes me is the existence of good, of divine paternity, I suppose."* [Mw]
Wilfred Monod [778]

---

[778] **Monod** Wilfred *(Paris, FRA: 1867-1943) Professor of theology.* Protestant.

Why did God, whose omnipotence, omniscience, wisdom, and will, allow evil to exist? Regardless of religious belief, no one can deny that the human race has experienced, and continues to endure many hardships and sufferings that faith alone cannot rationalize. In effect, the question of why God allows evil is as ancient as the creation of the world itself, and it continues to defy consensus, despite the undeniable existence of evil.

I find the perspective of Wilfred **Monod,** a provocative *Protestant Pastor*, particularly intriguing because it challenges us to delve deeper into its profound insights and grapple with its implications. Before delving into the heart of the matter, let's turn to his son **Théodore,** an esteemed Biologist and fellow *Protestant* (ref.579, p.185), who shed light on the subject. During an interview, Théodore stated:

> *"...it is the Christian revelation, evangelical,...... which introduced the notion of good into human history. The evil, that it is there, it is the continuation of its past"* [Md3].

Allow me to share an intriguing anecdote about W. **Monod**; who, according to Edmond Blattchen [Md3] once inscribed these powerful words on the walls of the *"Oeuvre de la Clairière"* in a poverty-stricken district of *"Les Halles"* in Paris:

> *"To believe anyway, to hope anyway, to love anyway."*

This inspired remark prompted his son Théodore to exclaim:

> *"That is the essence of faith. Despite all odds. It epitomizes the evangelical ideal. It's truly magnificent."*

The absence of Good, as W. **Monod** suggests is the primary origin of evil, often referred to as natural evil. This encompasses uncontrollable catastrophes like tsunamis, floods, volcanic eruptions, earthquakes, predators, and pandemics caused by lethal viruses or bacteria, which are consequences of the law of evolution. However, pandemics and their repercussions can also be attributed to human activities, such as war, terrorism, rape, and the misuse of scientific discoveries, ...this represents the second type of evil.

It's worth noting that for Hindu devotees (Hinduism is the world's third-largest religion after Christianity and Islam), evil is perceived as a blessing from the Lord, intended to liberate them from worldly materialism and focus on the spiritual realm. According to their beliefs, the origin of evil and suffering lies in the material world (cf. [Ba] T1-p. 113 & T3 pp. 307 & 308), emanating from *"external powers"* governed by three *"gunas"* (Virtue, Passion, and Ignorance), which shape our thoughts, actions and being:

> *"Material science born from human intellect, cannot alleviate our suffering, it is the superior inner power of Maya, under the direction of Krishna, the Supreme Lord that can truly set us free."*

Thus, our woes would have *"celestial, earthly, or bodily origins,"* and would be exacerbated by material progress. The stars are even held accountable for calamities like "*excessive heat and cold, abundant or insufficient rainfall, leading to famine, disease, poverty, and epidemics.... Causing physical and mental anguish.*"

But, how can we explain to skeptics, to those who endure suffering, why an all-powerful and Benevolent God allows evil to exist? Let's explore the perspectives of philosophers, theologians, and even believing scientists.

-   Among those who advocate for the necessity of evil, envisioning a future where Good will ultimately triumph, we can turn to **St. Paul the Apostle** (ref.67-p.21), who proclaimed during his visit to the Romans:

"For I consider that the sufferings of the present time are not worth comparing with the glory that is to be revealed to us..." [Rm 8: 18-30].

These words encapsulate the idea that present sufferings, while difficult and inexplicable, will pale in comparison to the magnificent future awaiting us.

Similarly, the teachings of Doctors of the Church like **St Augustin** (ref.79, p.24), and the *Dominican* **St Thomas Aquinas** (ref.127, p.36) shed light on the intricate relationship between good and evil, positing that the permission of evil serves a greater purpose and leads to the fulfillment of the good of the universe, respectively. We noted that **St Augustine** stated:

"...a hundred times God has permitted evil to derive a good, that is, a greater good."[Ag1]

and asserted that evil exists due to the misuse of human free will, while **St Thomas Aquinas** claimed that evil is a necessary consequence of the universe striving toward the ultimate good:

"...the permission of evil, tends to the good of the universe" [Ta].

According to him, virtue naturally leads man, enlightened by Reason, towards perfection. This idea was opposed by **Voltaire** (ref.187, p.57) in his tale "*Candide,*" which recounts the adventures of an optimistic young man despite the events he encounters. Note that a few years earlier he was of a contrary opinion stating:

"What is bad about you is good in the arrangement of things."

For Moses **Maimonides** (ref.114, p.33) a prominent *Jewish* philosopher and theologian, evil is not a creation from God, nor a supernatural force, but rather the absence of good; it is the consequence of the "*free will*" granted to man, who has the choice to do good or evil. He believed that the suffering experienced in the world serves as a test of human faithfulness and commitment to righteousness. The story of Job [Bi] in the Hebrew Bible exemplifies this notion, as Job's faith is put to the test through immense suffering, ultimately reaffirming his steadfast devotion to God. I have retained this quotation ([Ma2], Tome 2, chap.3):

"... for the end of everything is to become like him [God] in perfection......
no one can fail to recognize the action of a free will acting with intention
and not by necessity."

In **Islamic** theology, we find parallel ideas and a call for humans to demonstrate faith, devotion, and righteousness in the face of evil. The Qur'an speaks of suffering as a test, a trial designed to differentiate the righteous from the wicked, and an opportunity for individuals to strengthen their faith and connection with Allah.

For the mystic **Martines de Pasqually** (ref.555, p.174) God is neither responsible for the Good nor the Evil that affects us. It is not He who created Evil, but rebellious spirits opposed to His laws, precepts, and commandments. Similarly, good thoughts are birthed by loyal spirits. It is up to fallen humanity to choose its path through the gift of Free Will bestowed upon them. They can either align themselves with God's army, consisting of faithful angels, or with the rebels who initiated Evil. The good ones, those who remained faithful to God, serve as intermediaries between God and humanity. As Jesus told [Mt 26:52] one of His loyal followers who drew his word during His Passion when an armed crowd approached to seize Him: *"Put your sword back in its place.... Do you think that I cannot call on my Father, and He will at once put, at my disposal, more than twelve legions of angels?"*

One may wonder why God did not foresee some would rebel against His laws and why, following their transgression, He did not revoke the Free Will. Martines explains that if He had not endowed His creations with Free Will, the concepts of good and evil would have no place, and since His laws are eternal, He cannot revoke a right without contradicting Himself.

While these explanations may not fully satisfy those directly affected by suffering, they provide some philosophical and theological frameworks to understand the coexistence of evil and a benevolent God. Some individuals may question their faith, turning to agnosticism, or deism, while others find solace and strength in their religious beliefs despite their doubts. The resilience of faith in the face of suffering is a testament to the human spirit and the belief that a better world awaits those who persevere to believe in God, *the architect of the universe.* I have chosen as an example a person particularly affected, who was able to remain faithful to his religion, an extract from an interview with the 1986 **Nobel** Prize of Peace E. **Wiesel** (ref.774, p.294), a *Jewish* survivor of Auschwitz, by D. Jucaud of Paris Match [Ju]:

D.J.: *"What is your relationship with God at this time?"*

E.W.: *"Very tense since the war. I believe in God despite him, as I believe in man despite men. God never stops testing me. Since God is God, he is everywhere, in all acts, in all decisions, but then where was he during the Holocaust? We can consider ourselves before death as before God. Am I His victim? His partner? His adversary? I continue to ask questions that have no answers."*

I understand his doubts very well and admire the man who, despite all the difficulties of life, did not deny the *Judaic* religion in which he was brought up; even if at certain times, when he was interned in the concentration camp, he rebelled against God who *"had not kept his promises,"* namely, to protect *"his people."* In 1987, answering Michaël de Saint Cheron [Wi], he said clearly:

> *"...not having been abandoned by God nor abandoned God, it was his relationship with his faith that was shaken, but his faith was always present."*

His trust in God – who, according to *Jewish* tradition decides where one is born and the conditions in which one lives, it is He who personally watches over His *"treasures"* - remained intact. Wiesel provides an interesting answer to the question asked about the origin of suffering:

> *"The fact that there is so much suffering, agony, death, and casualties today shows that we - and all our contemporaries - have failed to match our actions to our abilities"* [Wi].

The *Jewish* philosopher, Emmanuel **Levinas,**[779] who was also a prisoner of war of the nazis, does not question whether God was present during the Holocaust. But he agrees with E. **Wiesel** in his analysis: it is up to humans to build a world where goodness would prevail. This creates a bridge with *Christian theologians* for whom God wants our Good but has left a place for Evil because he decided to leave us free will.

Anglican Archbishop Desmond **Tutu,**[780] winner of the 1984 **Nobel** Peace Prize, has never given up on the power of joy and redemption. Even in the face of apartheid, he endured and resisted the pains, sufferings, and injustices with grace and courage. He wisely said:

---

[779] **Levinas** Emmanuel *(Kaunas - Lithuania 1905-1995 Paris)* Jewish. *French philosopher of Lithuanian origin, came to France in 1923 to pursue his studies in philosophy at the University of Strasbourg. During a stay at the University of Freiburg, he was impressed by Martin Heidegger* one of the most important philosophers of the 20th century. Naturalized in 1939, Levinas joined the French army as a Russian-French interpreter. In 1940 his unit was captured by the Nazis and he was sent to a camp reserved for Jews, but his status as a prisoner of war preserved him from the Holocaust. However, his parents were exterminated in the camps. He gave lectures on the Talmud and taught at the Sorbonne. For 35 years he directed the Jewish Israelite Normal School. ** **Heidegger** Martin (DEU: Baden 1889-1976 Freiburg) *joined the Nazi party.* Protestant, *he converted to* Catholicism *at the end of his life.*
[780] **Tutu** Desmond *(ZAF: Klerksdorp 1931-2021 Cape Town). He began as a teacher but was outraged by the conditions of blacks and turned to theology. Ordained as* an Anglican priest *in 1961, he taught theology. Bishop of Lesotho in 1976, and Archbishop of Cape Town in 1986, he always passed on a message of peace and non-violence. In addition to the 1984* **Nobel** *Prize for Peace, he received the Schweitzer Prize for humanitarianism in 1986, the Gandhi Peace Prize in 2005, the 2013* **Templeton** *Prize, ...*

"I believe in God, who can respond to prayers, to whom we can give trust and without whom life on this earth would be without meaning (a tale told by an idiot). I believe that God has revealed Himself to us in many ways and through many men and women and that for us here in the West the clearest revelation is through Jesus and those that have followed him" [Tu].

While these courageous behaviors can be discussed, some mysterious aspects can only be explained by appealing to faith. Only faith in the existence of God and a better world, where justice reigns, can overcome these trials. It is, in my opinion, more difficult for those who suffer, with or without apparent reason than for those who have good fortune (some would say it is a gift from heaven, but it is difficult to admit for a true believer) not to depend on the whims of fate or the ignominy generated by some of their fellow human beings.

According to **Tihomir** Dimitrov,[781] the answer to a similar question asked to the 1993 **Nobel** Prize in Physics J.H. **Taylor,** *Quaker* (ref.377, p.112) was:

"We are active in the Religious Society of Friends, that is, the Quakers and it's been an important part of our lives, more so for my wife and me than for our children. My wife and I spend time with our faith group; it's a way for us to make connections with our philosophical views on life, why we are on the Earth, and what we can do for others…. The Quakers are a group of Christians who believe that there can be direct communication between an individual and the Spirit, which we may call God. By contemplation and deep inward looking one can effectively commune with the Spirit and learn things about oneself and about the way one should conduct oneself on the Earth."

This seems to have been the vision of Max **Planck** (ref.255, p.78), the father of quantum physics, the 1918 **Nobel** Prize winner in Physics, and a *Lutheran* who remained a practicing *Christian* despite the ordeals he underwent.[782] I have not found any writings by him on the subject nor for sir I. **Newton** (ref.185, p.59), a prominent *Anglican,* and his alter ego A. **Einstein** (ref.268, p.81), a *Pantheist.* Nevertheless, we know that Newton maintained correspondence with his friend and fellow Cambridge student, the *Anglican* philosopher and theologian Samuel **Clarke,** arguing that evil is a necessary consequence of *"free will,"* and the result of humans misusing it. He also believed that evil can lead to greater good, making

---

[781] **Tihomir** Dimiter. *M.Sc. in Psychology (1995), M.A. in Philosophy (1999) Compiler of "50 Nobel Laureates and Other Great Scientists who Believe in God" cf. Bibliography [Tr].*
[782] **Planck** Max, a Lutheran, was a churchwarden from 1920 until he died in 1947. *He lost his first wife in 1909 after 22 years of marriage, leaving him alone with two sons and two daughters. Karl, his eldest son was killed in 1916, and his daughters Margaret and Emma died in childbirth in 1917 and 1919 respectively. In 1944 his house was destroyed by bombing and to top it all off Erwin, his youngest son, was murdered by the Gestapo for having participated in a planned attack on Hitler.*

people overcome adversity and make positive changes in the world. Concerning A. **Einstein**, if he did not express his religious ideas on evil, we have no doubt [Ei1] that for him, anthropic evil is due to the misuse of technology and power by humanity, who foster greed, selfishness, and oppression.

The *Lutheran* polymath Gottfried **Leibniz** recalls that:

> "... among the ancients, the fall of Adam was called Felix culpa, a happy sin, because it had been repaired with immense advantage by the incarnation of the Son of God, who gave to the universe something nobler than anything that would otherwise have existed among creatures."

To summarize, there is a fairly broad consensus among non-scientist theologians, rabbis, imans, prominent scientists, and others that evil results from the "*Free will*" granted to Humanity by God with the idea that it will induce some positive effect on it. But we cannot conclude this paragraph without presenting the stances of some believing scientists for whom the answer is grounded in the theory of evolution and natural selection (cf. §5.4). Since the advent of Darwin's groundbreaking theory on the evolution of species, a remarkable quest has unfolded among distinguished scientists, seeking to harmonize the realms of science and faith within the intricate tapestry of pain and suffering.

Enter Peter[783] and Rosemary **Grant**[784] [Gr], two exceptional individuals whose *Christian* convictions have seamlessly coexisted with their transformative contributions to the fields of evolutionary zoology and biology.

Their enduring collaboration has not only left an indelible mark on our understanding of life's complexities but has also shaped the very fabric of our knowledge about Earth's diverse ecosystems.

---

[783] **Grant** Peter R. *(London, ENG 1936). His journey in the world of science started at the prestigious University of Cambridge, where he obtained in 1960 a BA (Hons). In 1964, he defended his thesis, at the University of British Columbia, it delved into the intricate relationship between ecology and evolution, showcasing their profound interconnection. Then, he joined Yale University for a post-doctoral fellowship, and McGill University as an Assistant Professor in 1965. He was elected Full Professor in 1975, after serving as an Associate Professor for five years. His dedication and contributions to the field were recognized when he was promoted to full Professor at Princeton University, NJ, where he served with distinction from 1985 to 2008. In that transformative year of 2008, Peter was honored with the title of Emeritus Professor. Throughout his illustrious career, Peter R. Grant has been a driving force in advancing our understanding of the natural world.*

[784] **Grant** Barbara Rosemary, born **Partington** (Arnside, ENG 1936). *After obtaining her BSc degree in zoology from the University of Edinburg in 1960, Barbara's academic pursuit led her to British Columbia. From 1964 to 1985, she held various esteemed positions as a Research Associate, contributing significantly to the field. Her dedication and unwavering commitment culminated in the completion of her thesis on Evolutionary Biology at Uppsala University, SWE. in 1985. This milestone was a testament to her profound scholarly achievements. In 1997, Barbara was honored with the position of Professor at Princeton University, where she continued her impactful work. In 2008, she received the well-deserved title of Emeritus professor, solidifying her legacy as a trailblazing figure in evolutionary biology.*

Picture this: Peter and Rosemary, born roughly 1,000 miles apart, shared an early and profound fascination with the wonders of the natural world during their formative years. Peter's insect collections and flower studies mirrored Rosemary's penchant for gathering plant fossils and comparing them to their living counterparts.

Their mutual curiosity laid the foundation for their future as scientists.

While Peter delved into the depths of zoology at the prestigious University of Cambridge, Rosemary Partington pursued her zoology BSc degree at the University of Edinburg. Fatefully, in the breathtaking landscapes of Vancouver, *Canada* at the University of British Columbia, Rosemary's academic journey intertwined with Peter's, and their kindred spirits aligned. Two years later, their marriage marked the beginning of a dynamic partnership in the realm of evolutionary zoology.

Their collective impact on our understanding of the natural world is immeasurable, continually inspiring generations of scientists worldwide. The Grant's extensive studies, particularly their four-decade exploration of finches[785] adaptation on the Galápagos Islands, have made significant contributions to our understanding of natural selection, speciation, and hybridization in these birds. Their work has illuminated the frequent occurrence of natural selection and, in some instances, the surprising speed of evolution. Their findings echo a liberating sentiment:

> *"A major burden was removed from the shoulders of believers when convincing evidence was advanced that the design of organisms need not be attributed to the immediate agency of the Creator, but rather is an outcome of natural processes."*

The Grant's illustrious careers have garnered them numerous prestigious medals, awards, and accolades from esteemed scientific Societies, including the Royal Society. In their journey, Peter and Rosemary Grant have not only advanced our scientific understanding but have also provided solace to those seeking to reconcile the wonders of creation with the tenets of faith.

An interesting stance is that one of the prominent biochemists Denis **Alexander** (ref.704, p.250) an apologetic *Baptist*. Although he acknowledges that pain and suffering are a reality of the natural world and that they can be caused

---

[785] Finches *are a group of about 18 species of passerine birds that live on the Galápagos Islands and Cocos Island. They are well known for their remarkable diversity, in beak form and function; each being adapted to different food sources. They are often considered an example of evolution by natural selection because they evolved from a single ancestor that arrived on the islands more than a million years ago. Charles Darwin collected some specimens of these birds during his voyage on the HMS Beagle. He did not realize their significance until later when he collaborated with an ornithologist named John Gould. Gould identified the different species and noticed the variations in their beaks. Darwin used this evidence to support his theory of evolution in his book On the Origin of Species².*

by various factors such as genetic mutations, environmental factors, and human actions, his credo, both in his numerous lectures, interviews, and books, is that:

> *"Whilst the cost of existence is extremely high, it can nonetheless be squared with the idea of a God of love whose ultimate purposes for humankind render that cost more comprehensible"* [Ar2].

His perspective is informed by both his scientific understanding of the natural world and his *Christian* faith As a biochemist analyzing the complex and dynamic system of interactions among living organisms and their environment he quotes various examples including that of the ambivalent carbon cycle. On the one hand, it is harmful to humanity through the emission of carbon dioxide (CO_2) resulting from the activities of volcanoes and mankind; on the other hand, it is essential to life on our planet. In effect, the carbon cycle is:

-   On one hand, responsible for greenhouse gas emissions such as CO_2 and methane (CH_4) into the atmosphere; causing global warming, affecting weather and climate events such as heat waves, droughts, floods, storms, tsunamis, sea level rise...... It is also the cause of the acidification of oceans and seas, resulting in negative effects on marine life: corals, mollusks, and plankton.....

-   On the other hand, it is essential for the sustainability of life on Earth, as it regulates the flow of carbon dioxide between the atmosphere and living organisms. Thanks to the photosynthesis process, green plants (containing chlorophyll) and some other organisms, absorb light. The light energy is converted into chemical energy allowing to convert water, CO_2, and minerals into Oxygen (O_2) and energy-rich organic compounds according to the following relation:

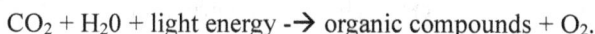

    $CO_2 + H_2O$ + light energy $\rightarrow$ organic compounds + O_2.

If photosynthesis ceased most organic matter, food, organisms, and gaseous oxygen would disappear, also I agree with his stance.

The *Christian* biochemist Francisco **Alaya** (ref.682, p.243), while studying theology in Salamanca, *Spain,* found that:

> *"Darwin was a much-welcomed friend"* [Dn2].

Upon his emigration to the USA in 1961, Alaya actively engaged in lectures, and debates with proponents of a literal interpretation of the Bible and Intelligent Design concepts. Concurrently, he pursued groundbreaking research in molecular biology, contributing to a novel understanding of species origin and advancements in preventing and treating diseases such as malaria.

In his publication: *"Darwin's Gift to Science and Religion,"* Alaya explains how natural selection solves the perennial problem of evil. He posits that floods, droughts, predators, parasites, dysfunctions, and diseases are not indicative of a flawed or malevolent design but rather inherent outcomes of life's evolutionary processes. According to Alaya, the features of organisms are not the result of deliberate design by the Creator, providing a unique perspective that bridges science and theology.

The common thread that runs through the views of R. and P. **Grant**, D. **Alexander**, and F. **Alaya**, whether through natural process or divine purpose, is an attempt to make sense of the complexity of life and the existence of pain and suffering.

Once is not custom, I give my perspective on the intriguing subjects of *"free will"* and the concepts of *"good and evil."* These reflections have been fortified by various readings and discussions, shedding light on these profound topics. First and foremost, the question arises: Did God bestow upon us the gift of free will? If our response leans towards negation, it would render our earthly existence devoid of purpose, reducing it to a mere predetermined script. I subscribe to the belief in a benevolent God, which leads me to attribute the existence of evil in its myriad forms to Satan. Consequently, my stance opposes the negation of free will because it would imply placing Satan on the same pedestal as God. This notion is incompatible with the fundamental concept of an Omnipotent God, particularly within the context of Abrahamic religions. For Christians like myself, this viewpoint is even more unfathomable, as the life and teachings of Jesus demonstrated His empathy towards human suffering and His ability to heal and triumph over Satan, the tempter. In light of these considerations, it becomes evident that God, by granting both nature and humanity the mantle of *"free will,"* opens up a realm of exploration. Among believing scientists, two hypotheses are often contemplated:

1st. The notion that God created the universe without a predetermined purpose. This perspective implies a *laissez-faire approach* as if God simply *"let it be"* or relinquished control of Satan. However, this view contradicts the very essence of God's anthropic project and His omnipotence. I firmly reject this thesis, as I have expounded upon in this essay.

2nd. The belief that God created the Best of All Possible Worlds (cf. §V.6) for the benefit of humanity.

This perspective encompasses the laws of nature we have observed, including the law of evolution and its corollary natural selection; an idea that remains irrefutable in contemporary times. However, for various reasons elucidated earlier, it is conceivable that God has imparted a direction to this law; a direction that can be likened to the *"Omega Point,"* as proposed by P. **Teilhard de Chardin** (ref.505, p.157) or the *"arrow of Time,"* as envisioned by I. **Prigogine** (ref.387, p.115). Nevertheless, this direction incorporates various waypoints, leading to oscillations between multiple possibilities, including those that give rise to adversity, a source of suffering that often appears unjust. Nonetheless, the law of natural selection ultimately prevails. This is one of the rationales for recognizing the *"hand of God"* in this intricate web of laws, which from the very outset, have governed evolution in all domains.

In conclusion, the interplay between free will, good, and evil remains a profound and ongoing inquiry, intertwined with the mysteries of existence and the divine. Although A. **McGrath** (ref.739, p.271) expert in the relationship between science and religion "*does not believe that humanity will ever fully explain suffering*" he shares the common opinion that:

"*There is more to life than the physical world and God urges us to find it*" [©€A].

These considerations invite us to ponder the nature of God's omnipotence and the intricate design of the universe, weaving together the fabric of our beliefs and understanding.

5.8. Is death an end in itself?

The question of whether death marks the end of our journey has intrigued humanity since the dawn of modern civilization. This enduring debate is evident in the ancient engravings found in various caves and archeological sites. While some pioneering thinkers like G. **Leibniz** (ref.181, p.53) have explored the concept of pre-established harmony, the voices of recognized scientists have largely been absent. Let's, for now, turn to the realm of philosophers and theologians, with a promise to revisit G. Leibniz later in our exploration of diverse beliefs in the afterlife.

Across different city-states and historical periods, **Hellenists'** beliefs about the afterlife have undergone significant evolution, shaped by philosophical and religious developments. A common thread emerges: souls, after death, are ushered into the Underworld, where they face judgment based on their earthly actions and deeds by Hades, the enigmatic ruler of this dark realm. To embark on this journey, souls must cross over the fabled River Styx, and the proper burial rites ensure their safe passage. Within the Underworld, distinct realms await:

- Elyseum, a realm of peace and joy reserved for heroes and those favored by the gods.

- The Fields of Asphodel where average souls reside, experiencing neither great suffering nor immense pleasure.

- Tartarus, the deepest and darkest part of the Underground, reserved for the most wicked souls.

Plato's Philosophical Insights. Among the variations, Plato (ref.54, p.15) stands out for his adaptation of Orphic beliefs[786] into a comprehensive theory of

---

[786] The cult of Orphism *was that of a mysterious Hellenistic sect founded by Orpheus, an ancient Greek legendary hero, son of Zeus. Its god was Dionysos also named Bacchus, son of Zeus. Its followers believed in reincarnation and original sin.*

the soul's destiny after death. Plato posited that the soul was immortal, originating from a divine source. Yet, he also proposed in Timeus that:

"Individual souls comprised residue of the universal soul substance, albeit of an inferior grade."

According to Plato, the soul could undergo reincarnation in different bodies, influenced by its earthly actions and virtues. He believed that through philosophical initiation and reminiscence, the soul could access knowledge of its origin and destiny.

Eternal Life in Harmony. For **Hinduism**, those who have detached themselves from material pursuits and devoted their lives to the Supreme Soul, God (Sri Krsna), will, upon death, merge into eternal life in perfect harmony with Krishna.

Common Threads in Abrahamic Religions. The three Abrahamic religions: Judaism, Christianity, and Islam possess unique beliefs regarding the afterlife while sharing some common tenets:

-   The soul separates from the physical body at the moment of death.

-   Judgment Day: God evaluates each soul based on deeds and faith during earthly life. Rewards await the righteous, while punishment awaits those who have rejected God and committed evil acts. Some theologians explore concepts like *Teshuvah* (repentance) in Judaism, *Purgatory* in Christianity, and the *Barzakh*, an intermediate state between death and judgment in Islam.

-   Resurrection. The body will be reconstituted and reunited with the soul.

As we circle back to our central theme, let's shift our focus to the perspectives of scientists. Blaise **Pascal**, a renowned scientist (ref.170, p.48), possessed a brilliant and inquisitive mind that constantly pondered the question:

*"What is man, this mortal, that you are mindful of him,
The son of Adam that you care for him?"* [PS 8:5]

In contemplating *"What is man in nature?"* Pascal answered, *"Nothing concerning the infinite, yet everything concerning nature."* According to him, we only comprehend life and death through Jesus Christ, who, by being crucified, buried, and resurrected, provides hope for an eternal spiritual life.

The distinguished *Lutheran* scientist, Gottfried **Leibniz,** the founder of the infinitesimal Calculus, and a rationalist philosopher, postulated that the soul is a monad (ref.182, p.53, and pp.55-56), a simple, indivisible substance characterized by perception and appetite. Each monad possesses a unique identity and nature, striving to increase its perception and happiness. Pre-established by God, monads play a vital role in the divine plan, harmonizing with their bodies and the universe, and collaborating with other monads to create the best of all possible worlds.

Leibniz envisioned the postmortem life of souls involving a metaphysical existence, with changing levels of perception and appetite, interacting with the world and other souls only through God's wisdom and goodness. Rewards and punishments are meted out based on moral actions, with varying levels of knowledge and bliss attainable.

Max **Planck** (refs.255&782, pp.78&303), a devout *Lutheran* and 1918 Nobel Prize winner in Physics, one of the Fathers of Quantum Mechanics, not only stood as one of the greatest scientists of our era, praised by A. Einstein, and L. de Broglie,... but also deeply permeated by a profound religious attitude [Pl 2 & Pl3]. For him, our soul, a part of God's Mind, being not bound by natural laws, is not subject to death but to judgment, and it is eternal. This perspective is not only found in some of his writings but also his behavior, especially in the face of his son's execution by the Gestapo. In a letter to his friend Anton Kippenberg, he expressed that his son's death had filled him with an inextinguishable grief, but he remained confident because he had always believed in a benevolent God with a purpose for his son in Heaven. This later is not a place but a state, a gift... Arno **Penzias** (ref.364, p.108), an orthodox *Jew*, shared a similar view to **Planck**'s. He expressed it during an interview with Rabbi J. O. Haberman [Hb]. Drawing a parallel to Mordecai's words of encouragement to his niece Ester, whom he had raised as his daughter, following her parents' passing (Esther 4:14), Penzias reaffirmed the idea that God has assigned each of us a unique place and purpose in this vast universe. Regardless of our religious affiliations, there's no reason to assume that everything comes to an end upon our earthly departure. Our actions, however inconspicuous they may seem, will leave an impact. In the worst-case scenario, our children may carry the torch forward.

I cannot conclude this exploration of death without reminding you that, long related to the realms of science fiction, the concept of a Multiverse or Multiple Universes has captured (cf. chapter IV.2.) the imagination of cosmologists since Hugh **Everett's** groundbreaking work in 1967 (ref.575, p.184). Pioneering minds such as Andrei **Linde** (ref.561, p.179), and Andreas **Albrecht** (ref.562 p.179), ... have since embraced this notion. The idea of a parallel universe, impervious to electromagnetic waves yet receptive to gravitational waves, as elucidated by the Ligo discovery in 2016 (chapter II.10), has gained traction, existing near perhaps mere millimeters away, and intertwined with our own. However, envisioning such possibilities wasn't the norm when Dominican Scholar Father Antonin-Dalmace **Sertillanges**,[787] who passed away in 1948, contemplated the mysteries of death. I would go so far as to characterize his ideas as visionary:

---

[787] **Sertillanges** Antonin-Dalmace *(FRA: Clermond-Ferrand 1863-1948 Sallanches). Born Antonin Gilbert, he was both a moral philosopher and a Dominican priest. His scholarly pursuits were dedicated to exploring the moral theory of St Thomas Aquinas. He founded the* "Revue Thomiste" *and assumed the role of professor of moral philosophy at the Institut Catholique de Paris. Additionally, he authored numerous philosophical essays.*

"Through death, the family is not destroyed; it transforms: a part of it goes into the invisible. We believe that death is an absence when it is a discreet presence. We believe it creates an infinite distance, when it eliminates all distance, bringing back to the spirit what was localized in the flesh... to die is to reunite, it's a paradox to assert that.....".

My perspective is bolded and further strengthened by reflections from Great Scientists and Nobel laureates. As a Roman *Catholic* who believes in the resurrection, I am inclined to trust that we are being tested, for we have a role to fulfill even after our earthly life has ended. Death is a doorway to eternal life. In the *Christian* faith, marked by the anticipation of the resurrection of the dead, earthly life's purpose is to earn the promised eternal life by following God's commandments, as stated by Christ. This perspective also applies to *Muslims*, for whom death unveils the invisible world that awaits. Although this idea remains unverifiable, some individuals who have neared death have reported similar experiences. As mentioned in the Qur'an (v.19 and 22, Surah 50):

"The agony of death brings forth the truth: this is what you were turning away from. You were indifferent to it. Well, We remove your veil; your sight is clear today."

Conclusion

"This conviction, linked to a deep feeling of a superior reason, revealing itself in the world of experience, translates for me the idea of God...I affirm that the religious feeling cosmic is the most powerful and noble motive of the scientific research."

A. Einstein [Ei]

"Religion has a most important role in pointing out that there is more to life than selfish materialism"

A. Hewish (ref. 371)

I have chosen the following epigraph for this essay from Gottfried **Leibniz** (ref.182, p.53): *"Why is there something rather than nothing?."* This question, posed by[788] a prominent religious scientist of Isaac **Newton**'s stature (ref.185, p.56), is a poignant inquiry addressed to atheists who adhere to materialism. He was not asking about a cause like all other causes, but asking about the root of all possible causes. **Leibniz** endeavored to provide a more profound answer to this question through his concept of *"pre-established harmony"* and the associated monads [Lei], which draw parallels with certain *Hinduism* theories. His unwavering belief in the existence of God is evident in his prolific writings. While I may not entirely align with his perspective, his *"metaphysical image of the world"* is a compelling subject for exploration.

It's worth noting that A. **Einstein** (ref.268, p.81) held a similar scientific viewpoint, as evident in his statement (cf. [Ei1], p.20):

"His (The scientist's) religiosity consists of astonishment and ecstasy at the harmony of the laws of nature, revealing an intelligence so superior that all human thought and ingenuity can only reveal their derisory nothingness in the face of it."

Over four centuries, immense progress has been achieved in science, leading to a deeper understanding of the universe's structure. While some *"mysteries"* have been unveiled, others have arisen, prompting Albert **Einstein**'s observation:

"What is incomprehensible is that the universe is comprehensible."

---

[788] *Others, but less known than G. **Leibniz**, have asked it.*

These discoveries have painted a picture of a universe more intricate than that envisioned by philosophers and exegetes who lived before the advent of modern science. Today, we comprehend that quarks and leptons, governed by the four fundamental forces that may have been unified at the moment of the Big Bang, constitute the universe's fundamental building blocks. Their properties are determined solely by "*initial conditions*" and approximately fifteen "*fundamental constants.*" These constants, not predicted by any theory, hold the fate of our universe in their delicate balance. As G. **Leibniz** aptly noted [Lz2]:

> "*...these laws* (in the discovery of which he participated as he points out) *do not rely on the principle of necessity but on the principle of wisdom's choice. This stands as one of the most compelling and nuanced arguments for the existence of God, particularly for those who delve into these matters.*"

Two centuries later, Max **Planck** (refs.255&782, Pp.78&302), the 1918 **Nobel** Prize in Physics, universally recognized as the father of Modern Physics, a devout *Protestant*, will state a similar viewpoint:

> "*The existence of these constants is a palpable proof of the existence in nature of something real and independent of every human measurement*"

Hence, there is no room for the concept of "*pure chance,*" as we have sought to elucidate. Paradoxically, this strengthens rather than undermines the position of theists, in contrast to the misconceptions of the 17th century during the dawn of experimental science. However, my intention was not to prove but to support the notion, that our faith in God, *the Creator*, should not be solely rooted in scientific discoveries. Regardless of one's belief system, we have observed that all the necessary conditions for life to exist have been meticulously put in place. Therefore, for believers, the answer to **Leibniz**'s question is succinct: had the universe not been created as God intended, there would be no fertile ground for humanity to flourish. This perspective aligns with the strong anthropic principle. However, my aim was not to scientifically establish the existence or nonexistence of God, as I mentioned in the introduction. Instead, I aimed to demonstrate the following points:

- Belief in the existence of God is not naive.

- Moderate scientific rationality can coexist with religion; as agnostics have testified.

- On the other hand, some materialists exhibit an overly rigid rationalism, that takes on religious characteristics, a phenomenon more prevalent in France than in the USA, for example,

- *Atheist negationists* lack valid arguments. John **Lennox**,[789] Professor Emeritus of Mathematics at Oxford University, ENG, and an *apologist for Christianity* [Lx], recently expounded on why atheists lack compelling arguments. It was during a conference in Haifa, ISR, attended by Christian students. Lennox, a seasoned defender of his faith in Christ since his days at Cambridge University, is well-known for his televised debates with prominent atheist scientists, including the eminent atheist ethologist Richard **Dawkins** (ref.698, p.247).

To advance toward my goals, I have engaged with some of the most prominent scientists of our era, recognizing that science is the cornerstone of materialist thinking. While many of us may be unaware of it, scientists themselves are well aware of the controversies within the field. Therefore, there's no inherent reason for these controversies to vanish when they delve into matters of faith. Interestingly, a significant number of Nobel Prize laureates,[790] a prestigious accolade established only in 1901, profess a belief in the existence of a divine being. In my explorations, I have encountered individuals from various faith traditions, including *Judaism, Christianity, Islam, Hinduism, Confucianism*, and beyond. Additionally, I have conversed with proponents of deism, some of whom grapple with the challenges of aligning with organized religions, which is understandable given their unique perspectives.

Also, our objective has been to objectively illustrate the evolution of thoughts:

- the "*ancient worldview*" held that the world was eternal, shaped by gods from a primordial, timeless substance. These gods, embodying Nature's faces and forces, were products of the primitive Universe.

- Subsequently, with "*Moses*," the *Hebrews*, guided by the Pentateuch, embraced a creative perspective on the world, centered around Yahweh-God, who fulfilled their spiritual needs. They honored Him with the title "*God of Israel*," a name passed down from Jacob, the third Patriarch following his father **Isaac**, and his grandfather **Abraham**.

---

[789] **Lennox** John *[Northern Ireland, GB 1943). Master of Arts and Ph.D. in Mathematics from Cambridge and Oxford Universities, Ph.D. in Science from Cardiff University, taught Mathematics and Science & Religion at Oxford. Professor Emeritus of Mathematics at Oxford, Philosopher of Science, and recipient of numerous scientific awards has given numerous conferences in America and Europe. Author of more than 70 scientific articles, and 2 monographs in Mathematics. Pastoral Advisor to Green Templeton College, Oxford. Apologist for Christianity.*

[790] *During a conference in Haifa (in front of Christian students), John* **Lennox** *declared that from 1901, the year of the creation of the Nobel Prize, to 2000: 70% of the Nobel Prize winners in Physics believe in the existence of God, which is higher than the percentage of individuals that one can interview at random in the street.*

- In "*Hellenic antiquity*," despite the prevalence of polytheism, scientific rationalism emerged alongside attempts at mechanistic interpretations, with a focus on finality. **Plato** and **Aristotle** stood as undisputed Masters of thought of this era. It's worth noting that for Aristotle and his disciples, a first cause (a god) was deemed necessary to "*shape*" the limited world in which they resided. This rationalism would leave a profound imprint on Western thought.

- The "*Middle Ages*" witnessed the Hellenization of the Arab-Islamic world, often referred to as the "*Islamic Golden Age*." The Arabic language served as a conduit for Aristotelian ideas, enriched by the thinkers of the Arab-Muslim empire, some of whom made significant contributions to the advancement of science, particularly in fields like medicine and optics. Until the end of the 16th century, the domain of science remained intertwined with theology.

- The **17th century**, marked a significant shift in the relationship between science and religion. *Catholic* scientists like René **Descartes** [De] (ref.152, p.43), Galileo **Galilei** (ref.144 to 146, p.41), and Blaise **Pascal** (ref.170, p.48), paved the way for a confrontation between mathematical models and empirical evidence. Philosophers with a keen interest in science were also vocal during this period. **Descartes'** assertion that: "*our senses do not reveal the true nature of things but only their utility or harm to us*" took on profound significance. "*Materialist scientists*" hoped to undermine religious beliefs, yet, as *Christian* cosmologist Stephen **Barr**,[791] **2006** Laureate of the prestigious **Templeton Prize**, observed this was not the case. Even "*The great Mr. Newton*" (refs.185&204, p.59&61), an *Anglican,* one of the most prominent scientists and *theologians* of our era, held a theistic view of God as the creator of the universe. His support for Laplacian determinism, albeit unintentionally, gave rise to confusion among some minds sharing the idea that God has given man Free Will. Similarly, Charles **Darwin**'s theory in **1850** (ref.232, p.70), though misinterpreted by literalists in

---

[791] **Barr** *Stephen (N-Y, USA 1953). Ph.D. in Particle Physics in 1978 at Princeton US-NJ, he held various research positions in the Universities of Pennsylvania, Washington, and Brookhaven National Laboratory before joining in 1987 the University of Delaware where he was appointed 2011 Director of the Bartol Research Institute and is Professor Emeritus. A recognized cosmologist, his name is associated with a diagram, the solution to a problem in particle physics, ...*
Catholic apologist, *he participates in many programs and conferences on* "Science & Religion." *Author of numerous scientific articles as well as popularization articles. Member of the Academy of Catholic Theology, he received in 2006 the prestigious* **Templeton** *Prize, in 2007 the Benemerenti medal from the hands of Pope* **Benedict XVI**...

their reading of the Scriptures, found favor among theistic evolutionists theists and sparked passionate debates.

- In the vibrant landscape of the **18th century Enlightenment**, a period teeming with intellectual ferment, it proves challenging to identify a stalwart defender of religion against the prevailing tide of freethinkers. However, in the illustrious figure of L. **Euler** (ref.210, p.63), hailed as the Master of mathematicians in his era, and revered as one of the greatest mathematicians of all time, we find a compelling voice. Euler, whose prolific writings spanned philosophy and theology, held a conviction that science and religion could not only coexist but harmonize in a complementary dance. In his seminal work *"Defense of the Divine Revelation against the Objections of the Freethinkers"* Euler articulates a perspective that transcends the simplistic notions of some of his contemporaries. He challenges the prevailing ideas:

"It is not sufficient to say with the blind philosophers, that God is the soul of the world, that he is the universal spirit, that he is the internal mover of everything. We must go further, and conceive him as independent of all material beings, as the source of all existence, and as the original cause of all that is beautiful, good, and perfect in the universe."

- In the mosaic of scientific progress realized during the **19th century**, **Pasteur**'s contributions (ref.240, p.73) shine as a beacon of innovation and compassion, transcending the boundaries of his era and leaving an enduring impact on the trajectory of human advancement. He aptly stated:[792]

"Little science estranges men from God, but much science leads them back to Him."

- In the **20th century**, the quest for answers regarding the origin of the Universe, underwent a profound transformation. No longer confined to theological explanations, this pursuit became the realm of a select group of visionary scientists proficient in the intricate realms of mathematics, navigating the uncharted waters of a four-dimensional, non-Euclidean space.

A pivotal moment in this intellectual journey was the revelation of electromagnetic waves, an unprecedented revelation that unfolded in the wake of James C. **Maxwell**'s groundbreaking discoveries (ref.234, p.71).

---

[792] *Extract from the book of René Vallery-Radot, Pasteur's son-in-law* (Life of Pasteur, 1902).

Maxwell, *a devout Presbyterian,* brought to light phenomena previously unimaginable. His revelations set the stage for the introduction in 1905 of **Einstein**'s transcendent general theory of relativity (ref.268, p.81); a Spinozian *Pantheist* whose paradigm-shifting work catapulted theoretical physics into a new epoch. Einstein's genius was underscored by the remarkable fact that many of his predictions found empirical confirmation through the ingenious experiments of his contemporaries.

As the torchbearers of science advanced so did the tapestry of knowledge, particularly in physics, astronomy, and biology. Yet, with each revelation, new enigmas surfaced. The advent of Quantum Mechanics, exemplified by W. **Heisenberg**'s principle of indeterminacy (ref.316, p.93), ushered in an era where the very fabric of reality was shown to possess inherent uncertainty. For example, the accurate trajectory of an isolated electron eluded definitive determination, while the collective behavior of electrons revealed a statistical dance of probabilities. Here we confront the limits of science and reason, hinting at a human indeterminism rooted in the boundaries of our logic and reason.

In contrast, the creation of the universe by a divine hand remains firmly deterministic, shrouded in the unknowable mystery of God. Naturally, we possess limited insights into the Divine, for neither God nor the realm of spirituality finds a place within the structured models of cosmologists. Nevertheless, the march of physics has illuminated the boundaries of human comprehension. A good example is our profound ignorance of dark matter and dark energy these elusive elements that make up a large portion of the cosmos. These mysterious entities defy categorization as atoms or molecules, shunning our scrutiny. Yet their undeniable influence upon gravity serves as a haunting reminder of the limits of human knowledge.

- The **21st century** has unequivocally underscored the chasm between the realm of science and the loftier world[793] **Plato** once alluded to. This profound distinction crystallized with the momentous discovery of gravitational waves in 2015 (cf. chap.2, p.135) and, subsequently, the unveiling of the long-anticipated image of a Black Hole (Sagittarius A*) at the heart of our galaxy in 2019. These breakthroughs birthed new mathematical models anchored in the Theory of General Relativity, delving into n-dimensional spaces, where n can extend to 12, and even more. The embrace of these cutting-edge theories concerning the universe's origins necessitates from non-specialists in cosmology, astronomy, and theoretical physics, a *faith in science and scientists,* perhaps akin to the devotion of the faithful who find solace in scriptures,

---

[793] In "The Banquet" ***Plato*** *imagines a dialogue with Antisthenes during which he evokes the visible world, an imperfect copy of the superior world that we do not see. To Antisthenes saying:* "I can see the horse but I cannot see the kaballik*" ***Plato*** *answers:* "Precisely to see the horse you need eyes and you have them, to see the kaballik you need intelligence and you don't have it." * *Kaballik = Caballéité: concept of horse.*

apostolic testimonies, and the sacrifices of saints in nurturing their faith in God.

Our journey through the millennia has shifted us from ancient cosmogony, where the wisdom of the ancients reigned supreme, to cosmology, an intricately specialized bastion of the sciences that has surged forward since the dawn of the 20th century. As our knowledge expands, rendering the universe more comprehensible to a select few, it also unfurls a cascade of new questions, potentially challenging previously established doctrines. Yet this complexity underscores the intricacies of celestial mechanics. Even the foremost cosmologists cannot definitively unveil the infinitesimal interval preceding Planck's time. This is the enigma that both believers and skeptics ponder. Some believe that the elusive Unified Theory (TOE) may liberate them from this initial singularity, enabling a scientific grasp of the universe's inception, but doubts persist.

In the face of these cosmic enigmas, it becomes evident that the universe occasionally withholds its secrets, waiting to be unlocked by the divine wisdom of a higher power. While physicists grapple with the unknown, we are reminded that science and reason have their bounds, and the cosmos' awe-inspiring complexity perpetually inspires humility and wonder in the hearts of those who seek to unravel its deepest mysteries. Although our discourse has centered on the Cosmos, delving into the realm of Life would have been equally intriguing. Therefore I defer to Francisco J. **Alaya** (ref.682, p.243) an eminent evolutionary biologist of global renown, a former *Benedictine Catholic*, and an eager defender of the theory of evolution. He aptly noted:

"Science and religion address different facets of reality, and one who believes that everything can be explained through a simplistic materialistic lens may be deemed naive."

This naivety echoes the sentiment of 1933 **Nobel** Prize-winning *Christian* physicist E. **Schrödinger** (ref.309, p.91), a philosopher, who wrote in *"Nature and the Greeks"* (1954):

"In short, I have not the slightest hesitation in declaring that to accept an existing material world as an explanation of the fact that we are all finally empirically in the same environment is metaphysical and mystical. It is convenient but a bit naive. By adopting this hypothesis, we lose a lot of things. But in any case, no one has the right to denounce other points of view as mystical and metaphysical, imagining that his own is devoid of these weaknesses."

In chapter 4 (*"The Arithmetical Paradox: The Oneness of Mind"*), Schrödinger, expresses, particularly his monistic view of reality, inspired by the Vedantic philosophy of India, according to which consciousness is singular and plurality is only in appearance (cf. chap.3, p.167). He also criticizes the materialistic conception of the world, which he considers naive and metaphysical, imagining that his own is empty of these weaknesses. He emphasizes the fact that the attempt to entirely eradicate theoretical Metaphysics, as Emmanuel **Kant** said in *"Critique and Judgment"* (1790),[794] would render:

> *"... the art and the science lifeless, deprived of soul, and incapable of the least ulterior development."*

Having showcased that numerous esteemed scientists personally embrace the concept of a supreme architect, whom they call God, and even lean towards theism,[795] it becomes evident that belief in God is not antithetical to contemporary scientific thought. This hope is that these pages provide the elements needed to confront skepticism and, with unwavering faith, whether any potential mockery from atheists or proponents or probabilistic theories that, if true, might fail to account for the creation of a masterpiece like Mozart's work.

To resume, exploration throughout the centuries has confirmed that it is impossible to scientifically disprove the existence of God, the Creator of *"Heaven and Earth,"* as affirmed by the three Abrahamic religions and others. Perhaps this is because, as Mahatma **Gandhi**[796] posited, religious sentiment resides within us, consciously or unconsciously, transcending the realm of the intellect. [Col2 p. 265].

For skeptics, I turn to Lord William **Kelvin** (ref.227, p.68), the progenitor of Thermodynamics, and a devout *Christian*, who asserted:[797]

---

[794] *In "Critique of Judgment" (1970), **Kant** discusses the difference between fine and mechanical art. He argues that fine art is not based on rules or concepts but on the artist's genius, who produces original and expressive works that inspire others. On the other hand, mechanical art is based on imitation and skill, lacking soul and spirit.*

[795] *Scientists are generally satisfied to report on the state of their research, leaving everyone free to interpret the facts in one way or another. I have traced the religions of some scientists; nevertheless, there may be some among those I have not given a label who are followers of a religion. The Freemasons, for example, when adhering to the REAA (the majority) although theists (condition N°1 to be accepted) must leave their political and religious opinions at the door of their meetings.*

[796] **Mahatma Gandhi** *(IND: Porbandar, Guraja 1869-1948 Delhi). The Indians consider him* "the father of India," *hence the name Mahatma (great soul) which he always refused to associate with his name. He was very much influenced by Indian philosophy and was against the caste system. After studying at the University College London, where he joined the vegetarian society, he became a lawyer. During that time he became interested in religions, which led him to embody the doctrine of non-violence that would guide his actions throughout his life. After a stay in South Africa from 1893 to 1915, he returned to India, founded the Sabarmati ashram*, and entered politics, being appointed head of the Congress party in 1921. He devoted himself to the struggle for the liberation of India from English colonization by boycotting foreign products and advocating non-violence.*

* Ashram*: hermitage, generally an isolated place in a forest or a mountain, where followers gather around a spiritual master.*

[797] *Wikipedia:* "Lord Kelvin, aphorisms and phrases in English."

"Do not be afraid to be free thinkers! If you think hard enough you will be forced by science to believe in God, the founder of all religions. You will find that science is not antagonistic to religion but an auxiliary to it."

"The deeper I delve into my scientific inquiries, the more I am inclined to believe that science and atheism are incompatible."

"... Atheism is an idea so insane that I cannot put it into words."

As expressed by the *Anglican* poet and painter Dante G. **Rossetti:** [798]

"The atheist's most challenging moment arises when gratitude swells within, and there is no one to thank."

It suggests, that believers while contemplating the beauty and perfection of nature, have cause to thank God for His divine craftsmanship.

I will let the word to Michal **Heller** (ref.700, p.248), a prominent contemporary cosmologist and *Catholic priest*, laureate of the prestigious 2008 **Templeton** Prize, a great admirer of **Pythagoras** (ref.55, p.16) and **Leibniz** (ref.182, p.53) said:

"Does the universe need to have cause? Causal explanations are a vital part of the scientific method. Various processes in the universe can be displayed as a succession of states, in such a way that the preceding state is a cause of the succeeding one. If we look deeper at such processes, we see that there is always a dynamic law prescribing how one state should generate another state. But dynamical laws are expressed in the form of mathematical equations, and if we ask about the cause of the universe we should ask about the cause of mathematical laws. By doing so we are back in the blueprint of God's thinking of the universe. The question on ultimate causality is translated into Leibniz's question: 'Why is there something rather than nothing?' When asking this question, we are not asking about a cause like all other causes. We are asking about the root of all possible causes,"

---

[798] **Rossetti** Dante Gabriel *(ENG: London 1828–1882 Birchington-on-Sea)*. Anglican *painter, poet, illustrator, and translator (English <-> Italian). He attended King's College School and founded with the* religious Anglican *painter* William H. **Hunt** *(London, ENG: 1827–1910) and the* pious Anglican *painter* Sir John E. **Millais** *(ENG: Southampton 1829-1896 London) et al. the Pre-Raphaelite Brotherhood (the group of 7 members who intended to revitalize art by highlighting details of nature. He was a reader of the Bible, W. Shakespeare, ..* J. E. Millais *was the youngest (18 years) student to be admitted to the Royal Academy of Arts, granted in 1885 the title of Baronet by Prime Minister* W. E **Gladstone***, and elected in 1896 President of the Royal Academy of Arts.*

and, to Dr. Frank **Wilczek** (ref.381, p.113), *Pantheist,* Laureate of the **Nobel** Prize in Physics in **2004**. In **2022**, the Templeton Prize committee recognized his exceptional use of mathematical tools to explore the profound questions that have intrigued humanity for centuries, including our place in the universe. This well deserved accolade affirmed his ability to peer into the mysteries of existence using equations, diagrams, calculations, and intellectual concepts, affirming the intrinsic link between the pursuit of understanding the cosmos and a deeper understanding of the divine. In accepting the **Templeton** Prize, Wilczek eloquently expressed his belief, saying:

> *"The central miracle of physics to me is that by playing with equations, drawing diagrams, doing calculations, and working within the world of mental concepts and manipulations, you are describing the real world. If you were looking to try to understand what God is by understanding God's work, that's it."*

This remarkable scientist continues to blaze new trails in physics, introducing groundbreaking concepts and theories that unveil hidden dimensions of reality, much in the spirit of Sir J. Templeton. With his unique blend of space and time, advanced mathematics, and profound philosophical ideas, Wilczek's work invites readers to witness a universe of awe-inspiring beauty and intellectual wonder. In doing so, he exemplifies the profound interconnectedness of science and spirituality offering a beacon of inspiration for all who seek to explore the cosmos and the divine in tandem.

Bibliography

[Ag1] St Augustin *"Les Confessions"* Essai Poche Flammarion 1964.

[Ag2] St Augustin *"La Cité de Dieu"* 3 tomes Essai 1994 ;

[Al] Alia David *"De l'infiniment petit à l'infiniment grand,"* **interview** (6/02/2004) voir Wikipédia *et dans Internet Lamed.fr du 3/11/2017.*

[Amp] An Daniel et al. *"Apparent evidence for Hawking points in CMB Sky"* archiv: 1808.01740v4 [Astroph.CO] March 2, 2020.

[An] Atlan Henri dans *"Le monde s'est-il créé tout seul ?"* *Entretiens avec Patrice Van Eersel et la collaboration de Sylvain Michelet* de Thuan T. X., Prigogine I., Jacquard A., de Rosnay J., Pelt J-M. & Atlan H. Albin Michel, Paris 2008.

[Ar] Alexander Denis *"Is Evolution a Chance Process?"* [Al] Scientia et Fides [online] October 30, 2020, T.8, nr 2, s. 15-41.

[Ar2] Alexander Denis *"Is There Purpose in Biology? The Coast of Existence and the God of Love."* Lion Hudson Ltd, Oxford, 2018, pp. 288; ISBN: 978-0857217141.

[At] Atlas Experiment *"Atlas illuminates the Higgs boson at 13 TeV "* CERNCOURIER Vol. 18, N°3, Av. 2018, p. 13.

[Au] Augier Henry *"Dieu à l'épreuve des sciences"* Ellébore, Paris 2013.

[Ba] Baillon Jean-François *"Isaac Newton - écrits sur la religion"* Gallimard, Paris 1996.

[Bar] Barr Stephen *"Modern Physics and Ancient Faith."* University of Notre Dame Press. Flanner Hall In, 2003.

[Be] Berkeley – *Cours de Physique – Mécanique Vol. 1.* by C. Kittle, W. D. Knight & M. A. Ruderman Armand Colin Paris 1972.

[Ben] Benoit XVI *"Lumière du monde,"* Bayard – Paris - 2010.

[Bh] Barish Barry C. *"An ultimate goal is the use of gravitational waves to study the Big Bang itself."* CERNCOURIER Vol. 57, N°10, Dec... 2017, pp. 20 - 21.

[Bi] Jerusalem Bible School *"La Sainte Bible"* translated in French under the direction of the Jesuralem Bible school - Edition Du Cerf, Paris, 1961.

[Bl] Blattchen Edmond *"Enfants de la Terre"* Ed. Alice–2016 collection: L'intégrale des entretiens d'Ed. Blattchen *(BEL: Ougrée, Liege)..*

[Bo] Boslough John *"Beyond the black hole - Stephen Hawking's Universe"* Fontana/Collins, Glasgow 1990.

[BP] Bowler Peter J. *"Reconciling science and religion: The Debate in Early-Twentieth-Century Britain"* Chicago University Press 2014.

[Br] Bridoux André *"Descartes Œuvres et Lettres."* La Pléiade, Gallimard, Paris 1953.

[Bro] de Broglie Louis *"Mécanique ondulatoire du photon et Théorie quantique des champs."* Gauthier-Villars, Paris – 1952.

[Bt] Babut Daniel *"la religion des philosophes grecs."* PUF 1974.

[Bw] Barrow John D. et al. *"The anthropic cosmologic Principe"* - Oxford University Press 1985

[By] Boyle Robert *"The Sceptical Chymist"* Eds. J.M. Dent & Sons London, Dutton & Co New-York 1661. Online on *Internet Archive.*

[Ca] Casadebaig Philippe *"Aristote ou le langage de l'être* "tiré de" *Les Philosophes de Platon à Sartre."* Articles rassemblés par Grateloup Philippe. Hachette, Paris-1985.

[Cl] Campbell John *"Rutherford, Transmutation and the Proton"* CERNCOURIER 2019 Vol. 59 N°3. Pp. 27-30

[CC1910] CERNCOURIER *"New gravitational-wave events"* Ed. Matthew Chalmers, Jan-Feb. 2019-p.7.

[CC1911] CERNCOURIER *"Colliders join the hunt for dark energy"* Ed. Matthew Chalmers, Jan-Feb. 2019p.12.

[CD] Chandler Daniel & Dixon Edward *"The Inaugural Distinguished Christian Scholar Lecture: An Interview with 2014 lecturer Dr. F. P. Brooks, Jr,"* christianityandscholarship.org.

[Ce] Cercignani C. *"Ludwig Boltzmann: The Man Who Trusted Atoms"* Oxford 2006. And J. of Physics Vol. 68, Pp 90-91.

[Co] Collognat Annie *"25 métamorphoses d'Ovide"* Librairie Générale Française 2007.

[Col] Collins Francis *"The Language of God - A Scientist Presents Evidence for Belief."* Ed. Free Press - NY, London, Toronto, Sydney, 2006.

[Col2] Collins Francis *"Belief: Readings on the Reason for Faith."* Ed. HarperOne San Francisco – 2010.

[Com] Combes Françoise *"La Matière Noire dans l'Univers"* Discours inaugural lors de son accession à la chaire *"Galaxies et Cosmologie"* au Collège de France. December 18, 2014.

[Cs] Cairns-Smith Graham *"l'énigme de la vie – Une enquête scientifique"* Odile Jacob Paris 1985.

[Da] Dahl Fasmer *"Histoire merveilleuse d'Albert Schweitzer"* Bibliothèque Rouge et Or *"Souveraine "* Ed. G.P. Paris – 1956.

[De] Debré Bernard *"Des savants et des dieux "* - Cherche-Midi, Paris 2013.

[De & Sa] Décote Georges et Sabbah Hélène, *"Itinéraires Littéraires - Le XVIIIème siècle"* Hatier, Pa

[Dn] Darwin Charles *"The Origin of Species"* Introduced and abridged by Philip Appleman, W.W. Norton & Company N-Y, London - 1975.

[Dn2] Darwin Charles *"Darwin's Gift to Science and Religion."* Joseph Henry Press, Washington, D.C

[Do] Dobjansky Theodosius, *in The American Biology Teacher* vol. 31 03/1973.

[DP] Deterre Philippe et Ploux Jean-Marie *"Un Dieu Créateur."* Ed. Salvator, Paris 2020.

[Du] Duhem Pierre *"La théorie physique. Son objet et sa structure."* M. Rivière & Cie. – Paris 1914.

[Dy] Dyson Freeman, *"Talk: Freeman Dyson-Wikiquote,"* in "wikiquote.org" according to Wikipedia biography, or *"Disturbing the Universe"* Basic Books, N-Y 1979.

[Ec] Eccles John, sir *"Comment la conscience contrôle le cerveau"* Fayard 1997 ou *"How the self controls the brain»* 1994. read pp. 153, 172 – 215 & 217. As well as in les Cahiers des Sciences du Figaro Magazine" *Le cerveau peut-il comprendre le cerveau"* 1982.

[Ec2] Eccles John, sir *"Le Miracle de l'Existence Humaine"* 3ème millénaire – Evreux, Mars 1989 N°12 Pp. 26-34.

[Ed] Eddington Arthur (sir) *"The Nature of the Physical World."* Cambridge University Press, Cambridge, 1928.

[Ei1] Einstein Albert *"Comment je vois le monde"* - Flammarion, Paris 1979.

[Ei2] Einstein Albert & Infeld Leopold *"L'évolution des idées en physique des premiers concepts aux théories de la relativité et des quantas"* Flammarion, Paris 1983.

[Ei3] Einstein Albert *"Conceptions Scientifiques, Morales et Sociales"* Flammarion, Paris 1952.

[EW] Josy Eisenberg et Elie Wiesel *"Job ou Dieu dans la tempête"* Fayard/Verdier, Paris 1986.

[Fa] Fannon Dominic *"Psychiatric Bulletin ,"* vol 30, N° 12 -2006.

[Fk] Falk Darrelle R. *"Coming to Peace with Religion: Bridging of the worlds between Faith and Biology"* InterVarsity Press, Downers Grove, IL.

[Fl] Flew Anthony *"There is a God: How the World's Most Notorious Atheist Changed His Mind"* HarperOne, N-Y, 2008.

[Fz] Fontez Mathilde *"Planètes océan,"* Sciences & Vie, Nov. 2018.

[Fo] Fourier Joseph *"Mémoires de l'Académie des Sciences de l'Institut de France."* Gauthier-Villars, T.6 pp. LXI-LXXXI, Paris 1823.

[Fr] Françis (Pope) Encyclic *"Laudato Si.'"* Parole et Silence 2015.

[Fr2] Françis (Pope) Apostolic Letter "Sublimitas Et Miseria Hominis," 2023 The Holy See https://www.vatican.va

[Gc] Gillen Alan L. and Cargill Michael *"The Signature of God in Medicine and Microbiology. An Apologetic Argument for Declarative Design in the Discoveries of Alexander Fleming."* Liberty University – Scholars *Crossing*, March 16, 2016. Cf. Wikipedia.

[Ge] Gershon Tim *"We need to talk about the Higgs,"* CERN Courier p.5. Apr. 2018.

[Gi] Giberson Karl W. & Collins Francis S. *"The Language of Science and Faith."* *IVP Books – InterVarsity* - Press – Downers Grove, IL – 2011.

[Gk] Grothendieck Alexander *"La Clef des Songes ou Dialogue avec le Bon Dieu"* http://www.cm2vivi2002.free.fr

[Go] Goodall Jane *"Reason and Hope: a Spiritual Journey"* London: Thorsons 1999.

[Gp] *"Les Philosophes de Platon à Sartre,"* under Léon-Louis Grateloup's direction– Hachette-Paris-1985.

[Gr] Greene Brian *"La Magie du Cosmos"* Laffont, Paris 2005.

[Gs] Graves Dan *"Scientists of Faith,"* Jenson Books Inc. - North Logan, UT-1996.

[Gt] Grant Peter and Rosemary *"Evolutionary Biology and Religion: Allies or Adversaries?"* in *"The Oxford Handbook of Religion and Science"* 2008.

[Gu] Guitton Jean, Bogdanov Grichka & Igor *"Dieu et la science – vers le métaréalisme"* Grasset, Paris 1991.

[Ha1] Hawking Stephen *"Une Belle Histoire du Temps"* – Flammarion, Paris (citation p. 163) 2005.

[Ha2] Hawking Stephen *"A brief history of time"* – Bantam Books, London - 1988.

[Hal] Haldane John B.S. *"The Inequality of Man"* Pelican - Ed. A 12 – 1939 - p. 114.

[Hat] *"Itinéraires Littéraires, XVIIème siècle,"* Hatier-Paris, 1988.

[Hb] Haberman Joshua O. *"The God I Believe in: Conversations about Judaïsm with...."* The Free Press, Maxwell Macmillan int. N-Y., Toronto, 1994.

[He] Heisenberg Werner *"La Partie et le Tout. Le monde de la physique atomique."* Albin Michel, Paris, 1969.

[He2] Heisenberg Werner *"La physique et la philosophie »,* Albin Michel, Paris, 1971.

[Ho] Sir J. Houghton *"The Search for God: Can Science Help?"* Publisher Oxford: Lion 1995.

[Hs] Hess Victor *"A Physicist's Faith"* by Society of Catholic Scientists 31/03/2021, replication on Internet of the 1946 original from V. F. Hess

[Hu1] Hutchinson Ian H. *"The Faith of Great Scientists. J. C. Maxwell"* MIT IAP Seminar Jan 1998, 2006, http://silas.psfc.mit.edu/Maxwell.

[Hu2] Hutchinson Ian H. *"Can a Scientist Believe in Miracles?: An MIT Professor Answers Questions on God and Science"* IVP Books - Nottingham 2018.

[Hy] Hailey Charles J., K. Mori, et al. *"A density cusp of quiescent X-ray binaries in the central parsec of the Galaxy."* Nature 556 (7699), pp. 70-73 Ap. 2018.

[Ir1] Irwin James *"To rule the night: the discovery voyage of astronaut J. Irwin."* Holman Bible Publishers, Nashville, TN, 1973.

[Ir2] Irwin James *"More than earthlings: An Astronaut's Thoughts for Christ-Centered Living."* Broadman Press, Nashville, TN, 1983.

[Ir3] Irwin James *"More than an Ark: Spiritual lessons learned while searching for Noah's Ark."* Broadman Press, Nashville, TN, 1985.

[Is] Isaacson Walter *"Leonardo Da Vinci"* Simon & Schuster Paperbacks New-York 2018.

[Ja] Jasper Karl *"La situation spirituelle de notre époque."* Cerf, Paris 1986.

[Ja1] Jastrow Robert: *"God and the astronomers"* W.W. Norton & Company 1978 (Commentaries on Catholics and Jewishes)

[Ja2] Jastrow Robert *"On God and the Big Bang"* YouTube 26 juin 2012.

[JP] Jean-Paul II *"Foi et raison"* Encyclic Letter *Fides et ratio* 14/09/1998 - Pierre TEQUI éditeur Paris.

[Ju] Jucaud Dany *"Quand Elie Wiesel répondait à Paris Match"* Paris Match Actualités 02/07/2016

[Ka] Kant Emmanuel *"Critique de la Raison Pure ,"* translated and proposed par A. Renault. G-F Flammarion, Paris 2006.

[Kd] Krsna-Dvaipäyana Vyäsa *"Le Srimad-Bhägavatam ,"* translated by his Divine Grace A.C. Bhaktivedanta Swami Prabhupäda, Editions Bhaktivedanta – Paris, 1976.

[Ko] Kole Merlin R. *"Black holes galore at the galactic core."* CERNCOURIER Vol. 58, N° 6, p.17 - Ju/Au 2018.

[Ku] Kulkarni Viraj *"What E. Schrödinger Said About the Upanishads"* from *science.therewire.in-* September 05,2020.

[La] de Lavigne *Elisabeth, traductrice de "25 savants confessent leur foi en Dieu" Alétia June 21, 2014.*

[Lam] Lamont Ann *"James Joule – The Great Experimenter Who Was Guided by God."* Creation 15, N° 2 (March 1993) PP. 47-50.

[Le] *Lelouch Claude & Reeves Hubert "les belles histoires de la belle histoire" Eds. Plume Paris 1992.*

[Lé] Léon XIII. Encyclique *"Providentissimus deus"* 1893.

[Li] Livingstone David N. *"Darwin's Forgotten Defenders. The Encounter between Evangelical Theology and Evolutionary Thought."* Regent College Publishing. Vancouver 1984.

[Lr] Leprince-Ringuet L. *"Le Grand Merdier ou l'espoir pour demain."* Flammarion, Paris 1978.

[Lr2] Leprince-Ringuet L. *"Foi de Physicien ," Fayard-Paris 1970.*

[Ls] Lewis C.S. *"Mere Christianity"* Kindle.

[Lt] Lederman Leon & Dick Teresi *"The God particle: if the universe is the answer, What is the question?"* Dell Publishing, NY, 1993.

[Lu] Henri de Lubac: *"Drame de l'humanisme athée"* – Cerf, Paris 1990.

[Lx] Lennox John: *"God's Undertaker: Has Science Buried God?"* – Zondervan 2011.

[Lz] Leibniz *Gottfried W. "La monadologie" Nouvelle édition par A. Bertrand. Eugène Belin, Paris 1886.*

[Lz2] Leibniz *Gottfried W. "Discours de Métaphysique"* 1686, one of *"8 Œuvres de Leibniz"* (French Edition), e-artnow, Édition du Kindle.

[Ma1] Maïmonide *Moïse: "Le guide des égarés" Tome 1. G.-P. Maisonneuve & Larose, Paris 1963.*

[Ma2] Maïmonide *Moïse: "Le guide des égarés" Tome 2. G.-P. Maisonneuve & Larose, Paris 1963.*

[Mb] McBrierty Vincent J.: *"Ernest Thomas Sinton Walton, The Irish Scientist, 1903-1995," Trinity College Dublin Press, 2003.*

[Mb2] McBrierty Vincent J.: *"Connected Christianity – Reflections of a Catholic Academic"* - Association of Papal Orders in Ireland. 2016.

[Md1] Monod Théodore: *"Le Chercheur d'Absolu ," Le Cherche Midi 1997.*

[Md2] Monod Théodore: *"Terre et ciel"* – Entretiens avec S. Estibal - Actes Sud 1998.

[Md3] Monod Théodore: *"Enfants de la Terre" Collection of television interviews of* Edmond Blattchen - *Alice* – 2016.

[Md4] Monod Théodore: *"Balthus Entretiens"* – *Thome Media 2017.*

[MdC] Monod Cyrille *"Les carnets de Th. Monod"* collected by C. Monod – Le Pré aux Clercs 2000:

[MdW] Monod Wilfred *"Le problème du bien: essai de théodicitée et journal d'un pasteur"* Ed. Felix AlcanParis 1934.

[Me] Mendel Gregor J. *(1865) translated in English in 1901 by Druery C. et al.: "Experiments on Plant Hybridization"* J. *of the Royal Horticultural Society"* 26: 1-32.

[Mg] McGrath Alister *"Science & Religion: A New Introduction"* 2nd Edition Wiley-Blackwell, Oxford – 2009.

[Mg2] McGrath Alister *"Dawkins' God: Genes, Memes, and the Meaning of Life"* 2nd Edition Wiley-Blackwell, Oxford – 2015.

[Mi] Khan Mashriqui Inayatullah *"An evening with sir James Jean"* Science & Faith, Wikipedia.

[ML] Margulis Lynn, Punset Eduardo *"Mind, Life, and Universe: Conversations with Great Scientists of Our Time"* Chelsea Green Publishing, 2007.

[Mo] Morris Simon C. *"Life's Solution: Inevitable Humans in a Lonely Universe"* Cambridge University Press – 2003. Voir aussi article de Rob Hengeveld *"Conway Morris S (2003) Life' Solution..."* Acta Bibliotheca Vol. 52, N° 3 pp 221-228.

[Mot1] Sir Mott Nevill F. *"A life in science,"* Taylor and Francis, 1986 London

[Mot2] Sir Mott Nevill F. *"Can Scientists Believe? Some examples of the attitude of scientists to religion,"* James & James London et al. 1991.

[MP] Martines de Pasqually *"Traité sur le réintégration des êtres dans leur première propriété, vertu et puissance spirituelle divine»,* présenté par R. Amadou. *Diffusion Martiniste, Le Trembley-Fr 2000.*

[Mt] Matura Humberto & Varela F; *"The Tree of Knowledge"* New Science Library, Boston 1986.

[Mr] Miller Kenneth *"Finding Darwin's God,"* HarperCollins, New York, 2007.

[Mu] Murray J. *"Calvin's Doctrine of Creation"* Westminster Theology Journal 17:31, 1954.

[Na] Navoni Marco *"Léonard de Vinci et les secrets du Codex Atlanticus"* National Geographic 2012.

[Ne1] Sir Newton Isaac *"Optique ,"* traduced from English by Jean-Paul Marat* in 1787. Christian Bourgeois Editor 1989.

[Ne2] Sir Newton Isaac: *"Principes Mathématiques de la Philosophie Naturelle."* Tomes 1&2. Traduction de la Marquise du Chatelet, commentaires de Clairaut. Albert Blanchard Paris, 1966.

[Op] Oparin Alexander I. (traduced by A. Synge, dir. G. Gavaudan & M. Guyot) *"L'origine de la vie sur la Terre."* Masson, Paris-1965.

[Or] Origène *"Traité des Principes, Vol. 4"* Cerf, Paris 1980. French translation by *"De Principiis"* edited in Latin in 220.

[Ov] Ovide *"Les métamorphoses,"* translated by Joseph Chamonard. GF-Flammarion, Paris 1966.

[Pa] Pascal Blaise *"Pensées."* Lefèvre, Paris, 1836.

[Par] Parker Steve *"Evolution – La grande histoire du vivant,"* Delachaux et Niestlé, Paris 2018.

[Pb] Peebles P.I.E. *"The Standard Cosmological Model,"* Les Rencontres de la Physique de la Vallée d'Aoste 1998. Ed. M. Greco.

[Pd] Peterson Daniel C. *"The religious faith of two Nobel laureate physicists."* The Patheos Religion Library, November 18th, 2019. Quoted in Wikipedia.

[Pe] Pelt Jean-Marie *"Dieu de l'univers, science et foi."* Fayard, Paris-1995.

[Pe2] Pelt Jean-Marie dans *"Le monde s'est-il créé tout seul ?"* Entretiens avec Patrice Van Eersel et la collaboration de Sylvain Michelet de Thuan T. X., Prigogine I., Jacquard A., de Rosnay J., Pelt J-M. & Atlan H. Albin Michel, Paris 2008.

[Pi] Picq Pascal *"Au commencement était l'homme – De Toumaï à Cro-Magnon."* Odile Jacob – Paris 2003.

[Pk] Pollack Robert *"Can Faith Broaden Reason"* - Trinity Church, Wall Street - Discovery Adult Education Program, May 22, 2016.

[Pk1] Pollack Robert *"The Faith of Biology and the Biology of Faith"* Columbia University Press –NY 2013.

[Pl1] Planck Max *"L'image du monde dans la physique moderne"* - Gonthier, Stuttgart-1963.

[Pl2] Planck Max *"A Scientific Autobiography, and other papers."* Williams and Norgate LTD-1950.

[Pl3] Planck Max *"Where is Science Going."* WW. Norton & Company, Inc. Publishers, New York -1932, on Wikipedia.

[Pla] Platon *"Plato's Timaeus»* Stanford Encyclopedia of Philosophy, Oct 25, 2005, reviewed May 13, 2022.

[Pn] Penrose Roger *"Cycles of Time: an extraordinary new vision of the universe"* The Bodley head 2010. Version read by Mann Bruce in 4CD. By BOT *(Box On Tape) 2011.*

[Pn2] Penrose Roger *"Les cycles du temps: une nouvelle vision de l'univers,"* Odile Jacob, Paris - 2013

[Po] Polkinghorne John *"The Faith of a Physicist, Reflections of a Bottom-Up Thinker».* The Gifford Lectures, 1993-94. Fortress Press Editions - 1996.

[Po2] Polkinghorne John *"The Science and Religion Debate: An Introduction."* Faraday Papers, no. 1 (2007) et http://www.faraday-institute.org

[Pr] Prigogine Iliya et al. *"Thermodynamics of cosmological matter creation."* PNAS October 1, 1988, 85 (20) 7428-7432.

[Pr2] Prigogine Iliya *"Le monde s'est-il créé tout seul ? (Did World created alone?)"* de Thuan T. X., Prigogine I., Jacquard A., de Rosnay J., Pelt J-M. & Atlan H. Albin Michel, Paris 2008.

[PS] Prigogine Iliya, Stengers Isabelle *"La Nouvelle Alliance"* Gallimard, Paris-1986.

[Pu] Pullman Bernard *"L'Atome, dans l'histoire de la pensée humaine"* Fayard-Le temps des sciences, Paris1995.

[Ra] Ranjbar Vahid Houston *"Boltzmann's Philosophy of Materialism and Abdul-Baha,"* https://vanidhoustonranjbar.medium.com February 26, 2020.

[Re] Reeves Hubert *"L'heure de s'enivrer - L'univers a-t-il un sens ?"* Le Seuil, Paris-1986.

[Re1] Reeves Hubert *"L'espace prend la forme de mon regard»* Myriam Solal, Paris-1995.

[Re2] Reeves Hubert *"Patience dans l'azur"* Le Seuil "Science ouverte," Paris-1981.

[Rg] Richards T. A. & S.L. Gomes *"How to build a microbial eye"* Nature 523, 166-167 (2015).

[Ro] de Rosnay Joël dans *"Le monde s'est-il créé tout seul ?"* Interviews with Patrice Van Eersel and the collaboration of Sylvain Michelet de Thuan T. X., Prigogine I., Jacquard A., de Rosnay J., Pelt J-M. & Atlan H. Albin Michel, Paris 2008.

[Rob] Robert Paul *"Le Grand Robert de la langue française"* 2ème édition, 9 + 5 vol. Paris 1985.

[Ru] Rousseau Jean-Jacques "Du *contrat social*" Flammarion, Paris 1992.

[Ry] Rey Alain: *"Le grand Robert de la langue française,"* 2ème édition en 12 volumes, Editeur Parmentier, Paris-1985.

[Sa] Sandage Allan *"Science finds God"* - News Week, July 20, 1998.

[Sc] Schrödinger Erwin *"Ma conception du Monde. Le Veda d'un physicien."* Mercure de France – 1982.

[Sc1] Schrödinger Erwin *"Mémoires de Mécanique Ondulatoire."* Alcan Paris – 1933.

[Sc2] Schrödinger Erwin *"What is life?."* Cambridge University Press – 1944.

[Sch] Schweitzer Albert *"Souvenirs de mon enfance."* Albin Michel – Paris – 1926.

[Sm] Salam Abdul *"The Art of the Physicist,"* New Scientist, Vol 35 p. 163 – July 20, 1967.

[Sc& Bi] Schrödinger Erwin *"L'ESPRIT ET LA MATIERE"* précédé de *"L'ELISION"* par Bitbol Michel. Le Seuil – Sources du Savoir – Paris 1990.

[S&V] Sari Akela et Greffoz Valérie, interview de Stephen Hawking pour *"Science et Vie"* n° 1058 - p. 114, 08/2008.

[Te] Teilhard de Chardin Pierre *"Hymne de l'Univers"* Seuil – Paris – 1961.

[Te2] Teilhard de Chardin Pierre *"Le Phénomène Humain"* France Loisirs, Paris 1990/1st Edition - Seuil, Paris 1955.

[Ta] St Thomas d'Aquin *"Somme théologique"* Kindle 2019.

[Th] Thonnard F-J. *"Précis d'histoire de la philosophie"* Société de St Jean l'Evangéliste - Desclée & Cie, Ed Pontifico Paris, Tournai, Roma 1945.

[Tr] Tihomir Dimiter *"50 Nobel Laureates and other Great Scientists who Believe in God"* http://nobelist.net.

[Ti] Tipler Frank *"The Physics of Immortality"* - Oxford University Press 1985.

[Tn] Thuan Trinh Xuan: dans *"Le monde s'est-il créé tout seul ?"* Entretiens avec Patrice Van Eersel et la collaboration de Sylvain Michelet de Thuan T. X., Prigogine I., Jacquard A., de Rosnay J., Pelt J-M. & Atlan H. Albin Michel, Paris 2008.

[To] Townes Charles H. *"The Convergence of Science and Religion"* in Thinks, Published by IBM March-April 1966, Vol 32-No. 2; and in Zygon J. of

Science and Religion- Vol. 1 N° 3 – Sept. 1966 p. 301-311 Wiley Online Library.

[Tu] Tutu Desmond *"God has a dream – A vision of Hope for Our Time."* Kindle Ed. 2005.

[Va1] Varela Francisco *"Tracer un chemin en marchant: le regard d'un biologiste sur la nouvelle biologie."* Pp. 40-48 de la Revue Le 3ème millénaire N°12 Eté 1989 - France.

[Vu1] Vasilescu Dan *"From Democritus to Schrödinger: A reflection on Quantum Modeling,"* Structural Chemistry, 21, 1289-1314.

[Vu2] Vasilescu Dan *"Approche Moléculaire et Thermodynamique de l'Origine de la Vie,"* contribution à *"Aux Origines de la Vie,"* coordonnateur Marcel Locquin, Fayard, Fondation Diderot, Paris-1987.

[Vi] Vischer Theodor *"Kritische Gänge, II,"* Verlag 2th ed. Leipzig 114.

[Vo] Voltaire *"Candide,"* Larousse, Evreux 1990.

[Wa1] Warfield Benjamin *"Evolution, Science, and Scripture: Selected Writings,"* Wipf and Stock Publishers, 2019. Reprint from Mark A. Noll, David N. Livingstone EDs.

[Wa2] Warfield Benjamin *"The Westminster Assembly and Its work,"* Princeton Theological Review 1904.

[Wh] White R. & McKenzie, D. *"Magmatism at rift zones. The generation of volcanic margins and flood basalts."* J. of Geophysical Research, 1994, Pp. 7685-7729.

[Wi] Wiesel Elie et Michaël de Saint-Cheron *"Le Mal Et L'exil. Dix ans après."* Rencontres – nouvelle cité Dame Press-2000.

Authors

NOM Prénom	Ref. - P.
Abdu'l-Baha *(Teheran, Persia 1844-1921 Haïfa) head of the* Baha'i faith.	14-5
Adams John C. *(GB: Laneast 1819–1892 Cambridge),* *astronomer,* Anglican.	230-70
Aetius Flavius *(Durostorum = Silistra, BGR 395–454 Ravenne, ITA), Roman Patrice.*	714-255
Ahmad Mirza *(Qadiyan, Punjab 1835-1908) founder of* Ahmadism.	394'-117
Alaya Pereda Francisco J. *(Madrid, ESP 1934), biologist.* Catholic.	682-243
Albrecht Andreas *(Ithaca, NY-US 1957), theoretician physicist.*	562-179
Aldrin Buzz *(Glen Ridge, NJ-US 1930), astronaut, Freemason,* Presbyterian.	761-283
Alexander Denis *(ENG 1945), molecular biologist,* Catholic.	704-250
Alhazen ou Ibn al-Haytham *(EGY: Bassora 965–1039 Cairo), experimental Scientist,* Chiite.	100-30
Alpher Ralph *(USA: Washington DC 1921-2001 Austin, TX), cosmologist* Agnostic Jew	350-105
Alter Harvey I. *(New York, USA 1935), 2020 **Nobel** Prize in Physiology or Medicine.* Jew.	406-120
Al-**Baghdahi** Abu'l Barakat *(IRQ: Balal 1080–1164/65 Bagdad), physician.* Jewish => Muslim.	105-31
Al-**Bayhaqî** Zahir al-Dîn (Sabzevar, IRN 1097–1169), *mathematician,* Muslim..	108-32
Al-Fârabi *(Fârâb, Turkestan 872-950 Damascus, SYR), philosopher, scientist.* Muslim.	97-29
Al-Fazari Muhammad *(Persian or Arab died in 796 or 806), an astronomer, and astrologer.* Muslim.	94-28
Al-**Ghazâlî** Abû Hamid Mohammed ibn Mohammed Ghazâlî *(Tus, IRN 1058-1111), philos.* Soufiste.	109-32
Al-**Khwarizmi** *(Khiva, OUZ 780–850 Bagdad, IRQ), mathematician, astronomer. geog.,* Muslim Imam.	91-27

Al **Kindi** *(IRQ: Koufa 801-873 Bagdad), mathematician, physicist scholar.* Muslim.	93-28
Al' **Mamûm** *(Bagdad, IRQ 786–833 Tarse, TUR)*, Calif, Muslim Theologian.	87-26
Al-**Razi** *(Rayy, IRN 865–925 or 935), polymath,* Muslim.	96-28
Al-**Tusi** Nasir al-Dîn *(Tus, IRN 1201-1274 Bagdad, IRQ) polymath,* Muslim.	120-34
Ampère André-Marie *(FR: Lyon 1775–1836 Marseille), physicist,* Catholic.	620-213
Anatole **France** *(FRA: Paris 1844–1924 St Cyr-sur-Loire), 1921 **Nobel** of Literature.* Free thinker.	625-216
Anaxagoras *(TUR - 500 to - 428 Lampsaque (Asie Mineure), astronomer, philosopher.*	631-220
Anaximander *(Miletus -610 to -546) astronomer, and philosopher.*	36-12
Anaximenes *(Miletus -550 to -480) astronomer, physicist, and philosopher.*	38-12
Anders William *(Hong Kong 1955), astronaut.*	758-282
Anderson Carl D. *(New York 1905–1991 San Francisco), 1936 **Nobel** Prize in physics.*	331-98
Aristarchus of Samos (Samos, GRC -310 to - 230), *astronomer, mathematician, philosopher.*	63-20
Aristotle *(Stagire nowadays Rtavro [GRC -384 to – 322 Chalcis, Eubea), polymath, and philosopher.*	59-18
Arius *(Cyrénaïque, LBY 256 – 336 Constantinople, TUR).* Priest, Christian Theologian.	208-62
Armstrong Niels *(OH-US: Wapakoneta 1930–2012 Cincinnati), Astronaut,* Freemason, Methodist.	760-282
Arrhenius Svante *SWE: Vik 1859-1927 Stockholm), 1903 **Nobel** Prize in Physics.* Lutheran	411-122
Ashkin *(USA: New York 1922-2020 Rumson, NJ)), 2018 **Nobel** Prize in Physics.* Jewish	400-118
Aspect Alain *(Agen, Fr 1947), 2022 **Nobel** Prize in Physics.*	324-96
Athanasius of Alexandria *(EGP: Damanhour 296/298 –373),* Christian Bishop, Father of the Church.	207-62
Atlan Henri *(Blida, ALG 1931), biologist,* Jew *Spinozian.*	175-51
Averroes *(Cordoba, ESP 1126–1198 Marrakech, MAR), physician.* Muslim.	111-32

Avicebron *(ESP: Malaga 1020 –1057 Valence), metaphysicist,* Rabbi.	104-31
Avicenna or **Ibn Sina** *(Afshana, UZB 980-1037 Hamadhan, IRN), physician,* Muslim.	101-30
Avogadro Amedeo *(ITA : Turin 1776 - 1856), a physical chemist. Catholic.*	219-66
Baha'u'llah *(Nur, Persia 1817-1892 Acre), funder of the* Baha'i faith.	15-5
Bainbridge John *(ENG: Asby-de-la-Zouch 1582-1643 Oxford) Astronomer,* Church of England Priest.	192'-58
Baltimore David *(New York 1938) 1975 **Nobel** Prize in Physiology and Medicine.*	639-224
Bardeen John *(USA: Madison, WI 1908–1991 Boston, MA). Physicist and Engineer. **Nobel** Prize in Physics in 1956 and 1972.*	265-80
Barish Barry Clark *(Omaha, NE-US 1936), 2017 **Nobel** Prize in Physics,* Jewish.	456-135
Barkla Charles G. *(Widnes, ENG 1877–1944 Edinburg, Scotland), 1901 **Nobel** in Physics.* Methodist.	329-98
Barr Stephen *(NY-US 1953), physicist, astrophysicist,* Catholic.	791-314
Barrow Isaac *(London, ENG 1630- 677), mathematician.* Anglican Theologian -> Catholic.	188-57
Barrow John D. *(London, ENG 1952), physicist, mathematician, astrophysicist.* Protestant.	696-247
Basov Nikolaï *(RUS: Ousman 1922–2001 Moscow), 1964 **Nobel** Prize in Physics.*	356-106
Baye Thomas *(ENG: London 1702-1761 Tunbridge Wells, Kent), mathematician.* Presbyterian Pastor.	220'-67
Beaud Richard *(Albeuve CHE 1942), philosopher, egyptologist,* Dominican Theologian	25-9
Beck Friedrich *(Wiesbaden, DEU 1927-2008), theoretical physicist.*	550-169
Becquerel Henri *(Fr: Paris 1852–1908 Le Croisic), 1903 **Nobel** Prize in Physics,* Catholic.	262-79
Behe Michael *(Altoona, PA-US 1952), biologist,* Catholic.	689-245
Bekenstein Jacob *(Mexico City 1947–2015 Helsinki, Finland), theoretical physicist.* Jewish.	385-114
Bell Burnell Jocelyn *(Lurgan, Ulster, IRL 1943), physicist.* Quaker.	370-110

Ben Abraham de Montpellier *Solomon, (first half of 13th century).* Rabbi	117-34
Ben Juda of **Ceuta**, *Magreb 1160-1226), physician, mathematician, astronomer, Rabbi.*	116-33
Ben Nachman Moses *(Girona 1194–1270 Acre, kingdom of Jerusalem), philosopher, physician,* Sephardic rabbi, Kabbalist	118-34
Bergson Henri *(Paris, Fr 1859-1941), Philos., **Nobel** Prize in Literature.* Jewish (*buried as* Catholic*).*	605-205
Bernard d'Espagnat *(FRA: Fourmagnac 1921-2015 Paris), theoretical physicist.* Christian.	740-272
Bernoulli Jacques (Bâle, CHE 1654–1705), *mathematician,* Protestant.	193-58
Birch Charles *(AUS: Melbourne1918-2009 Sydney), biologist, geneticist, and* Anglican theologian	541-165
Bohr Niels *(Copenhagen, DNK 1885–1962), 1922 **Nobel** Prize in Physics.* Atheist.	297-89
Boltzmann Ludwig *(Vienna, AUT 1844–1906 Duno, ITA), physicist,* Deist materialistic.	237-72
Borman Frank *(Gary, IN-US 1928), astronaut.* Christian.	757-282
Born Max *(Breslau-Empire allemand 1882-1970), 1954 **Nobel** Prize in Physics.* Jew -> Lutheran.	313-93
Boscovich Roger, J. *(Dubrovnik, HRV 1711-1787 Milan, ITA), mathematician,* Jesuit.	619-212
Bossuet Jacques-Bégnine *(Fr: Dijon 1627 – 1704 Paris),* Catholic Theologian.	183-56
Bothe Walther *(DEU: Oranienbourg 1891–1957 Heidelberg), 1939 **Nobel** Prize in Physics,* Lutheran.	320-95
Boyle Robert (Lismore, IRL 1627-1691 London, ENG), chemist, Anglican apologist.	196-59
Bradley James *(ENG: Sherborne 1693–1762 Chalford), astronomer,* Priest, Church of England.	211-64
Brailovsky Victor (Moscow, RUS 1935), *mathematician, informatician,* Orthodox Jewish.	364'-108
Brooks Frederick Jr. *(Durham, NC-US 1931), 1999 **Turing** Prize,* Evangelical apologist,	733-268
Brunet Michel *(Magné, Fr 1954), paleontologist.*	517-159
Brunschwig Jacques: *(FR Paris 1929-2010), philosopher.*	617-211
Buenanno Alessandra *(ITA), CERN collaborator.*	459-136

Cairns-Smith *(Kilmarnock, Scotland 1931-2016 Uplawmoore, ENG), biologist.*	630-219
Calvin Jean *(Noyon, Fr 1509-1564 Genève, CHE).* Protestant (Calvinist).	137-38
Campbell William C. *(Ramelton, North Ireland, UK 1930), biologist, parasitologist.* Protestant.	743-274
Cantor Georg *(St-Petersburg, RUS 1845-1918 Halle-German Empire), mathematician.* Lutheran.	174-51
Carrel Alexis *(FR: Sainte-Foye-lès-Lyon 1873–1944 Paris), 1912 Nobel Prize in Physio. or Med.* Catholic.	292-87
Carter Brandon *(AUS 1942), cosmologist* Theist.	648-229
Carver George W. *(Diamond US-MO 1864-1943 Tuskegee, AL), researcher, and inventor.* Christian.	752-279
Cauchy Augustin, Louis, baron *(FR: Paris 1789-1857 Sceaux), mathematician.* Catholic.	722-261
Cergnignani Carlo *(ITA: Teulada 1939-2010 Milano), mathematician*	767-286
Charpentier Emmanuelle *(Juvisy-sur-Orge 1968), 2020 Nobel Prize in Chemistry.* Catholic.	447-132
Chauvin Rémy *(Toulon 1913-2009 Sainte-Croix-aux-Mines), biologist.* Catholic.	680-243
Christine de **Lorraine** *(Nancy 1565-1637 Florence, ITA),* Catholic.	717-258
Cicero Marcus *(ITA: Arpinum -106 to -43 Gaeta) was a philosopher and politician.* Theist pagan.	80-24
Clark Ron *(South Africa 1944), archeologist.*	522-161
Clausius Rudolf *(Koszalin, POL 1822–1888 Bonn, DEU), physicist,* Protestant.	238-72
Clavius Christophorus *(Bamberg, DEU 1538–1612 Rome, ITA), astronomer,* Jesuit.	66'-21
Clinton Bill *(Hope, AR-US 1946), 42ème president des USA,* Baptist.	427-127
Cobb John B. *(Kobe, JPN 1925), environmentalist,* Christian.	776-296
Cohen-Tannoudji Claude *(Constantine, French Algeria 1933), 1997 Nobel Prize in Physics,* Jewish.	358-107
Collins Francis *(Staunton, VA-US 1950), geneticist, 2020 Templeton Prize,* Evangelical apologist.	433&435 -129
Combulazier Charles *(Fr: Marseille 1903–1991) Doctor in Sciences.* Catholic priest.	709-253

Compton Arthur *(US: Wooster, OH 1892–1962 Berkeley, CA)*, 1927 ***Nobel*** *in Physics*, Presbyterian.	326-98
Conklin Edwin G. *(USA: Waldo, OH 1863-1952 Princeton, NJ.)*, *Biologist*, Methodist.	633-220
Constantine IX *(TUR: Antioch 1000–1055 Constantinople)*, Christian *emperor.*	103-31
Copernicus Nicolaus *(Prussia: Thorn 1473-1543 Frauenberg)*, *astronomer*, Catholic canon.	135-38
Cory Carl *(Prague, CZK 1896–1984 Cambridge, US-MA)*, 1947 ***Nobel*** *Prize in Physiology.* Catholic.	340-101
Cory Gerty, born Radnitz *(Prague, CES 1898–1957 Glendale MI-US)*, 1947 ***Nobel*** *Prize in Physiology or Medicine* Jewish => Catholic	339-101
Costa de Beauregard *(Fr: Paris 1911–2007 Poitiers)*, *theoretical physicist*, Catholic.	729-265
Cousteau Jacques-Yves *(Fr: St André de Cubzac 1910–1997 Paris)*, *oceanographer*, Catholic.	637-223
Coyne *G*eorge V. *(Baltimore, MD-US 1933)*, *astronomer*, Catholic *Creationist.*	738-271
Crick Francis H. *(Northampton, ENG 1916-2004 San Diego US-CA)*, 1962 ***Nobel*** *in Physio.*, Atheist.	650-231
Curie Marie born **Sklodowska** *(Varsovie, POL 1867–1934 Passy, Fr)*. ***Nobel*** *in Physics (1903) & Chemistry (1911)*,	263&264 -79&80
Curie Pierre *(Fr: Paris 1859 – 1906). 1903* ***Nobel*** *Prize in Physics.*	241-74
Dallinger William H. *(ENG 1839–1990)*, *protozoologist*, Methodist minister.	662-238
Dalton John *(ENG: Eaglesfield 1766–1844 Manchester)*, *chemist*, Quaker.	217-65
Darwin Charles *(ENG: Shrewsbury 1809–1882 Down)*, *was a biologist.* Anglican he *became* Agnostic.	232-70
David de Posquières Abraham ben *(Fr: Narbonne 1120-1197 Posquières-Gard)*, Rabbi kabbalist.	598-202
Davies Paul *(London, ENG 1946)*, *theoretical physicist, cosmologist, and astrobiologist,* Christian.	554-173
Davisson Clinton J. (USA: *Bloomington, IL 1881–1958 Charlottesville, VA)*, 1907 ***Nobel*** *Prize in Physics.*	306-91
Dawkins Richard *(Nairobi, KEN 1941)*, *ethologist*, Atheist.	698-247
Decius *(Budalia, ALB 201-251 Roma, ITA) Emperor.*	70'-22

Democritus *(Abdera-Thrace –470 to –370 &* ***Leucippus****, co-inventor of atom concept.*	47-13
Dembski William *(Chicago, IL-US 1960), mathematician, informatician,* Evangelist Theologian.	691-246
Descartes René *(La Haye-en-Touraine, Fr 1596-1650 Stockholm, SWE), scientist, philos.* Catholic.	152-43
Deterre Philippe *(Troyes, FR 1952), biologist,* Catholic priest.	745-275
Deutsch David *(Haïfa, ISR 1953), Quantum Computer scientist,* Atheist.	576-184
de **Bérulle** *(Cérilly 1575–1629 Paris), Cardinal* Catholic.	155-44
de **Broglie** Louis V. *(Fr: Dieppe 1892– 1987 Louveciennes), 1929* ***Nobel*** *Prize in Physics,* Catholic.	303-90
de **Fermat** Pierre *(FRA: Beaumont-de-Lomagne 1607 - 1665 Castres), scientist.* Catholic.	161-45
de **Rochon** Alexis-Marie *(FRA: Brest 1741–1817 Paris), scientist,* Catholic priest.	194-58
de **Saint-Exupéry** Antoine *(FRA: Lyon 1900 – 1944 drowned in the Mediterranean Sea), writer.*	615-210
de **Vries** Hugo *(NLD: Haarlem - Ede), biologist, geneticist,* Anabaptist.	664-238
de **Waard** Cornelis *(Bergen-op-Zoom, NLD 1879–1963 Flessingue island, NLZ), was a scientific historian.*	167-47
delle **Colombe** *(ITA : 1565), scientist,* Catholic.	587-192
Dicke Robert *(USA: St Louis, MO 1916–1997 Princeton, NJ), scientist.*	365-108
Diogenes of (Sinope, *TUR - 413 to - 327 Corinth, GRC), philosopher, cynic school representative.*	39-12
Diophantus of (Alexandria, *EGY between 3rd and 2nd c. B.C. s.), arithmetician.*	92-28
Dirac Paul *(Bristol, ENG 1922–1984 Tallahassee, FL-US), 1933* ***Nobel*** *Prize in Physics.* Deist.	317-94
Dobjansky Theodosius *(Nemyriv-UKR 1900-75 San Jacinto, CA), genet.,* Russian Orthodox Christian.	539-165
Doudna Jennifer *(Washington DC-US 1964), 2020* ***Nobel*** *Prize in Chemistry.* Christian.	448-132
Duhem Pierre *(FRA: Paris 1861-1916 Cabrespine) physicist, chemist, and historian.* Catholic.	474-144
Dulbecco Renato *(Catanzaro, It 1914-2012 La Jolla, US-CA) 1975* ***Nobel*** *Prize in Physiology or Medicin.*	639'-224

Dumas Jean-Baptiste *(FRA : Alès 1800-1884 Cannes), chemist,* Catholic.	221-67
Dunston ^{Georgia M.} *(Norfolk, VA-US 1944)* microbiologist. Baptist.	436-130
Duve Christian *(Thames-Ditton, ENG 1917–2013 Nethen, BEL), biologist-geneticist,* Catholic.	640-224
Dyson Frank *(Measham, ENG. 1868-1939) Royal astronomer.* Baptist.	291- 87
Dyson Freeman *(Crowthorne, ENG. 1923), physicist. Christian.*	588-193
Eaves Lindon J. *(Walsall, ENG. 1944).* geneticist, Anglican priest.	736-269
Edison Thomas *(USA: Milan, OH 1847-1931 West Orange NJ), inventor.* Deist.	749-277
Einstein Albert *(Ulm, DEU 1879–1955 Princeton, NJ-US),* 1921 **Nobel** in Physics, Jew Pantheist.	268-81
Eisenberg Josy *(FR: Strasbourg 1933–2017)* Rabbi.	775-294
Ellis George *(Johannesburg, S-Africa 1939), cosmologist,* **2004 Templeton Prize,** Quaker.	771-289
Empedocles *(Agrigente, GRC~ - 495 à ~ - 435), engineer, physician, philosopher, poet...*	43-13
Englert François *(Etterbeek, BEL 1932)* Baron, 2013 **Nobel** Prize in Physics. Jewish -> Catholic.	450-133
Enns Peter *(Passaic, NJ-US 1961),* theologian Evangelist.	610-208
Epicurus *(GRC: Samos -341 to -270 Athens), philosopher, astronomer.*	61-20
Eratosthenes *(Cyrene, LBY -276 to -194 B.C. Alexandria, EGY) was a mathematician and astronomer.*	64-20
Euler Leonhard *(Basel, CHE 1707-1783 St Petersburg RUS), mathematician & physicist,* Calvinist.	210-63
Everett Hugh *(US: Washington DC 1930–1982 McLean, VA), mathematician, physicist, cosmologue.*	575-184
Fabre Jean-Henri *(FRA: Saint Léons 1823–1915 Sérignan-du-Comtat) was an ethologist.* Deist Christian.	253-77
Fackle Heino *(Cologne, DEU 1966), radio astronomer and physicist,* Protestant Pastor.	464-138
Falk Darrell R. *(Vancouver, CAN 1946), biologist,* Evangelist.	656-235
Faraday Michael *(ENG: Newington 1791-1867 Hampton Court), physicist.* Calvinist Presbyterian.	235-71
Farooqi Sadaf *(middle 1960), physician.* Muslim.	629-219

Fatio de **Duillier** Nicolas *(Basel, CHE 1664-1753 Worcester, ENG), mathemat., astrophysicist.,* Protestant.	202-61
Feit Carl *(Teaneck, NJ-US. 1945), microbiologist,* Rabbi.	764-285
Fermi Henrico *(Roma, ITA 1901–1954 Chicago, IL-US), 1938 **Nobel** Prize in Physics.*	268'-81
Fert Albert *(Carcassonne, FR 1938), 2007 **Nobel** Prize in Physics.*	440-131
Feynman Richard *(USA: Queens, NY 1918-1988 Los Angeles CA), 1965 **Nobel** Prize in Physics.* Free thinker of Jewish education.	361-108
Flavius Josephus *(Jerusalem, ISR 37/38-100 Roma, ITA), historiographer.*	29-10
Flew Anthony Garrard *(ENG: London 1923-2010 Reading), biologists,* Atheist -> Deist.	634-222
Fontenelle *(FRA: Rouen 1657–1757 Paris). Living dictionary, scientific populariser, cartesian.*	182'-53
Fourier Joseph Baron *(Auxerre 1768-1830 Paris), scientific,* Catholic.	596-198
Fra Mauro *(Venezia, ITA 1385–1460), cartographer,* Catholic Benedictin.	139-39
Franklin Rosalind *(ENG: Notting Hill 1920–1958 London), chemist, biologist, ...* Jewish.	432'-128
Fresnel Augustin *(FR: Broglie 1788-1827 Ville d'Avray), experiment. & theoretician. physicist.* Catholic	222-67
Friedmann Alexander *(RUS: St Petersburg 1888-1925 Leningrad), mathematician, and physicist.*	286-86
Friedmann Jerome I. *(Chicago, US-IL 1930), 1990 **Nobel** in Physics.* Atheist.	414-123
Fuller Richard B. *(USA: Milton, MA 1895–1983 Los Angeles, CA), architect, inventor,* Unitarian.	426-126
Funès José G. *(Cordoba, ARG 1963), astronomer,* Jesuit.	742-274
Galen Claudius *(Pergamon, TUR 129-216 Roma, ITA), physician,* Agnostic.	88'-27
Galilei Galileo *(ITA: Pisa 1564-1642 Arcetri) was an astronomer, and a physicist.* Catholic.	144 to 146-41
Galle Johan *(DEU: Radis 1812–1910), astronomer.*	229-69
Gamow Georges *(Odessa, UKR 1904-1968 Boulder, CO-US], theoretical physicist.*	349-105
Gandhi *(IND: Porbandar 1869-1948 Delhi) lawyer, politician, "Father of India."* Hinduist.	796-318

Gaon Saadja *(Medinet el-Fayoum, EGY 882-982 Bagdad, IRQ),* Rabbi.	98-29
Gassin dit **Gassendi** Pierre *(FRA: Champtercier (04) 1592- Paris 1655), astronomer.* Catholic canon.	177-52
Gaudi Antoni *(ESP: Reus 1852-1926 Barcelona), God's Architect,* Catholic.	710-254
Gauss Johann C *(DEU: Brunswick 1777–1885 Göttingen).* "*Prince of Mathematicians,*" Christian.	213-64
Geiger Hans *(DEU: Neustadt an der Weinstrabe, All. 1882– 1945 Potsdam) was a physicist.*	259-79
Gell-Mann Murray *(USA: New York 1929- 2019 Santa Fe, NM) 1969 **Nobel** Prize in Physics.*	417-124
Genzel Reinhard *(Bad Homburg vor der Höhe in Hesse 1952), 2020 **Nobel** Prize in Physics.*	466-138
Gerbert d'Aurillac *(Belliac FRA 945/950–1003 Roma, ITA), scientist,* **Pope Sylvester II**.	99-29
Germer Lester *(USA: Chicago, IL 1896–1971 Gardner, NY), physicist.*	306'-91
Ghez Andrea *(NY-US 1965) Astrophysicist, 2020 **Nobel** Prize in Physics.*	465-138
Giberson Karl W. *(Bath, Nouveau Brunswick, CAN 1957), physicist,* Evangelist.	699-247
Gilbert Walter *(Boston, US-MA 1932), 1980 **Nobel** Prize in Chemistry.*	439-131
Glashow Sheldon *(New York, USA 1933), theoretical physicist,* Jewish.	393-117
Gödel Kurt *(Brno, AUT 1906–1978 Princeton, NJ-US) mathematician, and philosopher.* Protestant.	184-56
Golshani Mehdi *(Isfahan, IRN 1939), physicist,* Muslim.	766-286
Goodall Jane *(London, Eng 1936) Primatologist, anthropologist, 2021 **Templeton** Prize,* Christian.	646-227
Goodenough John B. *(Jeina, DEU 1922), 2019 **Nobel** Prize in Physics,* Evangelist.	397-117
Goswami Amit (IND 1936) *physicist.* Hinduist,	325-97
Gould Stephen J. *(New York, USA 1941-2002), paleontologist, biologist,* Agnostic Jewish.	681-243
Grant Peter *(London, ENG. 1936), zoologist and evolutionary biologist.*	783-303
Grant Barbara Rosemary *(Amside, ENG. 1936), zoologist and evolutionary biologist.*	784-303

Grassé Pierre-Paul *(FRA: Périgueux 1895–1985 Carlux)*, *zoologist,* Catholic.	635-222
Gray ASA *(US: New York 1810-1888 Cambridge, MA)*, *physicist, botanist,* Presbyterian.	661-237
Green Michael, B. *(London, ENG 1946), theoretical physicist.*	403-119
Grimaldi Francesco *(Bologna, ITA 1618-1663, physicist,* Jesuit priest.	195-59
Gross David *(Washington-DC 1941) 1941* **Nobel** *Prize in Physics.*	381'-113
Grothendieck Alexandre *(Berlin, DEU 1928–2014 St Lizier, Fr), 1966 Medal* **Fields.** Atheist=>Deist.	593-196
Grünberg Peter *(Pilsen, CZ 1939-2018 Jülich, DE), 2007* **Nobel** *Prize in Physics,* Catholic.	441-131
Guillaume d'**Auvergne** *(FRA: Aurillac 1190– 1249 Paris)*, Catholic Theologian.	122-35
Guillaume de **Conches** *(Conches-en-Ouche 1080–1154)*, Catholic Theologian.	106-31
Gurzadyan Vahe *(Erevan, ARM 1955), was a physicist, mathematician, and cosmologist.*	577-184
Guth Allan *(New Brunswick, NJ-US 1947), cosmologist.*	391-116
Hahn Otto *(DEU: Frankfurt 1879–1968 Göttingen), 1945* **Nobel** *Prize of Chemistry,* Lutheran.	345-103
Haldane John B.S. *(Oxford, ENG 1892 – 1964 Bhubaneswar, IND), geneticist,* Atheist.	676-241
Haüy René Just *(FRA: St Just-en-Chaussée 1743–1822 Paris), mineralogist,* Catholic Priest.	243-74
Hawking Stephen *(ENG: Oxford 1942–2018 Cambridge), theoretical physicist.*	369-110
Hefner Philip *(Denver, CO-US 1932),* Lutheran Theologian.	707-252
Hegel Georg W. F. *(DEU: Stuttgart 1770-1831 Berlin), philosopher,* Christian.	154-44
Heidegger Martin *(DEU: Baden 1889–1976 Fribourg), philos., theologian.* Protestant => Catholic.	779'-301
Heisenberg Werner *(DEU: Wurtzburg, 1901-1976 Munich), 1932* **Nobel** *Prize of Physics,* Lutheran.	312-93
Heller Michael *(Tarnow, Pol. 1936), cosmologist,* **2008 Templeton Prize**, Catholic	700-248
Heraclitus (Ephesus, TUR -576 to -480), *philosopher and cosmologist.*	37-12

Herman Robert *(USA: New York 1914–1997 Austin, TX)*, *cosmologist.*	352-105
Hertz Heinrich R. *(Hambourg, German confederation 1857-1894 Berlin, DEU), was a physicist.*	249-75
Hess Victor F. *(Peggau, AUT 1883–1964 Mount Vernon, NY-US), 1936 **Nobel** in Physics,* Catholic.	332-99
Hewish Antony *(Fowey, UK 1924-2021), radio-astronomer, 1974 **Nobel** Prize in Physics,* Christian.	371-111
Higgs Peter *(Newcastle, ENG 1929), theoretical physicist. 2013 **Nobel** of Physics.*	449-133
Hipparchus *(Nicaea, TUR -190 to -120 Rhodes, Dodecanese), mathematician, astronomer ...*	63-20
Hippocrates *(GRC: Island of Cos -460 to -377 B.C. Larissa), physician.*	88'-27
Hodgson Peter *(London, ENG 1928-2008), nuclear physicist,* Catholic	732-267
Hofstadter *(New York 1915-1990 Stanford, CA), 1961 **Nobel** Prize in Physics.* Theist.	414'-123
Hooker Joseph *(UK; Halesworth 1817-1911 Berkshire), was a physician and an explorer.* Anglican.	658-236
Hubble Edwin P. *(USA: Marshfield, MO 1889–1953 San Marino, CA), astronomer.*	301-90
Hublin J-Jacques *(Mostaganem, Algeria, FRA 1953), paleo anthropologist.*	513-158
Hulse Russel A. *(New York 1950), astrophysicist, 1993 **Nobel** Prize in Physics.*	378-112
Hunt William H. *(London, ENG: 1827-1910), painter.* Anglican.	798"-319
Hutchinson Ian Horner *(ENG 1951), nuclear physicist,* Christian.	744-275
Huxley Thomas H. *(GB: Ealing 1825–1895 Eastbourne), biologist and anthropologist,* Agnostic.	660-237
Huyghens Christian *(La Haie, NL 1629–1696), was a mathematician, physicist, and astronomer.* Lutheran,	179-52
ibn'al-**Shatir** *(Damas, SYR 1304–1375), astronomer,* Muslim.	134-38
ibn **Firnas** Abbas *(ESP: Rond 810–887 Cordoba), physician, chemist, poet,* Muslim.	95-28
ibn **Ishaq** Hunayn *(Al-Hira, IRQ 808–877),* Nestorian Christian.	88-27
Ibn **Sahl** Sabur *(?? – 869), Persian physician,* Christian.	96'-29
ibn **Qurra** Thâbit *(Harran, TUR 826-901),* Sabean.	89-27

Irwin James B. *(USA: Pittsburg, PA 1930-1991 Glenwood Springs, CO), astronaut,* Baptist.	763-283
Isaacson Walter *(New Orleans, LA-US 1952) historian, and biographer.*	147-42
Jacobi Carl Gustav J. *(Kingdom of Prussia: Potsdam 1804-1851 Berlin), mathematician,* Anglican	214'-65
Jaki Stanley *(Györ, Hun 1924–2009 Madrid, Esp), astrophysicist, **Templeton Prize**,* Benedictin	731-267
Jasper Karl *(DEU: Oldenburg 1883-1969 Basel) existentialist philosopher.* Christian.	33-11
Jastrow Robert *(USA: New York 1925-2008 Arlington, VA), astrophysicist,* Agnostic.	367-109
Jefferson Thomas *(USA: Shadwell, VA 1743–1826 Monticello, VA). 3rd Pdt USA,* Christian or Deist.	429-127
Johnson Phillip *(1940), law professor,* Presbyterian.	688-245
Joliot Curie Frédéric *(Paris, FRA 1900-1950),1935 **Nobel** Prize in Chemistry*	264"-80
Joliot Curie Irène *(Paris, FRA 1897-1956),1935 **Nobel** Prize in Chemistry*	264'-80
Jordan Pascual *(DEU: Hanover 1902-1980 Hamburg), theoretical physicist.*	314-93
Joule P. James *(ENG: Salford 1818–1889 Sale), physicist, chemist, ...* Anglican.	224-68
Jouzel Jean *(Janzé, FRA 1947), glaciologist and climatologist, 2012 **Vetlesen** Prize,* Catholic.	410-122
Jung Carl G. *(CHE: Kesswil 1875–1961 Küsnacht) was a physician, and a psychiatrist.* Protestant.	677-242
Justinian 1st *(TUR: Tauresium 482–565 Constantinople), Roman emperor.*	84-26
Kant Emmanuel *(Könisberg-Prusse-Orientale 1724–1804), philosopher,* Agnostic.	2'–xi
Kastler Alfred *(FRA: Guebwiller 1902–1984 Bandol), 1966 **Nobel** in Physics,* Evangelical protestant.	622-214
Kendall Henry *(Boston, MA 1926). 1990 **Nobel** in Physics,* Theist.	415-123
Kendler Kenneth S. *(New York, USA 1950), geneticist,* Jewish.	718-258
Kennedy, John F. *(USA: Brooklyn, MA 1917–1963 Dallas, TX), 35th USA President,* Catholic.	762-283
Kepler Johannes *(DEU : Weil 1571–1630 Ratisbonne), astronomer,* Lutheran.	149&150 -42&43

Khan **Mashriqui** Inayatullah *(PAK: Amritsa 1888–1963 Lahore), mathematician, and logician.* Islamic scholar (Theistic Evolutionist).	754-280
Kocher Emil Th. *(Bern, CH 1841–1917), 1909* **Nobel** *of Physiology or Medicine.* Moravian Church,	748-277
Kronecker Léopold *(Legnica, POL 1823–1891 Berlin, DEU), mathematician,* Jewish => Christian.	216-65
Lambertini Prospéro L. *(ITA : Bologne 1675-1758 Rome),* Pope **Benedict XIV.**	212-64
Lamoureux Denis *(Alberta CAN 1954), biologist,* Catholic.	694-246
Landsteiner Karl *(Baden bei-Wien, AUT 1868-1943 US-NY), 1930* **Nobel** *in Physiol. Med.,* Catholic.	254-77
Laplace Pierre-Simon de, Marquis *(Beaumont-en-Auge 1749–1827 Paris), scientist,* Catholic**.**	220-66
Lavoisier Antoine de *(FR: Paris 1743–1794) Philosopher, Economist, and Chemist.* Catholic.	243'-74
Lawrence Ernest *(USA Canton, SD 1901-1958 Palo Alto, CA), 1939* **Nobel** *Prize in Physics,* Anglican.	341-102
Lederman Léon, M. *(USA: New York 1922-2018 Rexburg, ID),* 1988 **Nobel** Prize in Physics.	453-134
Leibniz Gottfried Wilhelm *(DEU : Leipzig 1646-1716 Hanover), polymath »,* Lutheran.	182-53
Leibowitz Yeshayahu *(Riga, LET 1909–1994 Jerusalem, ISR), biochemist,* Jewish.	719-259
Lejeune Jérôme *(FRA: Montrouge 1926–1994 Paris), geneticist,* Catholic.	642-226
Lemaître Georges *(BEL: Charleroi 1894-1966 Louvain) astronomer.* Catholic canon.	342-102
Lennox John *(IRL 1943) was a mathematician, and philosopher.* Christian apologist.	789-313
Leonardo da Vinci *(Toscana, ITA 1452–1519 Le Clos Lucé, Amboise, FRA), polymath,* Catholic.	148-42
Leprince-Ringuet Louis *(Alès 1901–2000 Paris) FRA: physicist,* Catholic.	597-201
Leucippus *(Miletus, Ionia –460 to –370) &* **Democritus**, *philosophers.*	46-13
Levinas Emmanuel *(Kaunas, LTU 1905–1995 Paris), philosopher,* Jewish.	779-301
Levi-Montalcini Rita *(ITA: Torino 1909-2012 Roma), 1986* **Nobel** *of Physiology or Medicine,* Sephardic Jew.5	721-259

Lewis Clive Staples *(Belfast, IRL 1898–1963 Oxford, ENG)*, *philosopher,* Catholic.	734-268
Lewis Merither *(USA: Charlottesville, VA 1774–1809 Nashville, TN), military explorator.*	428-127
Le Sage Georges-Louis *(Genève, CHE 1724–1803), physician.*	203-61
Le Verrier Urbain *(FRA: Saint-Lô 1811–1877 Paris) was an astronomer and mathematician.* Catholic.	228-69
Lincoln Abraham *(Comté de Hardin, KY 1809–1865 Washington-DC (murdered)). 16ème Pdt USA.*	472-143
Linde Andrei Dimitrievitch *(Moscow, RUS 1948), theoretical physicist.*	561-179
Lippershey Hans *(Wesel, DEU 1570-1619 Middelbourg, NLD), physicist.*	145-41
Lippmann Gabriel *(Bonnevoie, LUX 1845–1921 wrecked), 1908 **Nobel** Prize of Physics,* Jewish.	244-74
Littré Emile *(Paris1801–1881), physician,* Agnostic.	724-262
Lord Francis **Bacon** *(ENG: London 1561–1626 Highgate), was a scientist, philosopher,* and Anglican.	151-43
Lord Ernst B. **Chain** *(Berlin 1906–1979 Castlebar IRL), 1945 **Nobel** of Physiology or Medicine,* Jewish.	336-99
Lord John **Boyd-Orr** of Brechin, *(Scotland: Kilmaurs, 1880–1971 Edzell) was a scientific researcher in nutrition.* Free Church of Scotland.	337-100
Lord Howard **Florey** *(Adelaide, AUS 1898–1968 Oxford ENG), 1945 **Nobel** in Physiol.,* Agnostic.	335-99
Lord William **Kelvin** was *born* **Thomson** *(Belfast, IRL 1824-1907), Father of Thermod.,* Presbyterian.	227-68
Lord **Rayleigh** was *born* John W. **Strutt,** *(Essex, ENG: Langford Grove 1842–1919 Witham), 1904 **Nobel** Prize of Physics,* Anglican.	267-80
Lord Ernest **Rutherford,** Baron of Nelson *(Brightwater, NZL 1871–1937 Cambridge, ENG), 1908 **Nobel** Prize in Chemistry,* Anglican, near YEC	252-76
Lorentz Hendrik *(NLD : Arnhem 1863–1928 Haarlem), 1902 **Nobel** Prize in Physics,* Free thinker.	273-82
Lovell James *(Cleveland, OH-US 1928). Astronaut.*	759-282
Lucas Henry *(ENG 1610–1663),* Anglican priest, *sponsor of a scientific chair.*	189-57
Luther Martin *(Eisleben, Saxe 1483–1546),* Catholic -> Protestant Lutheran	136-38

Maïmonide Moses *(Cordoba, ESP 1135–1204 Fostat, EGY), physician, astronomer.* Jewish Theologian.	114-33
Maria da Novara Domenic*(Ferrara 1454-1504 Bologna), mathematician, astronomer, physician,* Catholic	135'-38
Marconi Guglielmo ITA: *(ITA: Bologna 1874–1937 Roma), 1909* **Nobel** *Prize of Physics,* Anglican.	250-76
Margenau Henry *(Bielefeld, DE 1901–1997 Hamden CT-US), was a physicist, and philosopher.* Christian.	471'-143
Mariotte Edme *(FRA: Dijon 1620–1684 Paris), chemist,* Catholic Priest.	197-59
Martines de Pasqually *(FRA: Grenoble 1727-1774 St Domingue), Freemason,* Mystic Catholic.	555-174
Mather John Cromwell *(Roanoke, VA-US 1946), 2006* **Nobel** *Prize in Physics.*	420-125
Maturana Humberto *(Santiago, CHL 1928), biologist,* Catholic.	685-244
Clerk **Maxwell** James (GB: *Edinburg 1831–1879 Cambridge), physicist,* Evangelical Presbyterian.	234-71
Mayor Michel *(Lausanne, CHE 1942), 2019* **Nobel** *Prize in Physics.*	421-125
McBrierty Vincent *(Belfast, IRL 1942, physicist,* Catholic.	756-282
McCready Price J. *(Havelock, CAN 1870-1963 Lorna Lind, US-CA),* Adventist, *Creationist Author;*	655-235
McCullough Matthew *(1985), CERN senior physicist.*	454-134
McGrath Alister *(Belfast, IRL 1953), molecular biophysicist.* Anglican Theologian.	739-271
Meissner Krzysztof *(Varsovie, POL 1961), theoretical physicist,* Catholic.	578-185
Meitner Lise *(Vienna, AUT 1878-1968 Cambridge, ENG), physicist,* Jewish -> Protestant.	347-104
Mendel Gregor *(CZE: Heinzendorf bei Odrau, Silesia 1822-1884 Brno), biologist,* Catholic monk.	657-236
Mercati Giovanni *(Reggio Emilia, ITA 1866–1957 Vatican),* Cardinal.	71-22
Mersenne Marin *(Oizé 1588–1648 Paris) was, a philosopher, scientist, and* Catholic priest.	166-47
Messier Charles *(FRA: Badonviller (31) 1739–1817 Paris), astronomer,* Christian.	463-137

Michell John *(ENG: 1724–1793 Yorkshire), physicist, and father of seismology.* Evangelist.	281-85
Michelson Albert *(Strzelno, POL 1852–1931 Pasadena, CA-US), 1907 **Nobel** of Physics..*	247-75
Miller Kenneth *(Rahway, NY-US 1950),* biologist, Catholic.	737-269
Millikan Robert *(USA: Morrison, IL1868–1953 San Marino, CA), 1923 **Nobel** in Physics,* Anglican.	727-264
Minkowski Hermann *(Alexotas, RUS 1864–1909 Göttingen, DEU), mathematician.*	313'-93
Mitchell Maria *(US-MA: Nantucket 1818 -1889 Lynn), astronomer,* Unitarian.	231-70
Monod Jacques *(FRA: Paris 1910–1976 Cannes), 1965 **Nobel** of Physiology or Medicine,* Atheist.	638-223
Monod Théodore *(FRA: Rouen 1902–2000 Versailles), biologist, explorer, humanist, ...* Protestant,	579-185
Monod Wilfred *(Paris, FRA: 1867-1943), professor of theology,* Protestant.	778-298
Morley Edward W. *(USA: Newark, NJ 1838–1923 West Hartford, CT), physicist.*	248-75
Morris Simon Conway *(Carshalton, ENG 1951), paleontologist,* Christian.	703-249
Moses *(16 – 17ème c. Before J-C),* founder of the Jewish religion.	29-10
Mössbauer Rudolf *(DEU: Munich 1929-2011 Grünwald), 1961 **Nobel** Prize in Physics,* Atheist.	414"-124
Munk Salomon *(Glogow, POL 1803 – 1867 Paris, FRA), librarian,* Jewish,	470-140
Muori Alysson *(Brazil ~ 1974), neurobiologist.*	537-164
Murray <u>Joseph E.</u> *(US-MA: Milford 1919-2012 Boston), 1969 **Nobel** of Physiology. or Medicine.,* Catholic**.**	412-123
Müller Walther *(Hanover, DEU 1905–1979 Walnut Creek, CA-US), physicist.*	259'-79
Nestorius *(Syrie 380 – 451 Kargeth),* Christian Patriarch *of Constantinople.*	85-26
Onsager Lars *(Oslo, DNK 1903–1976 Coral Gables, Fl-US), 1968 **Nobel** Prize in Chemistry.*	566-180
Oparin Alexandre I. *(RUS: Ouglicht 1894–1980 Moscou), biochemist,* Christian.	632-220
Oppenheimer Robert *(USA: New York 1904–1967 Princeton, NJ, theoretical physicist,* Mystic.	348-104
Origen *(Alexandria, EGY 185–253 Tyr, LYB),* Christian exeget.	70-22

Osborne Louis S. *(Roma, ITA: 1923–2012 Lexington, US-MA)* *« Gentle giant of physics,"* Christian.	419'-125
Ovid *(Sulmone, ITA -43 B.C. to 17 A.C.), layer, poet.*	627-218
Pääbo Svante *(Stockholm, Sweden 1955),* 2022 ***Nobel*** *Prize of Physiology or Medicine,* Agnostic?	442-131
Paley William *(ENG: Peterborough 1743–1805 Bishopwearmouth), naturalist,* Anglican apologist.	191-58
Papin Denis *(FRA: Blois 1647–1713), inventor,* Protestant.	181-53
Pardies *(FRA: Pau 1636–1674 Paris), physicist,* Jesuit.	180-53
Parmenides *(*of Elea, *Asia Menor -530 to -444), the Pythagorean, philosopher.*	40-13
Pascal Blaise *(FRA: Clermont-Ferrand 1623–1662 Paris), was a scientist, philosopher, and* Catholic.	170-48
Pasteur Louis *(FRA: Dôle 1822–1895 Villeneuve-l'Etang), pioneer in microbiology,* Catholic.	240-73
Pauli Wolfgang *(Vienna, AUT 1900–1958 Zurich, CHE),* 1945 ***Nobel*** *Prize in Physic,* Catholic.	310-92
Peacocke Arthur *(ENG: Watford 1924–2006 Oxford), biochemist,* Anglican panentheist.	651-233
Pecci Vicenzo => Pope **Leo XIII** *(ITA: Carpineto Romano 1810–1903 Vatican).*	76-23
Peebles Jim *(Winnipeg, CAN 1935), cosmologist,* 2019 ***Nobel*** *Prize in Physics.*	366-108
Pelt Jean-Marie *(FRA: Rodemack, Moselle 1933 – 2015 Metz), Life Sciences,* Catholic.	583-191
Penzias Arno Allan *(Munich, DEU 1933),* 1978 ***Nobel*** *Prize of Physics,* Conservative Jewish.	364-108
Perlmutter Saul *(Champaign-Urbana, IL-US 1959),* 2011 ***Nobel*** *Prize in Physics.*	443-132
Phillips William D. *(Wilkes-Barre, US-PA 1948),* 1997 **Nobel** *Prize in Physics.* Methodist.	359-107
Philo of (Alexandria, *EGY -20 to 45), philosopher.*	65-21
Philolaos of Crotone *(Metaponto, ITA −470 to −390 Thebes, GRC), scientist, philosopher, and politician.*	52-14
Pickford Martin *(Trowbridge, ENG 1943), palaeoprimatologist.*	519-160
Planck Max, Karl *(DEU: Kiel 1858–1947 Göttingen),* 1918 ***Nobel*** *Prize of Physics.* Lutheran.	255-78&782-302

Plantinga Alvin *(Ann Arbor, MI-US 1932), professor of Philosophy,* theologian protestant,.	570-182
Plato *(Athens, GRC -428/-427 to -348), real name:* **Aristocles,** *philosopher,* Theist pagan.	54-15
Plotinus *(Lycopolis, EGY 205–270 Naples, ITA), philosopher.*	78-23
Politzer David *(New York 1941), 1941* **Nobel** *Prize in Physics.*	381"-113
Polkinghorne John *(Weston-Super-Mare, ENG 1930), a theoretical physicist, theologian, and* Anglican priest. **2002 Templeton Prize.**	571-182
Pollack Robert E, *(Brooklyn 1940), biologist,* Jew.	741-273
Pope **Benedict XIV,** *born* **Lambertini** Prospéro L. *(ITA : Bologna 1675-1758 Roma).*	212-64
Pope **Benedict XVI,** *born* **Ratzinger** Joseph A. *(Marktl-Bavaria 1927).*	77-23
Pope **Grégoire XIII** *(ITA: Bologna 1502–1585 Roma).*	66-21
Pope **John-Paul II,** *born* **Wojtyla** Karol, J. *(Wadiwic, POL 1920–2005 Vatican).*	1–xi
Pope **Leo XIII** *born* Vincenzo **Pecci** *(Capiuto Romano, ITA 1810–1903).*	76-23
Pope **Pie XII,** *born* Eugenio, Maria, G. G. **Pavelli** *(ITA: Roma 1876–1958 Castel Gandolfo).*	679-243
Pope **Sylvester II** cf. Gerbert of Aurillac	99'-29
Popper Karl *(Vienna, AUT 1902–1994 London, ENG), philosopher of sciences,* Protestant.	540-165
Prigogine Iliya *(Moscow, RUS 1917–2003 Brussels, BEL), 1977* **Nobel** *Prize of Physics,* Jewish Atheist.	387-115
Prokhorov Alexander *(Atherton, AUS 1916–2002 Moscow, RUS), 1964* **Nobel** *in Physics,* Theist.	357-106
Psellos Michel *(Constantinople, TUR 1018–1078),* Christian monk.	102-31
Ptolemy *(EGY: upper Egypt around 90–168 Canopy), astronomer.*	68-21
Pullman Bernard *(Wloclawek, POL 1919–1996 Paris, FRA), biophysicist.*	45-13
Pythagoras *(Samos, Aegean island -580 to −495 Metaponte, ITA), philosopher, scientist.*	55-16
Queloz Didier (CHE 1966), *2019* **Nobel** *Prize in Physics.*	422-125
Ratzinger Joseph A. *(Marktl, DEU 1927)* => **Pope Benedict XVI.**	77-23

Ray John *(Black Notley, Essex, ENG: 1627-1705)*, naturalist, Anglican apologist.	190-58
Reeves Hubert *(Montréal, CAN 1932)*, scientist. *He said I am neither* an Atheist *nor* an Agnostic.	4–1
Ricci Mattéo *(Macerata, ITA 1552–1610 Beijing, CHN)*, scientist, Jesuit.	140-39
Rice Charles M. *(Sacramento CA-US 1952)* 2020 **Nobel** Prize *in Physiology or Medicine.*	408-120
Riemann (G. F.) Bernhard *(Breselenz, Kingdom of Hanover 1826–1866 Selasca, Kingdom of Italia)*. *Famous mathematician*, Christian.	214-65
Riess Adam *(Washington DC-US 1969)*, 2011 **Nobel** Prize in *Physics.*	445-132
Roberval « *The Mathematician* », born Gilles **Personne** *(FRA: Roberval 1602–1675 Paris)*, Catholic.	168-47
Roentgen Wilhem (DE: Lennep 1845-1923 Munich), 1901 **Nobel Prize** in Physics, Anglican *??*.	328-98
Römer Ole Ch. *(DNK : Aarhus 1644–1710 Copenhaguen)*, *astronomer.*	282-85
Rolston Holmes *(Staunton, VA-US 1932)*, scientist, **2003 Templeton Prize**, Presbyterian priest.	143-40
Rossetti Dante G. *(ENG: London 1828–1882 Birchington-on-Sea), painter, poet, illustrator,* Anglican	798-319
Rostand Jean *(FRA: Paris 1894–1977 Ville d'Avray), biologist*, Agnostic.	687-245
Rubens Heinrich *(DEU: Wiesbaden 1865–1922 Berlin), physicist.*	255'-78
Rubin Vera *(USA: Philadelphia, PA 1928–2016 Princeton, NJ), astronomer,* Jewish.	388-115
Ruffié Jacques *(FRA: Limoux 1921–2004), hematologist, geniticist*, Christian.	683-244
Salam Abdus *(Jhang Sadar, PAK 1926–1996 London, ENG). 1979 **Nobel** of Physics*, Ahmadist.	394-117
Sandage Allan *(USA: Iowa City, IA 1928–2010 San Gabriel, CA), Cosmologist,* Christian.	765-286
Sanger Frederick *(ENG: Rendcomb 1918-2013 Cambridge) 1958 & 1980 **Nobel** Prize in Chemistry*	438-130
Sapi Inès *Physicist CERN.*	119-34
Sartre Jean-Paul *(FRA: Paris 1905–1980), philosopher,* Atheist.	644-227

Schaefer Henry F. *(Grand Rapids, US-MI 1944), chemist,* Protestant.	692-246
Scharf Caleb *(ENG 1968). Researcher and writer in Astrobiology.*	491-150
Schawlow Arthur L. *(Mount Vernon, NY 1921-1999 Palo Alto, CA, 1981 **Nobel** in Physics,* Methodist.	354-106
Schmidt Brian *(Missoula, MT-US 1967). 2011 **Nobel** Prize in Physics.*	444-131
Schrödinger Erwin *(Vienna, AUT 1887–1961), 1956 **Nobel** Prize in Physics, Philosopher.* Anglican.	309-91
Schroeder Gérald *(USA 1940) theoretical physicist.* Orthodox Jewish.	592-196
Schwarz John, H. *(North Adams, MA-US 1941), theoretical physicist, 2013 Fundamental Physics Prize.*	402-119
Schwinger Julian *(USA: New York 1918–1994 Los-Ángeles, CA), theoretical physicist,* Jewish.	363-108
Septimius Severus *(Leptis Magna, LBY 146-211 Eboracum-Yorkshire, ENG),* pagan emperor.	70'-22
Sertillanges Antonin-Dalmasse *(FR: Clermond-Ferrand 1863-1948 Sallanges) Dominican philosopher.*	787-309
Sénut **Brigitte** *(Paris 1954) Paleopromatologist, paleoanthropologist.*	518-160
Sharansky Anatoli => Natan *(Donetz, UKR 1948), mathematician => Politician,* Atheist => Jewish.	730-266
Sharpless Karl B. *(Philadelphia, 1941),* **Nobel** *Prize in Chemistry in 2001 and 2022,* Quaker.	266-80
Sir William **Henry Bragg** *(ENG: Westward,1862-1942 London), 1915 **Nobel** Prize in Physics.* Anglican.	295-88
Sir William **Lawrence Bragg** *(Adelaide, S-Australia 1890-1971 Waldringfield, ENG) 1915 **Nobel** Prize in Physics.* Anglican.	296-88
Sir James **Chadwick** *(GB: Bollington 1891-1974 Cambridge), 1932 **Nobel** Prize in Physics.*	300-89
Sir John **Cockcroft** *(ENG: Todmorden 1897-1967 Cambridge), 1951 **Nobel** Prize in Physics,* Methodist.	343-103
Sir Frank **Dyson** *(Measham, ENG. 1868-1939 into the sea), Royal astronomer,* Baptist.	291-87
Sir John **Eccles** *(Melbourne, AUS 1903-1997 Locarno, CHE), 1963 **Nobel** of Physiology,* Deist.	164-46
Sir Arthur **Eddington** *(ENG: Kendal 1882-1844 Cambridge), Astronomer,* Quaker.	290-87

Sir Alexander **Fleming** *(UK: Darvel, Scotl. 1881–1955 London)*, 1945 **Nobel** Prize in Physiology, Presbyterian.	334-99
Sir John A. **Fleming** *(ENG: Lancaster 1849–1945 Sidmouth)*, *physicist, engineer,* Christian,	251-76
Sir William R. **Hamilton** *(Dublin, IRL: 1805–1865), scientist,* Anglican.	223-67
Sir William **Herschel** *(Hanover, DEU 1738-1822 Slough, ENG), astronomer,* Lutheran.	595-198
Sir John T. **Houghton** *(Wales: Dyserth 1931-2020 Dolgellau),* 1955 **Nobel** *Prize of Peace,* Evangelist.	409-121
Sir Michael **Houghton** *(UK 1949), 2020 **Nobel** Prize for Physiology or Medicine.*	407-120
Sir Fred **Hoyle** *(ENG: Bingley, Yorkshire 1915-2001 Bournemouth), cosmologist & Astron,* Atheist.	351-105
Sir James H. **Jeans** *(London 1877-1946 Dorking, Surrey) was a physicist, astronomer, and mathematician.* Anglican.	753-280
Sir Ernest **Marsden** *(ENG: Manchester 1889-1970 Wellington), was a physicist.*	260-79
Sir John **Millais** *(ENG: Southampton 1829-1896 London), painter.* Anglican.	798'-319
Sir Nevill F. **Mott** *(Leeds, ENG 1905), 1977 **Nobel** Prize in Physics,* Anglican.	589-194
Sir Isaac **Newton** *(ENG: Woolsthrope 12/1642 or 01/1643–1727 Kensington), physicist,* Anglican.	185&204 -56&61
Sir Robert **Penrose** *(Colchester, ENG 1931), mathematician, 2020 **Nobel** of Physics,* Agnostic Jewish.	368-109
Sir Martin **Ryle** *(ENG: Brighton 1918–1984 Cambridge), astronomer.*	373-111
Sir Charles **Sherrington** *(ENG: Islington 1857–1952 Eastbourne), **1932 Nobel** Prize in Physiology or Medicine,* Anglican.	164'-46
Sir George G. **Stokes** *(Comté de Sligo, IRL 1819–1903 Cambridge, ENG), Lucasian Pr.* Anglican.	226-68
Sir John **Templeton** *(Winchester, TN-US 1912–2008 Nassau, BHS), economist,* Presbyterian.	434-129
Sir George Paget **Thomson** (Cambridge, ENG: 1892–1975), 1937 **Nobel** Prize in Physics, Anglican.	307-91
Sir Joseph J. **Thomson** *(ENG: Manchester, 1856-1940 Cambridge), 1906 **Nobel** in Physics,* Anglican.	308-91

Smalley Richard *(USA: Akron, OH 1943-2005 Houston TX)*, *1996 **Nobel** Prize in Chemistry*, Christian.	424-126
Smoot George F. *(Yukon, FL-US 1945)*, 2006 **Nobel** Prize in Physics.	419-125
Socrates *(near Athens, GRC -469 to -399 B.C., sentenced to death), philosophers' Master.*	53-15
Soddy Frederick *(ENG: Eastbourne 1877–1956 Brighton)*, *1921 **Nobel** Prize in Chemistry.*	258-78
Sommerfeld Arnold *(Könisberg, Prussia 1868–1951 Munich, DEU), theoretical physicist.*	299-89
Sonier de Lubac Henri *(FRA: Cambrai 1896–1991 Paris)*, Jesuit Theologian Cardinal-diacre.	73-22
Spinoza Baruch *(NLD: Amsterdam 1632 – 1677 La Haye) was a philosopher.* Pantheistic Jew.	172-50
Spottiswoode William *(ENG, London 1825–1883)*, *mathematician, physicist,* Anglican.	659-237
Steinhardt Paul *(Washington D-C, USA 1952), cosmologist.*	563-179
St Albert the Great *(DEU: Lauinguen 1193-1280 Cologne)*, *scientist,* Dominican.	123-35
St Ambrose of Milano (Trier, DEU 340-397 Milano ITA), Doctor and Father of the Church	81-24
St **Anselme** de **Cantorbery** d'Aoste *(Aosta, ITA 1033-1109 Canterbury, ENG),* Catholic monk.	715-256
St Augustin of Hippo *(Thagaste [Souk Ahras, Roman ALG] 354-430 Hippo [Annaba])*, Christian, Doctor, and Father of the Church.	79-24
St Duns Scotus John *(Dun, ENG 1266–1308 Cologne, DEU)*, Franciscan theologian.	128-36
St Paul the Apostle *(Tarsus 0-65 Roma executed).*	67-21
St **Thomas** d'**Aquin** St *(ITA : Aquino-Lazio 1224–1274 Priverno-Lazio),* Dominican theologian, Doctor, and Father of the Church.	127-36
Strassmann Friedman *"Fritz" (DEU: Boppard 1902-1980 Mainz), chemist.*	346-104
Sudarshan George *(Pallam British India 1931–2018 Austin, TX-US), theoretical physicist,* Hinduist.	582-190
Swimme Brian *(Seattle, US-WA 1950), cosmologist, evolutionist* Theist.	705-251
Taylor John G. *(ENG 1931–2002), mathematician.*	384-114

Taylor Joseph H. *(Philadelphia, PA-US 1941), astrophysicist, 1993 Nobel Prize in Physics.* Quaker.	377-112
Taylor Richard E. *(Medicine Hat, CAN 1929-Stanford, US-CA 2018). 1990 Nobel Prize in Physics.*	416-123
Teilhard de **Chardin** Pierre *(Orcines, FRA 1881-1955 New York, US), paleontologist, geologist,* Jesuit.	505-155
Temin Howard M. *(USA: Philadelphia, PA 1934–1994 Madison, WI), 1975 Nobel in Physiology,* Jewish.	639-224
Thalès *(Milet, Ionia, end of beginning of Vth c. B.C.), initiator of Philosophy, polymath.*	35-12
Thompson Benjamin *(Woburg, US-MA 1753–1814 Auteuil, FRA), philanthropist,* Anglican	262'-79
Thomson James *(Ballinainch, IRL 1786–1849 Glasgow, Scotland), mathematician,* Presbyterian.	227'-69
Thomson James *(Belfast, IRL 1822–1892 Glasgow, Scotland), physicist,* Presbyterian.	227''-69
Thonnard François-J. *(BEL: Barvaud-sur-Ourthe 1896–1974 Saint-Gérard),* Catholic priest.	58'-18
Thorne Kip *(Logan, US-UH 1940), 2017 Nobel Prize in Physics.*	458-136
Thuan Trinh Xuan *(Hanoï, VNM 1948) was an astrophysicist.* Buddhist.	647-229
Tihomir (1995) *double Master's degree in Psychology and Philosophy.*	781-302
Tipler Franck *(Andalusia, AL-US 1953), mathematician* Theist, *adept of I.D. as an idea.*	695-247
Tolman Richard C. *(USA: West Newton, MA 1881–1948 Pasadena, CA), physicist, chemist,...*	573-183
Tomonaga Sin-Itiro *(Tokyo, JPN 1906–1979), theoretical physicist.*	362-108
Torrey John *(New York: 1796–1873), was a physician, botanist, and chemist.* Protestant.	661'-237
Torricelli Evangelista *(ITA: Faenza 1608–1647 Firenza), scientist,* Jesuit.	178-52
Townes Charles *(USA: Greenville, SC 1915–2015 Oakland, CA), 1964 Nobel of Physics,* Protestant.	355-106
Tutu Desmond *(ZAF: Klerksdorp 1931-2021 Cap), 1984 Nobel Prize for Peace,* Anglican *Archbishop.*	780-301
Varela Francisco (Santiago, CHL 1946–2001 Paris, FRA), *biologist,* Buddhist.	684-244

Vasilescu Dan *(FR: Paris 1937), biophysicist,* Agnostic.	567-180
Venter Craig (Salt Lake City, UT-US 1946), *biologist.*	431-128
Voltaire, born François Marie **Arouet** *(Paris 1694–1778), philosopher,* Deist.	187-57
Von **Braun** Wernher *(Wirsitz, Posnania 1912–1977 Alexandria, US-VA), inventor,* Lutheran.	624-215
Von **Laüe** Max *(DEU: Pfaffendorf 1879-1960 Berlin), was a physicist.* Catholic.	294-88
Walcott Charles D. *(USA: New York Mills, NY 1850–1927 Washington, DC), paleontologist,* Anglican.	663-238
Wallis John *(ENG: Ashford 1616–1703 Oxford), mathematician,* Presbyterian.	176-51
Walton Ernest *(IRL: Dungarvan 1903–1995 Belfast), 1951Nobel Prize of Physics,* Methodist.	344-103
Warfield Benjamin *(USA: Lexington, KY 1851–1921 Princeton, NJ),* Presbyterian *Bishop.*	665-238
Watson James D. *(Chicago, US-IL 1928), 1962 **Nobel** in Physiology or Medicine.*	432-128
Weinberg Steven *(New York, USA 1933), theoretical physicist.*	395-117
Weisley John *(ENG: Epworth 1703-1791 London).* Methodist *founder,* Pastor.	735-268
Weiss Rainer *(Berlin, DEU 1932), 2017 **Nobel** Prize in Physics,* Agnostic.	457-135
Wheeler John A. *(USA: Jacksonville, FL 1911–2008 Hightstown, NJ), nuclear physicist,* Unitarian.	280-84
Whewell William *(ENG: Lancaster 1794–1886 Cambridge) polymath, scientist;* Anglican *theologian.*	233-70
Whiston William *(ENG: Norton 1667–1752 Rutland), Mathematician.* Presbyterian Theologian.	209-63
White Robert *(ENG: 1952), geophysicist.*	706-252
Whitehead Alfred N. *(Ramsgate, ENG 1861-1947 Cambridge, US-MA), math., philos., metaph.* Theist.	542-166
Whittingham Stanley *(Nottingham 1941), 2019 **Nobel** Prize in Chemistry,* Christian.	398-118
Wiesel Elie (Sighetu Marmatiei, ROU 1928–2016 New York, USA), *1986 **Nobel** Prize for Peace,* Hassimid Jew.	774-294
Wilczek Frank *(New York, USA 1951), 2004 **Nobel** in Physics, 2022 Templeton Prize,* Pantheist.	381-113
Wilkins John *(ENG: Fawsley 1614–1672 London),* Anglican *bishop.*	192-58

William of Ockham (Ockham, ENG 1285–1347 Munich, DEU) *philosopher.* Franciscan *theologian.*	581-186
Wilson Robert Woodrow *(Houston, TX-US 1936),* 1978 ***Nobel** Prize in Physics,* Agnostic.	364'-108
Winchester Albert *(Waco, TX-US 1908–2007), biologist,* Baptist.	755-281
Witten Edward (Baltimore, MD-US 1951), *theoretical physicist, Medal **Fields**.* Jewish.	404-119
Wojtyla Karol, J. (Wadiwice, POL 1920–2005 Vatican) => **Pope St John-Paul II.**	1-xv
Xenophanes of (Colophon, *Ionia -570 to -475), philosopher.*	41-13
Xu Guangqi *(CHN: Shanghai 1562–1633 Beijing), scientist,* Jesuit.	141-39
Yahya Harun / **Oktan** Adnan *(Ankara, TR 1956), Interior designer, negationist,* Islamist *creationist.*	635-235
Yoshino Akira *(Suita, JPN 1948),* 2019 ***Nobel** Prize in Chemistry.*	399-118
Zeeman Pieter *(NLD : Zonnemaire 1865–1943 Amsterdam),* 1902 ***Nobel** Prize in Physics,* Free thinker.	273'-82
Zeno of Elea *(Elea -490 to -430), dialectician.*	42-13
Zoroaster *(Persia around 10th c. B.C.), founder of Zoroastrianism.*	5-2
Zucky Fritz *(Varna, BGR 1898–1974 Pasadena, CA-US), astronomer.*	390-116

Glossary

Accident: ref.469-p.141
Adventist: ref.654-p.236.
Ahmadism: ref.394–p.117
Amphibological: ref.585–p.194
Anagogical: ref.75–p.23.
Anthropic Principles: ref.648–p.231.
Anthropoid/Simiiform: ref.510-p.159.
Antiparticle: ref.318–p.94.
Arianist: ref.208-p.62.
Ashram: ref.796'-p.318.
Astronomical Unit = UA: ref.496–p.154.
Atomystic: ref.607–p.209.
Australopithecus: ref.520–p.162.
Big Crunch: p.86.
Big Rip: ref.551–p.172.
Big Splat: ref.574–p.185.
Black Hole: refs.279&380-Pp.84&113.
Bohr's Atom: ref.298–p.89.
Bohr's complementarity principle: 321–p.96.
Brahman / Brahmana: ref.44–p.13.
Brane: §4.2.1. p.184.
Brightest: ref.492–p.153.
Brownian motion: ref.288–p.86.
Cabbala/Kaballa: ref.603–p.206.
Calâm/Kalâm & Mu'tazilite: ref.90–p.27.
Catarrhinian: ref.511–p.159.
Church Father/Doctor: ref.9–p.3.

Clear idea: ref.158-44.
COBE (satellite): ref.418–p.124.
Compete: ref.594–p.199.
Contingent: ref.626&773–p.220&294.
Cosmogenese: ref.670–p.241.
Cosmological constant: ref.289–p.87.
Cromagnon man: ref.532-p.165.
Dark Matter: ref.390-p.116 & §1979-p.116.
Deduction: ref.157–p.44.
Deist vs Theist: ref.186–p.57.
Demiurge: ref.23–p.9.
Denisova man: ref.530-p.163.
DNA: Appendix 2 – p.142.
Duality: ref.304–p.90.
Eclectism: ref.80'–p.24.
EHT: ref.462-p.138.
Eleatic School: ref.8–p.2.
Ennead: ref.24–p.9.
Entropy: ref. 238'–p.72.
Epicurism / Epicurus: ref.61–p.20.
Epicycle: ref.69–p.22.
ESO: ref.467-p.139.
Eudemonism: ref.57–p.17.
Existentialism: ref.32–p.9.
Fiqh vs Hadith: ref. 112–p. 32.
Fullereness: 425-p.126.
Gaia: ref.501–p.156.
Gene / Genome: Appendix 2 – p.142.
Geodesic: ref.284–p.86.
Gödel's theorem: ref.613, p.212.

Rutherford's Atomic Model:
ref.261–p.79.
Samaritan: ref.31-p.10.
Sassanides: ref.10–p.3.
School of Chartres: ref.107–p.32.
Secular: ref.121-p.35.
Septuagint: ref.30–p.10.
Sinanthropus (Peking man):
ref.527–p.163.
Singularity: ref.479-p.149.
Space-Time: ref.278–p.84.
Spallation: ref.486–p.150.
Standard Model: ref.382–p.113.
Star: ref.488-p.151.
Strepsirrhinian: ref.508-p.159.
String Theory: ref.401–p.119.
Substance: ref.34–p.11 & 51-p.14.
Sufi: ref.110–p.32.
Supernova: ref.490-p.152.
Tarsiiform: 509-p.159.
Tartarus: ref.21-p.8.
Theist vs **Deist:** ref.186–p.57.
Theory of Special Relativity:
ref.269-p.82.
Thorah: ref.27–p.9.
Three Universe theory: ref.548-
p.170.
T.O.E. (Theory Of Everything):
ref.383–114.
Trinity dogma: ref.206–p.62.
Tropological: ref.74-p.23.
Universe's expansion: ref.338–
p.101.
Univocity: ref.129-p.37.
Upanishad: ref.6–p.2.
Vaisesika or **Vaiheshika-Sutra:**
ref.49–p.14.
Vedânta-Sutras: ref.6–p.2.
Verb (the) or the « **Word** »:
ref.711–p.255.
VIRGO/LIGO: ref.455–p.135.
W.M.A.P.: p.135, 2nd alinea.

YEC: Young Earth Creationist: ref.
653–p.236.